RATIONALITY IN ECONOMICS

The principal findings of experimental economics are that impersonal exchange in markets converges in repeated interaction to the equilibrium states implied by economic theory, under information conditions far weaker than specified in the theory. In personal, social, and economic exchange, as studied in two-person games, cooperation exceeds the prediction of traditional game theory. This book relates these two findings to field studies and applications and integrates them with the main themes of the Scottish Enlightenment and with the thoughts of F. A. Hayek: Through emergent socioeconomic institutions and cultural norms, people achieve ends that are unintended and poorly understood. In cultural changes, the role of constructivism, or reason, is to provide variation, and the role of ecological processes is to select the norms and institutions that serve the fitness needs of societies.

Vernon L. Smith was awarded the Nobel Prize in Economic Science in 2002 for having established laboratory experiments as a tool in empirical economic analysis, especially in the study of alternative market mechanisms. He is professor of economics and law at George Mason University and a research scholar in the Interdisciplinary Center for Economic Science in Arlington, Virginia. Professor Smith is the president of the International Foundation for Research in Experimental Economics, which he helped found in 1997 to support education and research in experimental economics. He has authored or coauthored more than 250 articles on capital theory, finance, natural resource economics, and experimental economics. Cambridge University Press published a collection of his essays, *Papers in Experimental Economics*, in 1991 and published a second collection of his more recent papers, *Bargaining and Market Behavior*, in 2000. Professor Smith is a Fellow of the Econometric Society, the American Association for the Advancement of Science, and the American Academy of Arts and Sciences; in 1995 he was elected a member of the National Academy of Science. He has received an honorary doctor of management degree from Purdue University and an honorary doctorate from George Mason University, as well as honors from universities in other countries. Professor Smith's current experimental research involves the joint emergence of specialization and exchange under different property right histories; antitrust implications of bundled pricing of commodities; compensation elections for protecting minorities in majority-rule democracies; and demand response and active retail choice in electricity pricing.

Rationality in Economics

Constructivist and Ecological Forms

VERNON L. SMITH

George Mason University

CAMBRIDGE
UNIVERSITY PRESS

CAMBRIDGE UNIVERSITY PRESS
Cambridge, New York, Melbourne, Madrid, Cape Town, Singapore,
São Paulo, Delhi, Dubai, Tokyo, Mexico City

Cambridge University Press
32 Avenue of the Americas, New York, NY 10013-2473, USA

www.cambridge.org
Information on this title: www.cambridge.org/9780521133388

First published 2008
Reprinted 2008, 2009
First paperback edition 2009
Reprinted 2010

A catalog record for this publication is available from the British Library.

Library of Congress Cataloging in Publication Data

Smith, Vernon L.
Rationality in economics : constructivist and ecological forms / Vernon L. Smith.
p.   cm.
Includes bibliographical references and index.
ISBN 978-0-521-87135-8 (hardback)
1. Experimental economics. 2. Economics – Methodology. 3. Constructivism (Philosophy)
4. Ecology. I. Title. II. Series.
HB131.S38   2007
330.01–dc22   2007002899

ISBN 978-0-521-87135-8 Hardback
ISBN 978-0-521-13338-8 Paperback

To my wife
Candace Cavanah Smith

When we leave our closet, and engage in the common affairs of life, [reason's] conclusions seem to vanish, like the phantoms of the night on the appearance of the morning; and 'tis difficult for us to retain even that conviction, which we had attained with difficulty. . . .

Hume (1739; 1985, p. 507)

All particulars become meaningless if we lose sight of the pattern they jointly constitute.

Polanyi (1962, p. 57)

We have become accustomed to the idea that a natural system like the human body or an ecosystem regulates itself. To explain the regulation, we look for feedback loops rather than a central planning and directing body. But somehow our intuitions about self-regulation do not carry over to the artificial systems of human society. [Thus] . . . the . . . disbelief always expressed by [my] architecture students [about] . . . medieval cities as marvelously patterned systems that had mostly just "grown" in response to myriads of individual decisions. To my students a pattern implied a planner. . . . The idea that a city could acquire its pattern as "naturally" as a snowflake was foreign to them.

Simon (1981; 1996, p. 33)

Contents

Preface

I began developing and applying experimental economics methods to the study of behavior and market performance in the 1950s and 1960s and started teaching a graduate course in experimental economics in 1963; these early research exercises continued and occasionally started to include experiments that had economic design and "policy" applications in the 1960s. Thus, laboratory experiments in the 1960s examining rules for auctioning U.S. Treasury securities, in confluence with other forces, helped motivate a field experiment by Treasury in the 1970s, consisting of sixteen bond auctions, and this led to changes in policy in the 1980s and 1990s.

At the University of Arizona, along with several of my remarkable students and colleagues, we started to do electronic trading experiments – "E_Commerce" in the lab – in 1976 (Williams, 1980). Primarily these were exercises testing and exploring theoretical and other hypotheses about the performance of markets under controlled laboratory conditions. In the 1980s, these efforts grew naturally through our incremental learning into using experimental economics more systematically as a framework for communication and interaction with business, legal, engineering, regulatory, and other practitioners, in addition to students, and as a test bed for market designs that are applied in the world and used for postimplementation dialogue in ongoing rule evaluation.

All these laboratory experiences changed the way many of us thought about economic analysis and action, as experimental methods took on a life of their own – a fact that I had no conscious awareness of initially, as I was still thoroughly imbued with the prevailing orthodox way of economic thinking. The transformation began in the 1960s, but progressed slowly. There are many reasons for the change, but of unique significance is the discovery that programming myself through the challenging exercises of designing and conducting experiments forced me to think through the process rules and procedures of institutional arrangements within which agents interact. Few are as skilled as was Albert Einstein in acquiring new understanding by the device of formulating detailed and imaginative mental experiments – the *Gedankenexperiment,* a concept introduced into German by Ernst Mach. Scientists need the challenge of real experiments to discipline their thinking in the required painstaking detail. This practice is what fuels the development of experimental knowledge in economics and all of science. That knowledge has a life of its own, whose traditions and techniques are distinct from the theory and the test hypotheses associated with each particular science.

Economic theory became, in my thinking, a framework for the prediction of equilibrium prices and allocations, implemented by the rules of extant trading institutions. Experiments provided a way of bridging the gap between equilibrium theory – pencil-and-paper thought models – and economic action by agents governed by market institutions that are complex to the participants but who do not approach their task by thinking about it the way we do as economists when we do economic theory. Experiments constituted a substitute for the missing dynamic process analysis that had not been part of the standard equilibrium tool kit, a kit that had focused only on what might be the equilibrium shadow cast ahead by any such process.

Also important was the early discovery, its replication, and ongoing generalization that humans could quickly learn to function in these private incomplete information environments using the action (property) right rules of extant institutions and their natural cognitive skills to explore exchange opportunities and achieve over time the efficient outcomes predicted by the modeler, alone armed with complete information. Humans functioned well in the heart of that rule-governed dynamic process but were not aware of the shadow ahead. Central to my new awareness was a growing and unsettling realization of the unsolved puzzle of *how* economic agents/subjects acquire the tacit knowledge that enabled them to function so well in socioeconomic environments – a knowledge-acquisition problem little recognized or studied and understood by economics and psychology. Agent actions, however, are not governed by the same mental processes we use to construct the theory. And their quick proficiency in repetitive markets with low asymmetric information is startling and awe inspiring. I think it says much about why these institutions have survived, grown, and daily beget new emergent variations in the communication age.

This was a humbling experience once I realized that in terms of formal modeling, none of us knows much beyond anecdotes about how either subjects in the lab or economic agents perform their task and nothing about how they process messages in time – and neither do they, as becomes evident if you interrogate them. This observation has nothing to do with theoretical sophistication; put theorists in the experiment, as I have done, and they cannot articulate an explanation of their own behavior interacting with others through an institution. Moreover, their behavior is no more or less efficacious than the typical student subject in these dispersed private information markets. Repetitive or real-time action in incomplete information environments is an operating skill different from modeling based on the "given" information postulated to drive the economic environment that one seeks to understand in the sense of equilibrium, optimality, and welfare. This decision skill is based on a deep human capacity to acquire tacit knowledge that defies all but fragmentary articulation in natural or written language.

The learning from the discovery and observation of this skill in the laboratory has provided the basis for a productive interaction with managers and policy makers in industry and government. These practitioners relate easily to the experimental framework through hands-on demonstrations followed by presentations and to become quickly immersed in a helpful dialogue from which all who are involved learn together; it's "we," not a group consisting of "us" and "them." To do this, no formal economic background is needed, especially in a design problem too complex to model in accustomed ways. Practitioners are into problem solving and do not relate naturally to discussions driven by economic theory and its "applications" to their world because they do not automatically relate it to their experience, but they can appreciate working models when they see and experience them and become an active part of the design-testing process for new markets and management systems. Experiments provide the means for defining a common language and experiential base for problem solving.

After a couple of decades of laboratory experimental investigations, I realized that static equilibrium theory was gradually taking on a new and more vibrant institution-specific life because standard theory omitted what it was most important for us to understand – how message and allocation rules can affect equilibrium formation in dispersed information environments. Equilibrium theory began with a preference/production framework to support market prices and derived efficiency, ad hoc stability, and distributional properties of that system. The theory, however, contained no price-discovery process based on an articulated message space of communication among agents, the rules governing message exchange, and rules specifying how contracts emerged from that message exchange. It was static equilibrium theory without process, and that mode of thinking continued to dominate with the important new contributions in asymmetric information modeling.

This book offers an account and development of the details that indicate how my thinking changed and led me to a new appreciation of the classical scholars and of F. A. Hayek. It is true that I had read Hayek (1945) long ago, and its theme even led me to write Smith (1982b), but his other works I had either not read or their significance had escaped me because my mind was not ready to comprehend the enormity of their full meaning. This changed dramatically less than a decade ago – I was surprised recently to find that the collection Smith (2000) contains no references to Hayek – when I "really discovered" Hayek, returned to the classics, and saw in a fresh new light Adam Smith's (1759; 1982, 1776; 1981) works, those of David Hume, and others in the incredible Scottish Enlightenment. Adam Smith's (1759; 1982) first book is particularly insightful in the light of contemporary developments in cognitive psychology, but it was his narrower work in economics that would command the most acclaim.

My participation in Liberty Fund conferences based on the classics and their subsequent forms helped me to make these important rediscoveries and new integrations through the lens of my previous experience in the laboratory. The change wrought in my thinking by a lifetime career in experimental economics now enabled me to better appreciate the great depth of the Hayek program and that of his Scottish predecessors, which somehow had been mislaid along the mainstream technical way.

My hope is that with more concrete examples and demonstrations illustrating what Hayek was talking about – he gave us precious few – and what the Scottish geniuses were trying to convey to us, the twenty-first century will be a century of reawakening, a deepening of this intellectual enlightenment, and new inquiries based on new tools of analysis.

Because I am particularly concerned with integrating the experiments and field examples that I examine in the text with the themes of constructivist and ecological rationality, many of the examples are not treated in depth. However, I have tried to provide references that enable the interested reader to pursue a deeper study. In writing the text, many auxiliary, related, or supplemental commentaries were of relevance; I follow the usual style of placing some of these in footnotes, but I have made many of them more accessible by including them directly as shaded text, making it easy for the reader to follow these asides or bypass them to continue with the main text.

In the years while writing this book, there has occurred an explosion of literature on topics relevant to its themes. I have tried in many cases to connect into that rapidly changing literature, but I also needed to invoke a stopping rule. Hence, I will not do justice to all those connections, but I urge the reader to investigate them more deeply in response to his or her own intellectual curiosity.

Arlington, Virginia

Anchorage, Alaska

Tucson, Arizona

Acknowledgments

The subtitle of this book was directly suggested to me by Joel Norman's (2002) paper, "Two Visual Systems and Two Theories of Perception: An Attempt to Reconcile the Constructivist and Ecological Approaches." The term "ecological rationality" has been used fittingly by Gigerenzer et al. (1999) for application to important discoveries captured in the concept of "fast and frugal decision making" by individuals: "A heuristic is ecologically rational to the degree that it is adapted to the structure of an environment" (p. 13). My application of the term is concerned with adaptations that occur within institutions, markets, management, social, and other associations governed by informal or formal rule systems – in fact, any of these terms might be used in place of "heuristic" and this definition works for me. My emphasis is entirely complementary to that of Gigerenzer et al., although I make no attempt herein to integrate the two perspectives. Friedrich Hayek prominently identified both kinds of rationality but did not attach a name to the second.

I am indebted to Sid Siegel for technical and conceptual inspiration in the early 1960s; to George Horwich, John Hughes, Stan Reiter, and the Purdue faculty from 1955 to 1967 for warm, tolerant support beginning with my first experiment; to John Dickhaut, Charles Holt, Charles Plott, Martin Shubik, Shyam Sunder, and others for many significant encounters over the decades on institutional and experimental issues; and to students, visitors, and the current Interdisciplinary Center for Economic Science (ICES) team at George Mason University. In particular, I have benefited from many engaging discussions with Bart Wilson on Hayek, David Hume, Adam Smith, Ludwig Wittgenstein, and Michael Polanyi in attempting to understand the roots of economic behavior in tacit knowing. My debt to my coauthors, who have also been valued colleagues, will be evident in how dependent I have been upon our joint product.

I also want to thank three anonymous referees, whose detailed comments on a draft of this book were not just encouraging but also a constructive guide to further revisions, and especially Andreas Ortmann, who graciously provided extensive commentary on an earlier draft of the manuscript and whose summary I have drawn upon in writing the introduction.

I have extensively revised and expanded my Nobel lecture, "Constructivist and Ecological Rationality in Economics," portions of which have survived in Chapters 1 through 4, 7, 9 through 12, and 14. Reproduction has been granted with the kind permission of the Nobel Foundation.

Chapter 6 combines and revises the following two papers:

Banks, J., Mark Olson, David Porter, Stephen Rassenti, and Vernon Smith. 2003. "Theory, Experiment and the Federal Communications Commission Spectrum Actions." *Journal of Economic Behavior and Organization* 51:303–50. Available online at http://www.sciencedirect.com. Reproduction has been granted with the kind permission of Elsevier.

Porter, D., S. Rasseti, and Vernon Smith. 2003. "Combinatorial Auction Design." *Proceedings of the National Academy of Sciences, September 16*, 100(19):11153–7. Available online at http://www.pnas.org/cgi/reprint/100/19/11153.

Chapter 7 is a revised and expanded version of my Herbert Simon lecture:

Smith, Vernon L. 2005. "Behavioral Economics Research and the Foundations of Economics." *Journal of Socio-Economics* 34(2):135–50. Available online at http://www.sciencedirect.com. Reproduction has been granted with the kind permission of Elsevier.

Chapter 8 includes in part sections of the following work:

Smith, V. L., and F. Szidarovszky. 2004. "Monetary Rewards and Decision Cost in Strategic Interactions." In M. Augier and J. March, eds., *Models of a Man: Essays in Memory of Herbert A. Simon*. Cambridge: MIT Press. Reproduction has been granted with the kind permission of MIT Press.

Chapter 13 is a revised and expanded version of the following:

Smith, Vernon L. 2002. "Method in Experiment: Rhetoric and Reality." *Experimental Economics* 2:91–110. Reproduction has been granted with the kind permission of Elsevier.

Introduction

Rules alone can unite an extended order. . . . Neither all ends pursued, nor all means used, are known or need be known to anybody, in order for them to be taken account of within a spontaneous order. Such an order forms of itself. . . .
Hayek (1988, pp. 19–20)

. . . the realist . . . turns his back on the whole he cannot grasp and busies himself with a fragment.
Gibran (1918; 2002, p. 55)

Experimental economics is good at measurement, testing, and discovery in studying the microeconomics of human behavior governed by the informal norms of social exchange and the more explicit rules of exchange in institutions. It has not been good at integration and interpretation within the broader context of human social and economic development. The learning from a half-century of experimental discovery will be particularly significant if we can find a way to leverage that learning into a broader understanding of the human career; otherwise, the rewards from the range of our research will be too narrowly drawn, fragmented, and of passing interest, as scholars move on to the intricate details of whatever is next. This book is an outgrowth of my struggle to obtain a larger vision of meaning in social and market economic behavior, and to communicate whatever value that process might contribute to a larger community. I know that others have similar concerns because we have shared them from time to time in passing and in depth. The picture I see is still blurred. Its outlines, however, are unmistakable; it remains for others to sharpen or change that picture even if most just pursue their business in their own way without it.

If we are to confront the challenge of meaning, we must begin by recognizing that the phenomena that underlie our subject matter arise from the remarkable capacity of human sociality and culture to discover forms

1

of interaction and organization that have enabled impressive expansions in human betterment. The situations we model and study emerged naturally from individual interactions, associations, businesses, and collectives. The agents active in this process were naïve in economic understanding, but had deep personal experiential knowledge that served them well (Polanyi, 1962, 1969).

These considerations have heightened my interest in F. A. Hayek's important distinction between two kinds of rationality. I shall try to relate all of this book's discussion and examples – experimental, field empirical, descriptive – to the following two concepts of rationality:

Constructivist rationality, applied to individuals or organizations, involves the deliberate use of reason to analyze and prescribe actions judged to be better than alternative feasible actions that might be chosen. When applied to institutions, constructivism involves the deliberate design of rule systems to achieve desirable performance. The latter include the "optimal design" of institutions, where the intention is to provide incentives for agents to choose better actions than would result from alternative arrangements.

Ecological rationality refers to emergent order in the form of the practices, norms, and evolving institutional rules governing action by individuals that are part of our cultural and biological heritage and are created by human interactions, but not by conscious human design.

The two concepts are not inherently in opposition; the issues are emphatically *not* about constructivist *versus* ecological rationality, as some might infer or prefer, and in fact the two can and do work together. For example, in evolutionary processes, constructivist cultural innovations can provide variations while ecological fitness processes do the work of selection. We will encounter many examples in which the two kinds of rationality coincide, and others in which they diverge or at least are still seeking convergence.

To illustrate, people were specializing through trade in markets with asymmetric information before the agricultural revolution. Where the problem was not too intractable, our forebears long ago also discovered and solved some common problems and found private arrangements enabling needed public goods to be built. They overcame defection incentives to cooperate effectively, developed effective auction systems before the Christian epoch, and in time extended them to selling everything from art to securities. All these remarkable developments occurred in the midst of negative reciprocity, inhumane forms of punishment and violence, and persistently sharp in-group versus out-group differentiation in moral practices. Although as economists we have articulated rational models of public goods problems, such ways of thinking were not necessary in the past for societies to create emergent solutions out of human interactions uninformed by formal economic analysis. Similarly, in hundreds of market experiments, economically unsophisticated and naïve but proficient individuals produce rational outcomes without in fact having any knowledge of the rationality and efficiency of the outcomes they produce. Their effectiveness is perhaps less surprising once we recognize that their human forbearers and contemporaries used their cultural and biological inheritance to create the institutional forms that we study in the experiments, but our neoclassical models (since the 1870s) failed to anticipate or even to appreciate this important development as we proceeded to construct the concept of an "institution free core" of economic analysis.

As theorists, our first cut at constructivist problem definition and "solution" leads quite appropriately to concerns about incentive failure, but abstract approaches to incentives may omit significant features. Consider the problem of public goods provision. Initially we thought and taught that public goods could not be produced efficiently by private means. Yet the canonical example, the lighthouse, emitting signals that all ships could observe at zero marginal cost, was privately financed before economics had become a well-defined profession. The problem of supplying incentives for private investments and aborting free riders was solved practically by lighthouse owners who contracted with port authorities to charge docking ships for lighthouse services (Coase, 1974). These contracts allowed the capital cost of lighthouses (a discrete variable cost before it is incurred) to be prorated among ship dockings, since dockings provided an effective and practical measure of lighthouse service utilization and value in consumption. For "efficiency," it is argued, the so-called "fixed" cost, once incurred, should not affect the price of lighthouse services. However, this argument is a fallacious nonstarter because it omits the inefficiency that results if the lighthouse is not built.

And the famous "tragedy of the commons" in grazing cattle was decidedly not necessarily a tragedy for the high alpine Swiss cheese makers who for each summer at least since 1224 A.D. pastured their cows on the commons. Entry to summer pastures was controlled by a property right rule that "no citizen could send more cows to the alp than he could feed during the winter" (Netting 1976, p. 139).[1] These economic design problems were solved by

---

[1] In the solution that Coase found for the lighthouse, note that one might paraphrase Netting that no shipping company could pass more of its ships past the lighthouse than it paid for as part of the ship docking charges.

people completely unschooled in free-rider theory, but experienced enough in their behavioral coordination problem to seek solutions that might work. Somehow they perfected them by trial-and-error "natural experiments" over time.

Constructivist analysis enables us to see that these were examples of excludable public goods, and in all such cases the question is whether there are feasible ways of limiting use to avoid or internalize external costs, or of assuring payments that cover investment cost. Not every such institutional design problem has a solution that people are able to fashion out of their experience. Ostrom (1990) examines a variety of different common property resource problems around the world and the emergent self-governing institutions that solved or failed to solve the governance issues that they addressed.[2] The solutions, as in the preceding examples by Netting and Coase, are often ingenious beyond the imagination of our pencil-and-paper theories, whose primary value is in enabling us to see why there is and were problems that require solution but did not facilitate solutions such as those that emerged in these examples.

We have achieved little comprehensive understanding of the processes that show how either divergence or convergence may exist between the two concepts. In particular, our professional tradition is not geared to modeling ecological processes that can enable us to better understand emergent social systems. How, for example, might stateless groups discover specialization, comparative advantage, exchange mechanisms, and the supporting property rights that enable wealth creation?[3]

To explicate further the two kinds of rationality, consider this description using the perspective of game theory: Our professional approach to any observed problem area is to write down an abstract game model analyzing the phenomenon in a particular situation or institution – such as the free-rider problems in the lighthouse or in the grazing commons – contrasting the equilibrium of the model with optimality. The models of the institution and of an optimal outcome are each exercises in constructivist rationality. But in these exercises we take as given the abstract game situation or observed phenomenon, as well as the social structure and rules of the governing arrangements – for example, contracting lighthouse companies in

[2] For experiments in the voluntary provision of public goods based on incentive rules, see the references on the topic in Smith (1991); also see Ostrom et al. (1994) for a treatment of the theory, institutional analysis, and experimental and field studies of common property resource problems.

[3] A first effort to create an experimental design to examine these most rudimentary of all questions has been reported by Crockett et al. (2006).

a market, or developing a governing commons management institution in the Swiss Alps. The latter are natural or spontaneous examples of ecological rationality in the economy – self-governing institutions that emerged out of human experience. Now go to the laboratory to test, for example, a model of common property resources (such as the institution described by Netting, 1976); laboratory studies in this vein have been reported (see, for example, Cesari and Plott, 2003). The subjects in the experiment interact under rules derived from the observed field situation. The important difference is that because the experimenter assigns all the private values and costs in the economic environment created for the experiment, we can determine the equilibrium predicted to obtain and evaluate its optimality or efficiency. Suppose that the subjects converge to the predicted equilibrium in dynamic interaction over time. This is a laboratory example of ecological rationality showing the capacity of motivated subjects to achieve the efficient static outcome over time by unknown dynamic mental and social processes that are not modeled in these or other studies.

I think that improved understanding of various forms of ecological rationality will be born of a far better appreciation that most of human knowledge of "how," as opposed to knowledge of "that," depends heavily on autonomic functions of the brain. Human sociality leads to much unconscious learning in which the rules and norms of our socioeconomic skills are learned with little specific instructions, much as we learn natural language; think of it as the developing "social brain" at work. This contrasts with explicit learning of a new skill like playing a piano piece or bidding in an auction, which requires attention, emulation, and adaptation resources initially, but then soon becomes as unconscious a practice as any routine mental process that is taken over and guided by the practiced brain. This is what Polanyi (1962) calls tacit knowledge, and it has its own dynamics of acquisition through intuitive (inarticulate able) processes.[4] We learn social exchange without the self-aware application of attention, emulation, and adaptation resources, but the acquired skills enable gains from personal exchange that reward and help perpetuate that learning. Humans are not "thinking machines" in the sense that we *always* rely on self-aware cognitive processes, which is why all such approaches to learning are inherently limited.

---

[4] This knowledge is an essential reason why the transferability of results between the laboratory and the field, from one set of field observations to another, and between two laboratory experiments is fundamentally an empirical proposition, and not a methodological question to be settled by argument. (See Smith, 1982a, for discussions of "parallelism" between field and laboratory; field versus laboratory experiments is a frequently visited issue most recently explored by Harrison and List, 2004).

This book is prominently motivated by the results and methods stemming from the study of behavior in experimental economics and in field tests and applications, but these results and methods do not inform the whole of the book's message nor are they the only observations that give its message coherence. The laboratory evidence is broadly interpreted as a window on the human career – its development, meaning, and change. Accordingly, throughout I have tried to relate laboratory discoveries to history, anthropology, archeology, ethnology, field empirical studies, psychology (including its important social and evolutionary branches), animal behavior, philosophy, methodology of science, neuroscience, the history of ideas, and, indeed, life experience.

The very association of the word "ecological" with experimental evidence may seem strange to those who think that experimental evidence is somehow artificial, whereas "ecological" is natural. But "ecological" is just another word for the occurrence of a rule-governed, self-organized order, and I want to avoid compartmentalizing observations and labeling them in separate boxes without seeking unifying themes. We seek coherence, and if we are to find meaning we should not reject the idea that all humans in all situations are intuitive, feeling, searching, and acting organisms who do not naturally compartmentalize knowledge – except in formal modeling exercises – when they join the task of deciding and choosing. One should not presume that the actions chosen by a laboratory subject in a market, or in an anonymous interaction with another subject in an extensive form game, yields no insight into the human enterprise – at least not without a larger penetration of the experimental evidence and its use in test bedding in economic design and industry/policy applications, or without a larger examination of social science learning.

There are five fundamental propositions that inform much of the content of this book:

- Wealth creation depends essentially on knowledge and skill specialization. This includes innovation and technological change, because these are central parts of the acquisition of the tacit knowledge of "how."
- Specialization is possible only through the sharing and exchange systems that derive both from personal human sociality and impersonal market institutions. Hence, specialization is not a phenomenon that depends only on markets, although that is most certainly the source of its large-scale success in the modern world of wealth creation – a central theorem in Adam Smith's second book. Specialization and exchange

are far older than markets, which underscores the importance of better understanding human sociality.

- The personal knowledge that underlies specialization and exchange at any time is dispersed, private, and therefore asymmetric in all social systems.
- Neoclassical and information economics enable impersonal social systems to be characterized by an equilibrium given the state of existing knowledge, whereas experimental economics has demonstrated the efficacy with which people operating through extant impersonal market institutions are able to discover equilibrium outcomes through repeated interaction over time.
- Missing or incompletely developed in economics are models of how people are able to discover equilibrium outcomes so effectively given only private information and the message space of surviving market institutions; how the study of personal exchange systems can enable us to better understand their role in early human discoveries of specialization; how specialization and exchange relate to innovation and technological change; and how institutions emerge and survive in human socioeconomic development.

The main themes of the book may be summarized as follows[5]:

- What is generally known as "Das Adam Smith problem" (that there is an inherent contradiction between the *Theory of Moral Sentiments* and the *Wealth of Nations*) is an artificial problem in that "the propensity to truck, barter, and exchange one thing for another" applies to both personal exchange (which, as I see it in retrospect, is central to but certainly not all of the content of Smith's first book, which dealt broadly with human sociality) and impersonal exchange in markets (the theme of his second book). (Also see North 1990, 2005.)
- Both Hayek and Adam Smith (and his contemporaries, including David Hume and Adam Ferguson) well understood the coexistence of the two rational orders: constructivist and ecological. Many contemporary economists do not have such an understanding (but some do; see in particular the treatments by Binmore 1994, 1997; also Nelson and Nelson, 2002). Part of this book provides the conceptual foundation for these two rational orders.

---

[5] This summary draws directly on that of Andreas Ortmann, who reviewed an earlier version of this manuscript. I am much indebted and grateful to him for a thorough and inspiring review that enabled me to make many valuable revisions and additions to the text, the references, and the style.

- Traditional economic theory has long chased the fiction that purposive human action requires deliberate calculation based on constructivist rationality. Hence, for over eighty years following W. S. Jevons, who wrote in 1871, theory failed to anticipate that individuals do not require complete information to achieve equilibrium market outcomes in repeat interaction, a finding long replicated experimentally across many different, even quite complex, economic environments.

- In the midst of our constructivist adventures, and separate from them, institutions have emerged that are "ecologically rational" and that economists would be hard put to improve on, even if such institutions had always been an integral part of economists' perceived task. Underlying this second "rational order" are, roughly, Darwinian selection arguments. In the same way that natural systems such as an ecosystem or the human body (itself a cellular ecosystem) can regulate itself, so can social institutions (such as villages, cities, markets, associations, and scientific communities, which are all supported by endogenous property rights systems that sometimes become externally codified). Generally, human institutions and decision making are not guided only or primarily by constructivism, which is much more important in generating variations – social and economic innovations – than in selecting which ones shall survive.

- In achieving efficient cooperative outcomes in market exchange experiments, individuals are observed to maximize their payoffs, based on the use of monetary rewards to induce value (cost) on outcome states. The underlying classical model of behavior, *homo economicus*, appears thereby to be strongly supported in these impersonal exchange environments. But these exchange institutions in the laboratory are supported by externally enforced (property) rights to act that prohibit taking without paying, and giving without being compensated. Hence, action in the strict self-interest does not conflict with joint social betterment. But it would be quite wrong to conclude from this observation that across the extended range of experimental studies we always observe *homo economicus* taking action for his immediate interest alone.

- Thus, in Part III of the book I turn to an examination of the world of personal socioeconomic exchange, mostly as it emerges in two-person extensive form game trees. Exchange in these economic environments cannot lead to cooperative joint maximization without an individual being exposed to defection by her paired counterpart, who will defect indeed if he is *homo economicus* and always chooses according to payoff dominance. But such defection is not the norm even in single-play games between anonymous players. If property rights emerge in these two-person exchanges, they must do so by mutual consent in the form of reciprocity and sharing customs that eschew either party taking without giving. In these situations, we observe other-regarding behavior that supports more cooperation than the standard model predicts.

- Many if not most scholars, in the belief that our experiments fully control for everything except preferences, have modeled these other-regarding behaviors as due to other-regarding preferences (utility) in the tradition of static equilibrium theory. This model confounds reputation-based reciprocal motives to cooperate through exchange with the notion that cooperation requires preferences to be altruistic. Consequently, I prefer not to refer to the *homo economicus* model but rather more generally to the standard social science model (SSSM), which may appeal to social preferences or other formalisms to explain the prediction failures of the static selfishness model (Barkow et al., 1992).

- There is an interesting parallelism in the way that our brains and the socioeconomic world evolve and function. Both the world and our brains have evolved problem solutions, essentially via forms of selection that are not a significant part of our formal reasoning efforts. Whereas in the world our social brains have evolved institutions to solve problems, the brain has evolved internal off-line parallel processing capacities that enable us to function in daily life without continuous monitoring and conscious control, an important adaptation to the emergent mind as a scarce resource. Our unawareness of these processes, and our egocentric tendency to believe that we are in control, lead naturally to what Hayek (1988) called the "'fatal conceit;' the idea that the ability to acquire skills stems from reason" (p. 21).

- Hayek's research program identified three complex emergent forms of order in the biological and cultural coevolution of the human career; all are prominent in the concept of ecological rationality: (1) the internal order of the mind; (2) the external order of social exchange; and (3) the extended order of markets. The first, an inquiry into the neuropsychology of perception, began in the 1920s and was completed and published by Hayek (1952). The second form of order concerns human sociality in small group interaction. In this book, this form is particularly relevant to the study of exchange behavior in two-person extensive form games. The third, concerned with market order and welfare, is Hayek's best-known legacy. All these themes will be retrospectively evident in this book, although they were not part of the motivation and development of experimental economics or my own learning from experiment.

- Experimental economics allows us to study and better understand ecological rationality and the manner in which constructivist and ecological rationality can inform each other. Specifically, experimental economics allows us to test propositions derived from rational reconstructions of processes driven by ecological rationality to test the validity of those reconstructed interpretations. For example, in Chapter 12 we test the proposition that the cooperation observed in two-person trust games arises from reciprocity (favors are rewarded with favors) rather than altruistic preferences. Experiments provide a relatively low-cost methodology for studying that which is not or might be. But to do that successfully and comprehensively, we also have to look beyond the lab to related field studies and applications.

- This development, along with the many new computer communication technologies, has led to the important new subfield of Economic Systems Design (ESD), which combines constructivist tools and learning from experience (ecological processes) to fashion new group decision-making institutions, testing them in the laboratory and in the field, and modifying them in the light of experience. Testing is crucial because our constructions may err by failing to model the correct elements, by building on inappropriate assumptions, by being infeasible to implement or impractical for the participants, and so on.

Both the constructivist and ecological themes in this book apply also to method in science and experiment including economics. I explore that development (see Part IV) and use it to explain why the falsificationist thinking of scientists defines neither what scientists do or exclusively what they should do, although it explains much of what they say about what they do. I will also treat the logical incompleteness of the methods of science (and mathematics) and why the failure of all attempts to construct a rational methodology of science is not cause for alarm or postmodern cynicism. What saves the day is human sociality, as it operates in our scientific communities and enables us to muddle through in spite of the rhetoric of falsification tests. In this respect, human success in science is not so dissimilar from human success in creating wealth through specialization supported by markets and other social innovations created by human ingenuity.

Finally, Chapter 14 briefly discusses the emerging new science of neuro-economics, although its themes and those generally of neuroscience appear time and again in context, where appropriate, throughout the book.

PART ONE

RATIONALITY, MARKETS,
AND INSTITUTIONS

Subsidiary awareness and focal awareness are mutually exclusive. If a pianist shifts his attention from the piece he is playing to the observation of what he is doing with his fingers while playing it, he gets confused and may have to stop. This happens generally if we shift our focal attention to particulars of which we had previously been aware only in their subsidiary role.... The arts of doing and knowing, the valuation and the understanding of meanings, are ... only different aspects of extending our person into the subsidiary awareness of particulars which compose the whole.

Polanyi (1962, pp. 56, 65)

As we have no immediate experience of what other men feel, we can form no idea of the manner in which they are affected, but by conceiving what we ourselves would feel in a like situation.

Smith (1759; 1982, p. 9)

This division of labor ... is not originally the effect of any human wisdom, which foresees and intends that general opulence to which it gives occasion. It is the necessary, though very slow and gradual, consequence of a certain propensity in human nature which has in view no such extensive utility; the propensity to truck, barter, and exchange one thing for another.... It is not from the benevolence of the butcher, the brewer, or the baker, that we expect our dinner, but from their regard to their own self-interest. We address ourselves, not to their humanity but to their self-love, and never talk to them of our own necessities but of their advantages.

Smith (1776; 1981, pp. 25–7)

ONE

Rediscovering the Scottish Philosophers

... scientific discovery is notoriously guided by many unidentified clues, and so is to an important extent our ultimate decision of accepting the claims of a scientific discovery as valid.

Polanyi (1969, p. 184)

Exchange in Social and Economic Order

Historically, a recurrent theme in economics is that the values to which people respond are not confined to those one would expect based on the narrowly defined canons of rationality. These roots go back to Adam Smith, who examined the moral sympathies that characterize natural human sociality in *The Theory of Moral Sentiments* and the causal foundations of human economic welfare in *The Wealth of Nations*. Economists are largely untouched by Smith's first great work, which was eclipsed by *The Wealth of Nations*. It is telling that one of the economic profession's most highly respected historians of economic thought "found these two works in some measure basically inconsistent" (Viner, 1991, p. 250). But Viner's interpretation has been corrected by many reexaminations of Smith's work (see, for example, Montes, 2003; Meardon and Ortmann, 1996; Smith 1998).

These two works are not inconsistent if we recognize that a universal propensity for social exchange is a fundamental distinguishing feature of Homo sapiens, and that it finds expression in two distinct forms: personal exchange in small-group social transactions, and impersonal trade through markets. Thus, Smith was to some extent relying on one behavioral axiom, "the propensity to truck, barter, and exchange one thing for another," where the objects of trade are interpreted to include not only commercial goods and services but also gifts, assistance, and reciprocal favors out of sympathy, as "Kindness is the parent of kindness" (Smith, 1759/1982, p. 225). By interpreting personal local interactions and market participation as different expressions of the universal human propensity for engaging in exchange, we bring a unity of meaning to otherwise apparently contradictory forms of behavior. Smith's treatment of sympathy includes the modern idea of "mind reading," and is remarkably insightful and penetrating. As can be seen in the ethnographic record, daily life, and laboratory experiments, whether it is goods or favors that are exchanged, they bestow gains from trade that humans seek relentlessly in all social interactions. Thus, Smith's single axiom, broadly interpreted to include the social exchange of goods and favors across time, as well as the simultaneous trade of goods for money or other goods, is sufficient to characterize a major portion of the human social and cultural enterprise. It explains why human nature as expressed in behavior appears to be simultaneously self-regarding and other-regarding without recourse to an arbitrary expansion in the arguments of individual utility functions. It may also provide an understanding of the origin and ultimate foundation of human rights to act – "property rights."

What is traditionally called a "property right" is a guarantee allowing actions to be chosen within the guidelines defined by the right. We automatically look to the state as the guarantor against reprisal when rights are exercised, but we also know that the state can often be as much a part of the problem of human rights violation as its solution. Who is to monitor the monitor? But property rights predate nation-states. This is because social exchange within stateless tribes, and trade between such tribes, predate the agricultural revolution a mere eleven to twelve thousand years ago – little more than an eye blink in the time scale for the emergence of humanity, and less than 10 percent of the period since the emergence of our immediate Homo sapiens ancestors. Both social exchange and trade implicitly recognize mutual consensual rights to act, which are conveyed in what we commonly refer to as "property rights."

In what sense are such rights "natural" or emergent? The answer, I think, is to be found in the universality, spontaneity, and evolutionary fitness value of reciprocity in social exchange. Reciprocity in human nature – and prominently in our closest primate relative, the chimpanzee, but also in other animals (de Waal, 1989, 1997, 2005) – is the foundation of our uniqueness as creatures of social exchange, which we extended to include trade with non-kin and nontribal members long – perhaps very long – before we adopted herder and farmer life styles.

Political activists sometimes juxtapose property rights and "human rights" as mutually exclusive phenomena. But "property" is that over which an individual human, or association of humans, exercises some specified priority of action with respect to other humans. Only humans (and perhaps a few other animal species, notably chimps), but not property, can be recognized as allowed to act without reprisal from others. The emotional appeal of the slogan, "human rights, not property rights," appears to stem predominantly from an egalitarian ethic that seeks to dispossess those who are "propertied." Yet the essence of property rights is the right to the product of one's own labor and to the further productive yield generated by the savings from that product.

Property rights mean that (1) if I plant corn, I have the right to harvest the yield of that corn, and therefore the right to prevent passersby from harvesting it; and (2) if I save some of the income from the sale of that harvest and invest in more land, then I have the right to plant and harvest from that additional land. To be "propertied" is to have the right to accumulate. To accumulate is *not* to consume all that my labor and previous savings investment has produced. This allows all my accumulation to remain at work in society at large and for all others to benefit from my capital investment. This is the basis for all endogenously sustainable (that is, devoid of perpetual external transfers) net wealth accumulation in society. *There can be no other basis.* If there is any abridgement of an individual's rights to so harvest and accumulate, then there is a direct abridgement of the rights of all others to share in these external benefits and achieve a corresponding improvement in their welfare – benefits unintended by the investor-saver who seeks only his own security.

Adam Smith understood that consumption by the rich has an insignificant effect on the welfare of the poor, because most of the income of the rich is invested in the tools and knowledge of production (or "improvements") that provides external benefits for every consumer: "The rich only select from the heap what is most precious and agreeable ... though they mean only their own conveniency ... [and] ... the gratification of their own vain and insatiable desires, they divide with the poor the produce of all their improvements" (Smith, 1759/1982, p. 184).

This is of course one of the reasons why, as Bernard Mandeville (see the following section) humorously tells us that "the very Poor Liv'd better than the Rich Before...."

"Property rights" in this sense are essential to self-sustaining economic development and the reduction of poverty. Self-sustaining reductions in poverty must depend not on enforced redistribution of the yield from individual effort, but in finding the means of empowering and motivating the self development of the skills and knowledge that allow every individual to discover and obtain compensation for the products and services that enables him or her to deliver value to others through exchange. Empowerment in this sense has significant positive psychological underpinnings in which giving has been accompanied by the pride of receiving that enables escape from a residual sense of one-way dependence that can destroy the exchange relationship.

Lessons from Scotland

Contrary to the outpouring of vulgar representations, in Smith's view, each individual defined and pursued his own interest in his own way, and this form of "individualism" has been mischaracterized by the metaphor of the selfish "economic man" (cf. Hayek, 1976; 1991, p. 120). This unhistorical scholarship fails to recognize the key proposition articulated by the Scottish philosophers: To do good for others does *not require* deliberate action to further the perceived interest of others. As Mandeville so succinctly put it, "The worst of all the multitude did something for the common good" (Mandeville, 1705/2005, Oxford edition, pp. 17–37).[1] Smith correctly and perceptively saw a much richer tapestry in human sociality than the vanity and hypocrisy satirically exposed by Mandeville. But there was incredible insight in Mandeville, whose idea that unintended good for others could flow from doing well for yourself, as well as that the division of labor served as an engine of wealth creation, influenced those who followed, including Smith, who might have conceded Mandeville greater poetic license. It is not evident that Smith had much of a sense of humor, at least not Mandeville's. Thus, we have the observation concerning Smith that:

He was the most absent man in company that I ever saw. . . . Moving his Lips and talking to himself, and Smiling, in the midst of large Companys. If you awak'd him from his reverie, and made him attend to the Subject of Conversation, he immediately began a Harangue and never stop'd till he told you all he knew about

---

[1] This is not to suggest that Smith was in agreement with Mandeville, whose system – Smith called it "pernicious" – satirically derived all acts of virtue from human vanity. See Smith, 1759/1982, section VII.ii.4, p. 313, for his critique of Mandeville and his qualification.

it, with the utmost philosophical ingenuity. He knew nothing of characters (from Alexander Carlyle, *Autobiography*, quoted in Buchan, 2004, p. 134).[2]

Mandeville's (1705; 2005) influential and brilliant satire[3] deserves a longer excerpt, since he articulated a pre-Smithian statement (by fifty-four years) of specialization and gains from trade:

A Spacious Hive well stockt with Bees,
That liv'd in Luxury and Ease,...
Vast Numbers throng'd the fruitful Hive;

---

[2] Buchan adds the comment that, "*The Theory of Moral Sentiments*, when it appeared in April 1759, showed precisely what Smith had been up to during his reveries. Never was there a more fascinated observer of his own mental state, and of the curiosities and customs of a provincial society." Further, such reveries led to his much more popular, but in significant ways less penetrating work, *The Wealth of Nations* in 1776. (Smith, judging from Carlyle's account, may have had some of the earmarks of high-functioning autism or Asperger's Syndrome.)

[3] In Mandeville's time there were two conceptions of virtue: ascetic transcendence of the corruptions of human nature and dispassionate reason that eschewed all emotional impulse. But in his poem there is no such virtue; no actions based on emotionless reason or that are free from narrow selfishness. Private vices yield public virtues. The original poem appeared in 1705 as a pamphlet, became an underground classic that was severely criticized, and was republished in 1714 with a prose explanation and defense, and again in 1723 along with two other essays by Mandeville. It was reprinted many times in the next century, and translated into French and German. Of particular interest for the themes of this book, in the preface to the 1714 edition, Mandeville significantly and meaningfully recognizes an inherent conflict of perception between our personal social exchange characteristics and the invisible benefits that human behavior generates through the external order of markets, for he explains to his humorless critics:

> ... the main Design of the Fable ... is to shew the Impossibility of enjoying all the most elegant Comforts of Life that are to be met with in an industrious, wealthy and powerful Nation, and at the same time be bless'd with all the Virtue and Innocence that can be wish'd for in a Golden Age; from thence to expose the Unreasonableness and Folly of those, that desirous of being an opulent and flourishing People, and wonderfully greedy after all the Benefits they can receive as such, are yet always murmuring at and exclaiming against those Vices and Inconveniences, that from the Beginning of the World to this present Day, have been inseparable from all Kingdoms and States that ever were fam'd for Strength, Riches, and Politeness, at the same time ... why I have done all this, cui bono? and what Good these Notions will produce? truly, besides the Reader's Diversion, I believe none at all; but if I was ask'd, what Naturally ought to be expected from 'em, I wou'd answer, That in the first Place the People, who continually find fault with others, by reading them, would be taught to look at home, and examining their own Consciences, be made asham'd of always railing at what they are more or less guilty of themselves; and that in the next, those who are so fond of the Ease and Comforts, and reap all the Benefits that are the Consequence of a great and flourishing Nation, would learn more patiently to submit to those Inconveniences, which no Government upon Earth can remedy, when they should see the Impossibility of enjoying any great share of the first, without partaking likewise of the latter" (Mandeville, 1705/2005, preface).

Yet those vast Numbers made 'em thrive,
Millions endeavoring to supply

Each other's Lust and Vanity...
Some with vast Stocks and little Pains
Jump'd into Business of great Gains...
And all those, that in Emnity,
With downright Working, cunningly
Convert to their own Use the Labour
Of their good-natur'd heedless Neigbour.
These were call'd Knaves, but bar the Name,
The grave Industrious were the same;
All Trades and Places knew some Cheat,
No Calling was without Deceit...
Thus every Part was full of Vice,
Yet the whole Mass a Paradise;...
Such were the Blessings of that State;
Their Crimes conspir'd to make them Great;
And Virtue, who from Politicks

Had learn'd a Thousand Cunning Tricks,
Was, by their happy Influence,
Made Friends with Vice: And ever since,
The worst of all the Multitude
Did something for the Common Good...
Thus Vice nurs'd Ingenuity,

Which join'd with Time and Industry,
Had carry'd Life's Conveniences,
It's reap Pleasures, Comforts, Ease,
To such a Height, the very Poor
Liv'd better than the Rich Before,
And nothing could be added more...
Without great Vices, is a vain
EUTOPIA seated in the brain,
Fraud, Luxury and Pride must live,
While we the benefits receive...

Bare Virtue can't make Nations live
In Splendor; they, that would revive
A Golden Age, must be as free,
For Acorns as for Honesty.

Many contemporary scholars and not only popular writers have reversed Mandeville's (and Smith's wealth-of-nations') proposition, and argued that the standard socioeconomic science model (hereafter, the SSSM; see Cosmides and Tooby, 1992) requires, justifies, and promotes selfish behavior. That A implies B in no sense allows the reverse statement. But why would people, including economists, confuse necessary with sufficient conditions?

The answer is provided in this book's title text quotation from Hume: No one can consistently apply rational logical principles to everything he or she does.[4] As theorists, we live by proving theorems, and when in this mode we rarely make such errors. If perchance we slip, another will be quick to correct us.

If enforceable rights can never cover every margin of decision, then – contrary to the notion that markets depend on selfishness, and to the clever poetry of Mandeville – opportunism in all relational contracting and exchange across time are *costs, not benefits*, in achieving long-term value from trade; an ideology of honesty[5] means that people play the game of "trade" rather than "steal," although crime may frequently pay the rational lawbreaker who routinely chooses dominant strategies. Nor does nonselfish behavior in conjunction and parallel with ordinary market transactions prevent those transactions from promoting specialization and creating wealth. There is no inherent contradiction between self-regarding and other-regarding behavior, and as we shall see, the latter well serves each individual under the common cultural norms of reciprocity sharing as it derives from repeat interaction – other-regarding behavior does not require other-regarding utility.

Cultures that have evolved markets have enormously expanded resource specialization, created commensurate gains from exchange, and are wealthier than those that have not (see Scully, 1988; Demmert and Klein, 2003; Gwartney and Lawson, 2003; and the numerous references they contain). This proposition says nothing about the necessity of human selfishness for the attainment of economic betterment – the increased wealth of particular individuals can be used by them for consumption and investment; payment of taxes; Macarthur Fellows; and gifts to the symphony, the Smithsonian, or the poor. Markets economize on the need for virtue, but do not eliminate it and indeed depend on it to avoid a crushing burden of monitoring and enforcement cost. If every explicit or implicit contract required external policing resources to ensure efficient performance, the efficiency gains from specialization and exchange would be in danger of being gobbled up by these support costs. In this sense, the informal property right rules or norms of moral social engagement – thou shalt not kill, steal, bear false witness, commit adultery, or covet the possessions

---

[4] Our powers of analysis fall short even at our best. A missing chapter in the study of bounded rationality is its application to understanding and accepting with a little humility the severe limitations it imposes on our own professional development of economic theory.

[5] North (1990; 2005) has emphasized the importance of ideology in promoting economic growth by lowering transactions and enforcement costs.

of thy neighbor – strongly support wealth creation through the increased specialization made possible by personal social exchange and the extended order of markets.

Research in economic psychology[6] has prominently reported examples where "fairness" and other considerations are said to contradict the rationality assumptions of the SSSM. But experimental economists have reported mixed results on subject conformance with "rational models." In achieving gains for themselves and others, people often are better, in concordance with, or worse than constructive rational analysis predicts:

- Better in many two-person anonymous interactions (see Part III)
- As predicted in rapidly convergent repetitive-flow supply-and-demand markets (see Part II)
- Worse in certain asset-trading markets, although these still slowly converge in time (Smith et al., 1988; Porter and Smith, 1994)

---

[6] I will use the term "economic psychology" generally to refer to cognitive psychology as it has been applied to economic questions, and to a third subfield of experimental methods in economics recently product-differentiated as "behavioral economics" (Mullainathan and Thaler, 2001), and further differentiated into "behavioral game theory" (Camerer, 2003). The original foundations were laid by Ward Edwards, Daniel Kahneman, Anatol Rappoport, Sidney Siegel, Paul Slovic, and Amos Tversky, to name only a few of the most prominent. Behavioral economists have made a cottage industry of showing that the SSSM assumptions seem to apply almost nowhere to decisions, hypothetical or real. This is because their research program has been a candidly deliberate search for "Identifying the ways in which behavior differs from the standard model..." (Mullainathan and Thaler, 2001, p. 2), a search in what may be the tails of distributions. Lopes (1991), in a survey of the judgment and decision literature, suggests that it was after 1970 that the search for anomalies becomes evident. Others who have searched more generally in the distributions of behavior relative to model predictions have found both differences and confirmations; in some cases, the differences were accounted for by improved experiment or by refining the theory; in some cases the confirmations were not robust, in others quite robust; in still other cases the behavior could be said to be more "rational" than the models that were tested. Although this book is critical of much of standard theory and its conceptual foundations, it reports cases that both differ and are congruent with many traditional models. Even in the latter case there is usually plenty of room for improved understanding and in getting the details and dynamics right. I would wish that we could find solutions, not just fault, as there is more than enough of the latter to go around. Weber and Camerer (2006, p. 187) have sought to correct the "searching for anomalies" methodological stance of behavioral economics with a new definition based on "... using evidence and constructs from neighboring social sciences...." Experimental economics from its inception has been importantly interdisciplinary and influenced by experimental psychologists, as is plain in the references in this book. It has long focused on individual, group, but also rule-governed (institutional) behavior. Behavioral economics is not yet about new and fundamental foundations, but about the particular questions and results emphasized.

Patterns in these contradictions and confirmations can provide important clues to the implicit rules or norms that people may follow, and can motivate new theoretical hypotheses for examination in both the field and the laboratory. The pattern of results greatly modifies the prevailing and misguided rational SSSM, and richly modernizes the unadulterated message of the Scottish philosophers.

TWO

On Two Forms of Rationality

They're made out of meat.
Meat?...
There's no doubt about it. They're completely meat.
That's impossible. What about the radio signals? The messages to the stars?
... The signals come from machines.
Who made the machines? That's what we want to contact.
They made the machines... Meat made the machines.
That's ridiculous. How can meat make a machine? You're asking me to believe in sentient meat.
I'm not asking you, I'm telling you. These are the only sentient race in the sector and they're made out of meat...
No brain?
Oh, there's a brain all right. It's just that the brain is made out of meat.
So... what does the thinking?
You're not understanding, ... the brain does the thinking. The meat.
Thinking meat!
You are asking me to believe in thinking meat!
Yes, thinking meat! Conscious meat! Loving meat. Dreaming meat! The meat is the whole deal.

Bisson (1995)

Humans are built to see what they are expecting to see, and it's hard to *expect* to see something you've never seen.

Grandin and Johnson (2005, p. 51)

Introduction

The organizing principle throughout this work is the simultaneous existence of the two rational orders: constructivist and ecological. These two orders interact daily in ordinary human interaction, but that interaction is almost entirely invisible to our conscious experience. Theory is intensively

24

constructivist. What enables theory to do useful work (work is "accomplishment") is through testing processes in the laboratory, observations in the field, or both. I shall try to make the case that both orders are distinguishing characteristics of how we have developed as social creatures; *both* are essential to understanding and unifying a large body of experience from socioeconomic life, the experimental laboratory, and field applications after test-bedding, and to charting relevant new directions for economic theory as well as experimental-empirical programs.

Roughly, we associate constructivism with attempts to model, formally or informally, rational individual action and to invent or design social systems, and link ecological rationality with adaptive human decision and with group processes of discovery in natural social systems. As we shall see, the two need not be mutually exclusive, opposed, or incompatible: We can apply reason to understand and model emergent order in human cultures and to evaluate the intelligence and function of such order. Again, individuals and groups invent products, ideas, policies, and such, but whether they endure or are copied is subject to forces of selection and filtering that are well beyond the control of the initiators. Every business decision is someone's constructivist idea of a best or appropriate action, but whether that decision is ecologically fit is up to socioeconomic forces far beyond the originator. Ecological rationality, however, always has an empirical, evolutionary, and/or historical basis; constructivist rationality need have little, and where its specific abstract propositions lead to some form of implementation, it must survive tests of acceptability, fitness, and/or modification. The sooner a constructivist program is subjected to ecological filtering, the sooner people learn how it will fare in surviving.

In our time it was Hayek who articulated forcefully the idea that there are two kinds of rationality. He did this with characteristic insight in several lectures and articles. I am not aware, however, that this important idea has had any significant influence on economic thinking, certainly nothing like the influence of his recognition that the market pricing system serves as an information and coordinating system. In fact, the opposite has occurred: His message is sometimes garbled in secondary sources by competent people who do not take the time to try to understand, let alone read, what Hayek is saying. If my assessment is correct, why has the idea not been influential? I would conjecture that a critical element in understanding this proposition is to be found in human self-perception. We naturally recognize only one rational order because it is so firmly a part of the humanness of our reason and our mind's anthropocentric need to think it is in control. Emergent rational orders are not plainly visible to our perception, intellectual curiosity, and reason and therefore are not a natural part of our theoretical and professional investigations (see Giocoli, 2003). Neither are such orders visible to economists committed to developing analytical constructs of what it means to behave rationally, which is plainly seen as a deliberate self-consciously aware form of action.

Constructivist Rationality

The first concept of a rational order derives from the SSSM going back to the seventeenth century. The SSSM is an example of what Hayek has called constructivist rationality (or "constructivism"), which stems particularly from René Descartes (also Sir Francis Bacon and Thomas Hobbes), who believed and argued that all worthwhile social institutions were and should be created by conscious deductive processes of human reason. Thus, " . . . Descartes contended that all the useful human institutions were and ought to be deliberate creation(s) of conscious reason . . . a capacity of the mind to arrive at the truth by a deductive process from a few obvious and undoubtable premises" (Hayek, 1967, p. 85). In the nineteenth century, Jeremy Bentham and John Stuart Mill were among the leading constructivists. Bentham (and the utilitarians) sought to " . . . remake the whole of . . . [British] . . . law and institutions on rational principles" (Hayek, 1960, p 174).

Mill introduced the much-abused constructivist concept (but not the name) of "natural monopoly." To Mill it was transparently wasteful and duplicative to have two or more mail carriers operating on the same route. He is the intellectual father of the U.S. and other postal monopolies around the world, their resistance to innovation, and their demise in the face of the privatization movement in some countries and the growth of superior substitutes in others. Mill also could not imagine that it would be efficient for two cities to be connected by two parallel railroad tracks (Mill, 1848/1900, Vol. 1, pp. 131, 141–2, Vol. 2, p. 463). Mill died in 1873. I would conjecture that by the time of his death, men with little formal education were poised to follow the road to riches constructing the first parallel-route railroads. Emergent contradictions to constructivist natural monopoly theory are examples of ecological rationality. But much of economic policy, planning, regulation, antitrust, and economic theory is still in the grip of Mill's static analysis of natural monopoly.[1]

---

[1] Much of antitrust economics is mired in irrelevant static analysis based on a competitive pricing model of existing products that misses much of the dynamics of business change. You see this in the Microsoft case. And politically, with little support from economists,

As Hayek (1973) put it, the first constructivist view of rationality

... gives us a sense of unlimited power to realize our wishes... [and] ... holds that human institutions will serve human purposes only if they have been deliberately designed for these purposes.... [An institution's existence] ... is evidence of its having been created for a purpose, and always that we should so re-design society and its institutions that all our actions will be guided by known purposes.... This view is rooted ... in a ... propensity of primitive thought to interpret all regularity ... as the result of the design of a thinking mind" (pp. 8–9).

In economics, the SSSM leads to rational predictive models of decision that motivated research hypotheses that experimentalists have been testing in the laboratory since the mid-twentieth century. Although the test results often tend to be confirming in impersonal market exchange, the results are mixed and famously recalcitrant in personal exchange, notably in a great variety of two-person extensive form games where some half of the people appear to violate their own self-interest in risking defection by their bargaining counterpart, attempt to cooperate, and often succeed, even when anonymously paired. The results from some of these games (ultimatum, dictator, and trust) have motivated constructivist extensions of game theory based on other-regarding, in addition to own-regarding, preferences (e.g., Bolton, 1991; Rabin, 1993; Fehr and Schmidt, 1999; Bolton and Ockenfels, 2000; Charness and Rabin, 2003; Engelmann and Strobel, 2004; Sobel, 2005), but these models perform poorly in competition with exchange models when applied to wider classes of games (see Part III).

The results from repeated games have motivated extensions based on "learning" – the idea that the predictions of the SSSM might be approached over time by trial-and-error adaptation processes (Erev and Roth, 1998; Camerer, 2003; Camerer et al., 2004). These are models of short-run change

this is seen in the recurring popular complaint to the Federal Trade Commission that the oil companies are ripping off the consumer when prices respond to war-related or other concerns about oil supply (see Chapter 4 and Deck and Wilson, 2004). The latter are at least comprehensible under shocks in supply expectations. But in the Microsoft case it is easy to focus on the trees and not look at the forest. Thus, competition among operating systems with substantial freedom to make tie-in sales is an admissible economic alternative to standard models. It simply raises the stakes for an incumbent system and for its displacement by an innovating entrant. Operating systems do get displaced. Recall that in the 1960s it was IBM that was the evil monopolist that could not be displaced and whose dominance in the computer world required antitrust intervention. Yet IBM's DOS operating system was buried by Microsoft, and the latter appears to be racing to hold its historic edge over competing entrants who are making inroads. What changes from decade to decade are the names of the principals, not the static antitrust equilibrium principles that are argued. Linux, partly a voluntary-contribution operating system, is considered by many to be the emergent galloping runner-up in the current competition among operating systems.

over time, and do not address deeper questions of what of a strategic nature is being learned that is *transferable from one context to another*. This latter problem has been addressed by Cooper and Kagel (2003) in simple games with some encouraging but mixed results. These exercises, however, still await extension to the well-established behavioral differences between the extensive and normal form of games where comparison tests have so far failed to yield support for the behavioral equivalence of these standard game forms. Except for offering some limited discussion and tests of their equivalence, Part III of this book is concerned almost entirely with behavior in extensive form games.

An alternative and perhaps complementary explanation of some of these contradictions to theory is that people may use social-grown norms of trust and reciprocity[2] (including equity, meaning to each according to his or her perceived justly earned desert, that is, equality of opportunity, not outcome) to achieve cooperative states superior to individually rational defection outcomes. (Also see Binmore 1994, 1997, on social contract.) This book will report some experimental tests designed to separate competing preference and reciprocity theories of behavior in personal exchange. Although reciprocity seems to be a leader in the comparisons, I will summarize, its strength is not uniform across all tests, and much remains to be learned about the hidden recesses of meaning in human behavior and the circumstances in which cooperative or noncooperative behavior is manifest.[3] Technically, the issue can be posed as one of asking how most productively to model agent "types" by extending game theory so that types are an integral part of its *predictive* content, rather than merely imported as an ex post technical explanation of experimental results. For example, moves can signal types,

---

[2] Dissatisfied with the utilitarian approach because its predictions fail to account for the observed importance of instructions and procedures, we began investigating the reciprocity hypothesis in Hoffman et al. (1994). Mechanically, utilities can serve as intermediate placeholders for reciprocal trust, but as surface indicators serve poorly to generate new hypotheses designed to understand interactive processes. Good theory must be an engine for generating testable hypotheses, and utility theory runs out of fuel quickly. Utility values are seen as providing the ultimate "given" data of economics, and traditionally, as soon as utility is invoked as an explanation of something, the conversation stops.

[3] I am reminded of a department head from Hewlett-Packard (HP) visiting our lab. I naively assumed that he would be most interested in demonstrations of some of our market experiments. Not so. He was more interested in the "trust" experiments. Why? He saw the HP management problem as one of getting teams to cooperate internally by building trust and trustworthiness while being vigorous competitors externally. Could the trust games serve as a measurement and teaching tool for helping to solve this problem? This nicely illustrates the tension in Hayek's two-worlds quote in the opening text for Part III.

and effect decision, which explains why, as Chapter 12 demonstrates, game form (extensive versus normal from) matters, and why opportunity costs – payoffs available but not chosen – can impact observed outcomes. These considerations must be part of the internal structure of the theory such that outcomes become predictions conditional on the elementary characteristics of players who read each other's intentions. If such a program were successful, many of the basic results in game theory would become irrelevant or special cases of the extended theory. (In this regard, see Camerer et al., 2004, and their review of game-theoretic approaches.)

What is not methodologically satisfactory is to claim that any particular behavior – cooperative, positive reciprocity, punishment of defection, etc. – is consistent with game theory given that particular player type. Given to whom? From whence do such types come? What sustains them in any population and across successive generations? What are the observable defining characteristics of an agent type that enables us to predict the type's behavior in different game contexts? Game-theoretic strategic optimization mechanics can state what is optimal given the types and conditions, but such mechanics leaves all the action to be explained by external models of types and conditions, and its predictive content is sterile.

In this work, I will interpret reciprocity as the basis of personal social exchange and offer the hypothesis that it enabled the earliest specialization, and that barter, trade, and market exchange with strangers evolved from reciprocity. But reciprocity had a much broader function in tribal hunter-gatherer societies: Reciprocity was a gift exchange across time in which, particularly in ceremonial contexts – potlatch, kula, moka, abutu – it was essential to respond to a gift with a perceived greater gift. Reciprocity in this sense was at the foundation of more comprehensive alliance building between stateless societies. It was the means by which groups built pacts of mutual defense, provided aid during natural disasters, granted unmolested reciprocal access to trade routes, and established an environment to facilitate private trades – in short, gift exchange was the cement of reciprocal social constructions that enabled the provision of essential public goods (Dalton, 1979).

In market experiments – where cooperation can occur through the coordination function of prices produced by, but simultaneously resulting from, interaction with individual choice behavior – the results are more commonly in accord with standard competitive models that maximize group welfare. This professional victory is severely hollowed, however, by the failure of standard theory to predict the "surprisingly"[4] weak conditions under which the results obtain. Standard theory has also had limited success in modeling or explaining the emergence of the institutions that we have copied from the world of practice into the laboratory.

I want to acknowledge correspondence with Charles Plott and add his comment: "Although this is a giant victory for the economic theory of markets it simultaneously demonstrates that the theory is incomplete. The unexpectedly weak conditions under which the results obtain are good news for market performance, but not such good news for the scientific community because it demonstrates that we do not understand why markets work as they do." You do not have to have large numbers of agents, each an insignificant part of the whole – a few buyers and as many sellers (see Figure 4.1, with one active seller) are entirely adequate in a wide range of economic environments. They do not have to have complete or perfect or common information – each can have only private information – nor is it required that individuals make decisions systematically or be economically sophisticated. (Also see Selten, 1973; Huck et al., 2004.)

Many economists are either baffled that equilibrium theory works in these private information environments, or continue to think, speak, and write as if nothing had changed, that equilibrium theory requires complete or perfect information; others avoid confronting the question of what assumptions underlie the operational content of equilibrium theory. Consequently, the discoveries have yet to galvanize many programs of theory modification as applied to market equilibration processes.

Thus, for tractability, Cartesian rationalism provisionally assumes or "requires" agents to possess complete payoff and other information – far more than could ever be given to one mind. In economic analysis, the resulting exercises are believed to sharpen economic thinking, as "if-then" parables. Yet, these assumptions about the economic environment are unlikely to approximate the level of ignorance that has conditioned either individual behavior or our evolved institutions as abstract norms or rules independent of particular parameters that have survived as part of the world of experience. The temptation is to ignore this reality because it is poorly understood and does not yield to our familiar but grossly inadequate static modeling tools,

---

[4] Wilson (1992, p. 256) discusses an efficiency theorem, and suggests that the phenomenon is "perhaps unsurprising." It is nowadays, thanks to replication, but few have changed their thinking. Theory has lagged well behind the evidence, and yields inadequate testable insight into the process dynamics operating in different institutions.

and to proceed in the implicit belief that our parables capture what is most essential about what we observe. Having sharpened our understanding on Cartesian complete information parables, we carry these tools into the world for application without all the necessary caveats that reflect the tractability constraints imposed by our bounded professional cognitive capacities as theorists.

Throughout this book, I will use the term "environment" to mean the collection of agent characteristics; in the neoclassical model, these consist of agent values (preferences), endowments, and knowledge that define the gains from trade. The term "institution" will refer to the language (messages), rules of message exchange and contract in a market, or other group decision process. Finally, "behavior" refers to agent message choice given the environment and conditional on the institution (Smith, 1976, 1982a).[5] If X refers to agent outcomes, and M to agent messages, then the institution, I, consists of rules that define a mapping of messages into outcomes, $I: M \longrightarrow X$. Agents choose messages given their environment, E, conditional on the rules, I, and these choices constitute agent behavior, $B(E \mid I) \longrightarrow M$. In this characterization, think of institutions as algorithms whose property right rules define outcomes, given messages, whereas agent behavior is represented by decision algorithms for choosing messages, given the agent's environment and the institution. The roots of this schema were first articulated in the "decentralization" literature by Reiter (1959/1981) and Hurwicz (1960), who was much influenced by Hayek, and in the auction literature by Vickrey (1961); also see Smith (1982a) for a summary based on the experimental literature. This schema, in combination with laboratory investigations, created a whole new way of thinking about market and nonmarket economic systems whose full impact on professional economics is still being worked out.

In summary, constructivism uses reason deliberately to create rules of action, and to design human socioeconomic institutions that yield outcomes deemed preferable, given particular circumstances, to those produced by alternative arrangements. Although constructivism is one of the crowning achievements of the human intellect, it is important to remain sensitive to the fact that human institutions and most decision making are not guided only

---

[5] Many of the references herein to my own and coauthored works have been reprinted in Smith (1991, 2000).

or primarily by constructivism. Emergent arrangements, even if initially constructivist in form, must have fitness properties that take account of opportunity costs and environmental challenges invisible to our modeling efforts: What *is* depends vitally on what is *not*.

Limitations and Distractions of Constructivist Rationality

Since our theories and thought processes about social systems involve the conscious and deliberate use of reason, it is necessary to remind ourselves constantly that human activity is diffused and dominated by unconscious, autonomic, neuropsychological systems that enable people to function effectively without always calling upon the brain's scarcest resource: attention and self-aware reasoning circuitry. This is an important economizing property of how the brain works. If it were otherwise, no one could get through the day under the burden of the self-conscious monitoring and planning of every trivial action in detail. Thus, "If we stopped doing everything for which we do not know the reason, or for which we cannot provide a justification . . . we would probably soon be dead" (Hayek, 1988, p. 68).[6] Hayek has also emphasized the truth that no one can express in thoughts, let alone words, all that he or she knows, and does not know but might call upon or need to discover, for some purposive action. An individual's personal preferences are not instantly available but require recall or discovery in relevant contexts. Imagine the strain on the brain's resources if at the supermarket a shopper were required to evaluate his or her preferences explicitly for every combination of the tens of thousands of grocery items that are feasible for a given budget. Such mental processes are enormously opportunity-costly, and implicitly our brain knows – even if our conscious, planning, modeling mind does not know – that we must avoid incurring opportunity costs that are not worth the benefit. Rather, the challenge of any unfamiliar action or problem appears first to trigger a search by the brain to bring to the conscious mind that which is related to the decision context – the mind Googles the brain for context-relevant input. Context triggers autobiographic experiential memory, which explains why context surfaces as a nontrivial treatment, particularly in small-group experiments; it also explains why, over and over again, it has been shown that social context treatments effect decisions in one-shot games that allegedly control carefully for repeat interaction effects (see Chapters 10 through 12).

---

[6] See Rizzello (1999) for a treatment of the literature beyond Hayek's contribution to the workings of the mind in economic perception and decision.

The brain (including the entire neurophysiological system) takes over directly in the case of familiar, mastered tasks, and plays the equivalent of lightning chess when the "expert" trades, plays Beethoven's *Fifth Piano Concerto*, or connects with a ninety-five-mile-per-hour fastball – all without self-aware "thinking" and real-time management by the mind. Not part of our experience and awareness are the selective fitness processes of the past that have enabled the emergence of this enormous range of human competence: None of us living today has a single ancestor who did not survive to reach adult reproductive age. None died of a viral or bacterial disease; an infection; an inattentive hunting, climbing, or fishing accident; a social pathological deficiency; a defect of the senses; a critical natural language deficiency; and so on. Yet that long coevolutionary chain of cultural and biological adaptive success accounts for our current range of physical, mental, and cultural competence. But the vast majority of the thousands of contemporaries of our ancestors failed in some important aspect where our ancestors did not. The genes but also the cultural characteristics and institutions that survive are the ones that have created biological or cultural descendants who become "ancestors" who live and transmit this information (see Pinker, 1994, 2002).

We fail utterly to possess natural mechanisms for reminding ourselves of the brain's offline activities and accomplishments. This important proposition has led Gazzaniga (1998) to ask why the brain fools the mind into believing it is in control:

By the time we think we know something – it is part of our conscious experience – the brain has already done its work. It is old news to the brain, but fresh to "us." Systems built into the brain do their work automatically and largely outside of our conscious awareness. The brain finishes the work half a second before the information it processes reaches our consciousness. . . . We are clueless about how all this works and gets effected. We don't plan or articulate these actions. We simply observe the output. . . . The brain begins to cover for this "done deal" aspect of its functioning by creating in us the illusion that the events we are experiencing are happening in real time – not before our conscious experience of deciding to do something (pp. 63–4).

As stated by Libet (2004), whose pioneering contributions to the study of brain function in conscious mental activity dates from the 1960s:

. . . experimental evidence does show that certain stimuli in the sensory pathway, even when inadequate to produce any conscious experience, can nevertheless be usefully detected by the human subject. The important inference is, then, that neural activities inadequate to produce any subjective awareness can nevertheless help to mediate functions without awareness. Indeed much of our brain activities are of that nature (p. 28).

And to Hayek, who had a thorough grasp of these propositions, but without the advantage afforded by recent neuroscience understanding, what was the "fatal conceit"? "The idea that the ability to acquire skills stems from reason." The constructivist mind makes a fatal "error," blinding itself to understanding, as we are warned, "one should never suppose that our reason is in the higher critical position and that only those moral rules are valid that reason endorses" (Hayek, 1988, p. 21). But the anthropocentric mind routinely repeats this significant error.

That the brain is capable of off-line subconscious learning is shown by experiments with amnesiacs who are taught a new task. They learn to perform well, but memory of having learned the task escapes them (Knowlton et al., 1996). Also, there are numerous anecdotal reports of people seeking a solution to a problem who abandon it unsolved, but awaken in the morning with a solution. Such reports involve famous composers, poets, and many scientists (Mazzarello, 2000). For example, Dmitri Mendeleyev reported that the critical rule underlying his periodic table of the elements came in a dream after unsuccessful efforts to gain insight while awake. Poincaré (1913) reports having given up on solving a particularly difficult mathematical problem only to have the solution suddenly appear to him on a trip to Lyon.

I should note that the other side of the phenomenon of subconscious problem solving is the brain's ability to shut out all sense of self-awareness in the process of carrying out a difficult task – the sense of being "lost in thought." This has been documented in new brain-imaging studies of those regions of the brain that are involved in introspection and sensory perception; these functions, although well connected, are segregated through automatic switching mechanisms that allow the brain to concentrate its resources on the problem at hand (Goldberg et al., 2006).

It is known from many studies that neuronal patterns associated with task stimulus responses when awake are reactivated subsequently during sleep (Maquet, 2001). It's like the brain is "practicing" while asleep. Sleep appears to have a consolidating effect on memory, but may also restructure previous representations and yield insight. Thus Wagner et al. (2004) provide evidence that sleep has a facilitating role in subjects' discovery of a hidden abstract rule that abruptly improved their response speed in a mental task using two simple, explicit rules. More than twice as many subjects experienced this "insight" following sleep as after wakefulness, independent of time of day.

Libet (2004) reports that, "For many years, I have been keeping a pad and pencil by my bedside. When I wake during the night with a novel idea, I make notes for possible daytime action. A number of interesting solutions and explanations for research problems have appeared from that source" (p. 97).

New work in neuroscience has focused on the hypothesis that sleep strengthens procedural learning – how to do things, such as the ability to solve logical problems, and improve our musical or dramatic skills, rather than simply acquire knowledge of facts: the difference between knowing "how" and knowing "that." Researchers have long studied rapid eye movement (REM) sleep, which occurs periodically while we sleep and coincides with episodes of dreaming, because it seemed to be the proactive period in sleep. But studies attempting to link REM sleep with procedural learning have led to many contradictory findings. Recent studies of "slow-wave sleep" (non-REM periods) appear more promising in linking it to task learning, including different stages of slow-wave sleep with specific categories of tasks. (For a summary and references, see Nelson, 2004.)

A long history of studies identifies neural activity in the hippocampus with navigation learning in rats (O'Keefe and Nadel, 1978). The current view is that, "Memories develop in several stages. After the initial encoding of new information during learning, memories are consolidated 'off-line,' seemingly while not being actively thought about through a cascade of events that is not well understood. In humans and other mammals, such an enhancement of recent memory may occur during sleep" (Colgin and Moser, 2006, p. 615). What happens during slow-wave sleep in rats is a reactivation of the same encoded, temporally sequenced memories from earlier maze-running experience. In a new study of navigation learning in rats, Foster and Wilson (2006) show evidence of memory consolidation during rest periods while awake. Moreover, the sequences have a unique property: The temporal order of the experience is reversed – like "rewinding the tape of experience" – during rest periods, so that during rest the animal is recalling its most recent experience first, a phenomenon that is more readily observed in new environments than in familiar ones.

Most of our operating knowledge we do not remember learning. Hayek (1967) notes that "... modern English usage does not permit generally to employ the verb 'can' (in the sense of the German *können*) to describe all those instances in which an individual merely 'knows how' to do a thing... [including] ... the capacity to act according to rules which we may be able to discover but which we need not be able to state in order to obey them" (p. 44). Natural language is the most prominent example of unaware learning, but also music and virtually everything that constitutes our developmental socialization. We learn the rules of a language and of efficient social intercourse without explicit instruction, simply by exposure to family and extended-family social networks (Kagan and Lamb, 1987; Fiske, 1991; Kagan, 1994; Pinker, 1994). For a penetrating examination of the "cognitive and cultural conditions that make ... possible" the incredible range of human achievement based on "know-how," see Nelson and Nelson (2002); see also Polanyi (1962), particularly chapter 4.

Perhaps most of our individual mental activities and accomplishments are not accessible to our conscious awareness; all of them depend on mental processes that are inaccessible to that awareness. Similarly, people are not aware of a great range of socioeconomic phenomena, such as the productivity of social exchange systems and the external order of markets that underlie the creation of social and economic wealth.

Ecological Rationality

These considerations lead to the second concept of a rational social order: an ecological system, designed by no one mind, that emerges out of cultural and biological evolutionary processes[7] – home-grown principles of action, norms, traditions, and "morality." Paraphrasing Gigerenzer et al. ("A heuristic is ecologically rational to the degree that it is adapted to the structure of an environment" [1999, p. 13]), ecological rationality as it applies across the spectrum that I examine here can be defined as follows: The behavior of an individual, a market, an institution, or other social system involving collectives of individuals is ecologically rational to the degree that it is adapted to the structure of its environment.

---

[7] Many recognize that evolutionary processes are necessarily coevolutionary, but there is also deep denial of this, and bias that all is due to "culture" (which is even more poorly understood than biology), leading Pinker (2002) to investigate why such biases are so firmly entrenched. That these processes are coevolutionary is evident in the study of twins (Segal, 1999). Deconstructive reports argue that twin studies exhibit many of the usual data and statistical identification problems (Goldberger, 1979), but the need is for positive revisionist analysis. Data are always deficient, whether they tend to support heritable or environmental effects on observed characteristics. A prevailing view in neuroscience has been that a heritable abstract function can become dormant, atrophied, or malfunctional in the absence of initializing and maintenance input on a developmental time schedule for the brain's vision, language, and socialization circuitry, but in visual development this view has been recently challenged by the partial recovery of vision in a twenty-nine-year-old man in New Delhi, who continues to improve although blind from birth in a rare case of congenital aphakia in which the eyeball develops without a lens (Mandeville, 2006, pp. 271–2).

A significant part of the adaptation of individuals and their mental equipment includes the social structures that emerge from cultural and biological change. As interpreted by Hayek (1973), this concept of rationality "leads to the insight that there are limitations to what we can deliberately bring about, and...that that orderliness of society which greatly increased the effectiveness of individual action was...largely due to a process...in which practices...were preserved because they enabled the group in which they had arisen to prevail over others" (pp. 8–9).

Constructivism, however, can use reason in the form of rational reconstruction: to examine the behavior of individuals based on their experience and folk knowledge, who are "naïve" in their ability to apply constructivist tools to the decisions they make; to understand the emergent order in human cultures; to discover the possible intelligence embodied in the rules, norms, and institutions of our cultural and biological heritage that are created from human interactions but not by deliberate human design. People follow rules – morality, to David Hume – without being able to articulate them, but they may nevertheless be discoverable. This is the intellectual heritage of the Scottish philosophers and Hayek, who described and interpreted the social and economic order they observed and its ability to achieve desirable outcomes. The experimental laboratory provides a tool for testing hypotheses derived from models of emergent order.

"Morality" refers to any maxim of cohesive social behavior that survives the test of time, and is prominently represented by the great "shalt not" prohibitions of the leading world religions. Thou shalt not (1) steal, (2) covet the possessions of others, (3) commit murder, (4) commit adultery, or (5) bear false witness. The first two define and defend human property rights in the product of one's labor, and all resources accumulated by such labor, enabling the emergence of the extended order of mind through markets. The last three commandments protect the sanctity of social exchange – the external order of the mind. These modest exclusionary constraints leave an immense scope for freedom within their bounds. Corollaries, such as the Buddhist live-and-let-live version of the Golden Rule, are explicit in this respect: "Do not unto others as you would have them not do unto you."

Hume, an eighteenth-century precursor of Herbert Simon, was concerned with the limits of reason and the bounds on human understanding, and with scaling back the exaggerated claims and pretensions of Cartesian constructivism. Both Hume and Smith argued that the order in life and society follows from emergent norms and learning born of experience far more than the constructivist designs of reason. To Hume, rationality was phenomena that reason discovers in human institutions and practices. Thus, "the rules of morality . . . are not conclusions of our reason" (Hume, 1739/1985, p. 509). Smith developed the idea of emergent order for economics. Truth is discovered in the form of the intelligence embodied in rules and traditions that have formed, inscrutably, out of the ancient history of human social interactions. This is the antithesis of the anthropocentric belief that if an observed social mechanism is functional, somebody, somewhere, somehow in the unrecorded past surely must have used reason consciously to create it to serve its perceived intended purposes. This is the default folk belief in the historic origins of any legacy that is functional. But in cultural and bio-logical coevolution, order arises from mechanisms for generating variation to which is applied mechanisms for selection. Reason is good at providing variation, but poor at selection; that is, constructivism is a powerful engine for generating variation, but it is far too narrowly limited and inflexible in its ability to comprehend and apply all the relevant facts in order to serve the process of selection, which is better left to ecological processes that implicitly weights more versus less important influences.[8]

As I have already noted, one of the great achievements of human reason as applied to logic and mathematics is the distinction drawn between necessary and sufficient conditions. Yet this distinction is not part of our intuition about the causes or origins of human social phenomena. Consider any social order, institution, norm, or practice, **P**: natural language, money, reciprocity, trade, a market, an auction, and so on. In *some* examples, one can imagine that some one mind or group might have deliberately invented it. Thus **I** (a deliberate invention process) implies **P**, or **P** is necessary for **I**, that is, if **I**, then **P**. But what does not follow from this statement is that the observation of **P** is sufficient for **I**, that **P** implies **I**, that is, if **P**, then **I**. Humans (even mathematicians and logicians), when thinking casually about observed social phenomena, commit the error of converting "**I** implies **P**" into the belief that "**P** implies **I**." Such beliefs become a rigid view of the world, and we miss opportunities for deeper understanding of the social mind.[9]

In experimental economics, the eighteenth-century Scottish tradition is revealed in the observation of emergent order in numerous studies of

---

[8] I am indebted to Todd Zywicki for the suggestion that reason is good at providing variation but not selection.

[9] The famous quotation sometimes attributed to Frédéric Bastiat that "If goods don't cross borders, soldiers will," provides a somewhat different example brought recently to my attention by Mike Intrilligator. He noted in correspondence that the two greatest wars of the twentieth century, the two World Wars, involved combatants that had extensive trade

existing market institutions such as the continuous double auction (CDA). To paraphrase Smith, people in these experiments are led to promote group welfare, enhancing social ends that are not part of their intention. This principle is supported by hundreds of experiments whose environments and institutions (sealed bid, posted offer, and others besides CDA) may exceed the capacity of formal game-theoretic analysis to articulate predictive models. But they do not exceed the functional capacity of collectives of incompletely informed human decision makers, whose autonomous mental algorithms coordinate behavior through the rules of the institution – social algorithms – to generate high levels of measured performance.

What experimentalists have (quite unintentionally) brought to the table is a methodology for objectively testing the Scottish-Hayekian hypotheses under stronger controls than are otherwise available. This answers the question that Milton Friedman is said to have raised concerning the validity of Hayek's theory/reasoning: "How would you know?"[10] Hayek skated on the very edge of recognizing what experiments could do to test his proposition that prices in the market system serve to coordinate an equilibrium among agents with dispersed private information, then dismissed this potential insight (Hayek, 1978; 1984, p. 255; see Smith, 2002; Chapter 13 in this book). Economic historian Douglass North (1990) and political economist Elinor Ostrom (1982, 1990) have long explored the intelligence and efficacy embodied in emergent socioeconomic institutions that solve, or fail to solve, problems of growth and resource management. They study "natural" ecological experiments, from which we have learned immeasurably, and have encouraged important new experiments linking laboratory with field research. Acknowledging and investigating the workings of unseen processes are essential to the growth of our understanding of social phenomena, and enable us to probe beyond the anthropocentric limitations of constructivism.

The distinction between ecological and constructivist rationality is in principle related to Simon's distinction between subjective and objective rationality, procedural and substantive rationality, and between people making "good-enough" satisfactory decisions and making optimal decisions

---

relations. Intrilligator also noted that Norman Angell predicted an end to all war just before the start of World War I given the extensive trade among Britain, France, and Germany – one of the greatest prediction failures in history. Notice that this prediction failure is an example that negates a different proposition – namely, that if goods cross borders, soldiers will not.

[10] I am unable to find and provide a citation.

(Simon, 1955, 1956). Simon's conceptual distinctions do not rule out the discovery that the procedural rationality and "satisficing" moves of subjects can converge over time to the substantively rational competitive outcome of equilibrium theory. (For a formal example, see Lucas, 1986.) This was the significant contribution emerging from experimental economics, and it is clear from the title text quotation for this book that Simon was well aware that there were natural processes that yielded emergent social systems as well structured as those that produce a snowflake. Although this plainly is not the Simon who is remembered (see the essays in Augier and March, 2004), it was part of his conceptual framework. Simon's concern for the adaptation of decision making to the environment certainly did not rule out institutional change as part of the adaptation.

Both kinds of rationality have influenced the design and interpretation of experiments in economics. Thus, if people in certain contexts make choices that contradict our formal theory of rationality, rather than conclude that they are irrational, some ask why we should not instead reexamine maintained hypotheses, including all aspects of the experiments (procedures, payoffs, context, instructions, and so on) and inquire as to what new concepts and experimental designs can help us to better understand the behavior. What is the subject's perception of the problem that he or she is trying to solve? This is the legacy of Simon that is prominent in the methodology of experimental economics, the pioneering early work of the psychologist Sidney Siegel,[11] and today in the many contributions of Gerd Gigerenzer and his coworkers (Gigerenzer et al., 1999).

Finally, understanding decision making requires knowledge beyond the traditional bounds of economics,[12] a challenge to which Hume and Smith were not strangers. Thus, "an economist who is only an economist cannot be a good economist" (Hayek, 1956). This is manifest in the recent studies of the neural correlates of strategic interaction (now known as neuroeconomics) using functional magnetic imaging (fMRI) and other brain-imaging technologies. That research explores the neural correlates of intentions or "mind reading" and other hypotheses about information, choice, and one's own payoff versus others in determining interactive behavior.

---

[11] For a review of some of this early history see Smith (1992).

[12] I importune students to read narrowly within economics, but widely in science. Within economics there is essentially only one model to be adapted to every application: optimization subject to constraints due to resource limitations, institutional rules, and/or the behavior of others, as in a Cournot-Nash equilibrium. The economic literature is not the best place to find new inspiration beyond these traditional technical methods of modeling.

Hume and Smith were part of a broader "Scottish enlightenment" centered in Edinburgh that blossomed in the period from 1745 to 1789. It embraced philosophy, political and social systems, economics, psychology and the mind, science, and poetry and literature (Robert Burns and others). It was Smith who wrote an eighteenth-century "History of Astronomy" widely cited in the philosophy and history of science literature. James Hutton did for geology what Smith did for economics: His *The Theory of the Earth* was based on his study of rock formations and fossils in Scotland. The theme of evolutionary, unplanned, rule-governed natural order permeated the Enlightenment. Thus, Hutton wrote, "When we trace the parts of which this terrestrial system is composed ... the whole presents a machine of a peculiar construction by which it is adapted to a certain end. .... We are thus led to see a circulation in the matter of this globe, and a system of beautiful economy in the works of nature. This earth, like the body of an animal, is wasted at the same time that it is repaired ... " (quoted in Buchan, 2004, pp. 298–9). And as Adam Ferguson (friend to both Adam Smith and Hutton) wrote in *Essays on Civil Society*, "Every step and every movement of the multitude, even in what are termed enlightened ages, are made with equal blindness to the future; and nations stumble upon establishments, which are indeed the result of human action but not the execution of any human design ... " (quoted in ibid., p. 223). China's Great Leap Forward was a constructivist plan, blind to the accumulated intelligence and efficacy of ancient Chinese cultural traditions, that led to an economic and social disaster, now in the process of agonizing reappraisal and enterprising change driven by global trade.

Implications

This chapter has developed the main themes of contrast between two kinds of rationality – constructivist and ecological – and the potential for both experimental discoveries and field observations to provide a window on forms of human competence that are not part of our traditional constructivist modeling. These themes will be illustrated and discussed in a wide variety of examples drawn from economics, law, experimental economics, psychology, and the methodology of science. I will turn first in Chapters 3 through 8 to impersonal exchange through markets, drawing on learning from experiments and field observations to illustrate how the two kinds of rationality inform thematic learning from observation. Then I will examine personal exchange in Chapters 9 through 12, particularly in the context of two-person extensive form games, asking why constructivist models are of limited success in predicting behavior in single-play games, even when subjects are anonymously matched in accordance with the game-theoretic requirement that the actors be "strangers." Here I offer an exchange interpretation of behavior linking Chapters 9 through 12 to Chapters 3 through 8. In Chapter 13 I also address method in experimental science and its rationality, and in Chapter 14 I briefly discuss the unfolding possibilities for neuroscience to deepen and enlighten some of these issues.

PART TWO

IMPERSONAL EXCHANGE: THE EXTENDED ORDER OF THE MARKET

It appears more natural to regard action in a market ... as being action under low rather than high information conditions.... If each player knows only his own costs ... then neither non-cooperative game theory nor the theory of competition give conditions which are satisfactory in general to describe the attainment of an equilibrium.

Shubik (1959, p. 172)

... the aim of a skilful performance is achieved by the observance of a set of rules which are not known as such to the person following them.

Polanyi (1962, p. 49)

THREE

Relating the Two Concepts of a Rational Order

It is because every individual knows so little ... and we rarely know which of us knows best that we trust the independent and competitive efforts of many to induce the emergence of what we shall want when we see it.
                              Hayek (1960, p. 29)

Introduction

Constructivism takes as given the social structures generated by emergent institutions that we observe in the world and proceeds to model them formally. An example would be the Dutch auction or its alleged isomorphic equivalent, the sealed bid auction (Vickery, 1961; Milgrom and Weber, 1982). Constructivist models need not ask why or how an auction institution arose; or what were the ecological conditions that created it; or why there are so many distinct auction institutions. In some cases, it is the other way around. Thus, the revenue equivalence theorem in auction theory shows that the standard auctions generate identical expected seller revenues, which, if taken literally, leaves no modeled economic reason for choosing between them. But society chooses between them in particular applications. It is important to ask how and why, and to avoid the Cartesian error of dismissing such social developments as irrational.

More generally, using rational theory, one represents an observed socioeconomic situation with an abstract interactive game tree. Contrarily, the ecological concept of rationality asks certain prior questions: From whence came the structure captured by the tree? Why did this social practice occur, from which we can abstract a particular game, and not another practice with a different game-theoretic representation? Were there other practices and associated game trees that lacked fitness properties and were successfully invaded by what we observe? The end result of an unknown process is taken as the given to which constructivist game-theoretic tools are applied. There is a sense in which ecological systems, whether cultural or biological, must necessarily be, if they are not already, in the process of becoming rational: They serve the fitness needs of those who unintentionally created them through the history of their interactive experience. But was the adaptation formed to fit agents whose behavior is well represented by game-theoretic assumptions in contrast to any of the other forces that shaped the adaptation?

Constructivist mental models are based on assumptions about behavior, structure, and the value-knowledge environment. These assumptions might be correct, incorrect, or irrelevant, and the models may or may not lead to rational action in the sense of serving well the needs of those to whom the models are supposed to apply. The professional charge for which we, as theorists, are paid is to formulate and prove theorems. A theorem is a mapping from assumptions into testable or observable implications. The demands of tractability loom large in this exercise, and to get anything much in the way of results, it is necessary to consider both the assumptions and their implications as variables. Surely, few game theorists building on the assumption that agents always choose dominant strategies believed this to characterize the behavior of all agents in all situations, although it might be hoped that it applied to some. Hence, we see the nearly universal justification of theory as an exercise in "understanding." The temptation is to believe, however, that our "castles in the sky" (as W. "Buz" Brock would say) have direct meaning in our world of experience and to impose them where it may not be ecologically rational to do so. We need to remember that there is little guarantee that we can obtain empirically satisfactory characterizations of human social behavior with pencil and paper exercises that have not been tested and modified in the light of evidence from a large range of environmental and institutional contexts.

To understand what *is* – the surviving tip of the "can-do" knowledge iceberg – requires understanding of a great deal more that *is not*. This is because of the rich variety of alternatives that society may have tried, but that have failed. Nor is there any assurance that arrangements fit for one economic and social environment may be fit for another. In the laboratory, we not only can rationally reconstruct counterfactuals, as in a theoretical exercise in economic history, but also use experiments to test and examine their properties. Let us look at two contemporary nonexperimental examples.

Airline Route Deregulation

Airline route deregulation brought an unanticipated reorganization of the network, called the hub-and-spoke system (see, e.g., Donahue, 2002). I interpret this as an ecologically rational response, apparently anticipated by none of the constructivist arguments for deregulation, and predicted by none of the principals involved in the deregulation movement.[1] Nor could it have been uncovered, I submit, in the 1970s by opinion surveys of airline managers, nor by marketing surveys of airline customers. Unknown to both managers and customers was the subsequently revealed decision preferences of customers for frequency of daily departure and arrival times – a preference that had to be discovered through market trial-and-error experimentation. Nonstop service between secondary cities was simply not sustainable in a deregulated world of free choice. The only way simultaneously to achieve efficiency, the demand for frequency of service, and profitable load factors among secondary cities was for the flights to connect through hubs. A key factor in the equation was the condition that almost all of the total cost of a flight was avoidable fixed cost independent of number of passengers.

---

[1] Fred Smith, founder of Federal Express in 1971, did anticipate and then build a hub-and-spoke system for his highly successful and famous air cargo company. Here is how he describes his motivation and thinking, especially with respect to creating an efficient network for an air and ground supply chain:

When I came back from Vietnam in 1971, I saw that all the predictions I had made in my Yale paper in the mid-'60s had come true. Computers were replacing human functions, but the dependability of those products and the delivery systems for repairs were not up to par. I thought, Okay, now how can I fix this?

My solution was to create a delivery system that operates essentially the way a bank clearinghouse does: Put all points on a network and connect them through a central hub. If you take any individual transaction, that kind of system seems absurd – it means making at least one extra stop. But if you look at the network as a whole, it's an efficient way to create an enormous number of connections. If, for instance, you want to connect 100 markets with one another and if you do it all with direct point-to-point deliveries, it will take 100 times 99 – or 9,900 – direct deliveries. But if you go through a single clearing system, it will take at most 100 deliveries. So you're looking at a system that is about 100 times as efficient.... It comes from a mathematical science called topology. (Quoted in *Fortune* at www.fortune.com/fortune/fsb/specials/innovators/Smith.html; the same idea was subsequently employed in the airline industry.)

Here was a constructivist business model formulated by Smith, who implemented it and found financing to perform his entrepreneurial experiment. It was a success because it was discovered to be ecologically rational, exhibiting high fitness characteristics, but that was not a foregone conclusion – it had to be tested. Most such experiments in the business world are failures.

Survival in this brutal price-cutting environment, in which the last seat sold is the most profitable, does not tolerate partially filled planes (Van Boening and Wilcox, 1996; Durham et al., 2004, have reported experimental studies of posted price competition in harsh, avoidable, fixed-cost environments). Hence the hypothesis that a rational ecological equilibrium emerged to dominate repeated constructivist attempts by new business entrants and start-ups to satisfy an incompatible set of constraints provided by the microstructure of demand, traffic density, cost, profitability, and technology.

Might it have been otherwise if airport runway rights, or "slots," had been an integral part of the deregulation of airline routes, and the time-of-day spot pricing of slots had emerged to reflect hub congestion costs (Rassenti et al., 1982)? We do not know, but the effect of this hypothetical counterfactual on the viability of hub bypass could be assessed in laboratory experiments. As in all studies of what is not, the challenge is to estimate the parameters that would implement the appropriate economic environment, or to examine a range of parameters with the objective of bracketing the set of conditions that would provide varying degrees of support for hub bypass.

There was no paucity of constructivist business models that attempted to find profitable ways of providing nonstop service between secondary cities. The logic of these models seemed impeccable: You avoid the cost and delays of an intermediate stop. Thus, in the subnetwork route connecting Albuquerque, Tucson, and San Diego, service was tried and abandoned first by the major airlines and sequentially by two new start-up airlines – Sun and, subsequently, Arizona Airways – convinced that the service could be provided profitably. Sun ended in bankruptcy; later, Arizona Airways was formed to specialize in this three-city service operating smaller propeller-driven planes with lower break-even load factors, but this idea also ended in bankruptcy. Very recently, helped by growth in all three cities, Southwest Airlines (SWA) has finally managed a limited number of successful nonstop schedules as a supplement to service through hubs. This example appears to constitute a case study in how constructivist business models can provide variation, backed by a willingness to incur investment risk, and how a complex ecological environment can provide selection in a market's ongoing evolution.

Network topological efficiency, however, is yet to translate into profitable operations for the current number of competing U.S. airlines. Further consolidation and experimentation with fewer firms and/or different supply-pricing structures might or might not be in order given that the revenue of every major air passenger carrier is inadequate to cover cost. There should be no presumption that competition requires many firms in the airline industry where the current small number of majors is transparently more than adequate for ensuring prices that reflect efficient costs. The standard cost model of the firm is irrelevant. There is, of course, a profitable niche, now filled by Southwest Airlines, but this does not imply that the SWA niche can be replicated profitably by the few majors that face bankruptcy for the first, second, or third time. Since becoming bigger through consolidations has not provided the answer, it may be that larger organizations mean larger losses in this industry.

Antitrust policy is still too dominated by standard models of the simplistic marginal versus sunk fixed-cost dichotomy to deal effectively with the problem – very similar for every airline flying the same route network – that every flight incurs an avoidable fixed cost largely independent of load. In that environment, for a given price, the last passenger seat is the most profitable, and prices reflect bruising competition for that seat even with a small number of firms.

Deregulation also precipitated numerous air passenger pricing experiments and associated scheduling adjustments, as firms sought to find those arrangements that would better utilize their flight capacity. The resulting constructivist array of fare discounts that emerged – baffling both customers and agents – ultimately gave way to a simple and ecologically fit discount rule based on the Saturday night stay-over. This pricing rule provided incentives for flexible-schedule customers to depart on Saturday or return on Sunday. The rule became standard, and was easy to understand and implement. Consequently, the many empty seats on weekend flights under the old regulatory route and fare structure were much reduced and aircraft and crews were much more efficiently utilized in supporting the traffic flows that emerged.

An early pricing and promotional scheme that emerged and failed miserably was the idea of individual airlines promoting travel by giving each purchaser of a ticket a $50 or $100 voucher good for that specified cash discount on the purchase of any subsequent new ticket. A flurry of airlines started to copy this marketing ploy. Then some airline, followed soon by others, neutered it by offering to accept *all other airline vouchers* as cash discounts on their own tickets. Why bother to print vouchers for your company? Accept the vouchers of all others, and free-ride on their new voucher-endowed passengers. So the scheme unraveled as the airlines started to abandon their issuance of these promotional vouchers. But before that occurred, for a short time there emerged in busy hubs people who made a market in vouchers, since now they were effectively transferable and therefore liquid. It's called specialization, made possible by a new market. The original, seemingly well-hatched rational constructivist scheme was abandoned after a cascade of emergent adaptations; it was not robust in that economic environment because it was not ecologically rational.[2]

In economic theory, a rationally constructed model based on assumptions that enable a nice theorem to be proved may or may not be robust; hence the importance of testing it not only under the assumptions on the economic environment giving rise to the theorem, but under variations on those assumptions to test the breadth of its ecological rationality. Theorems are far more narrowly construed than either the laboratory or the world.

The California Energy Crisis

A second, and very troubling, example is the circumstances leading to the California energy crisis. As in other regions of the country and the world, deregulation was effected as a planned transition with numerous political and stakeholder compromises. In California, liberalization took the form of deregulating wholesale markets and prices while continuing to regulate retail prices at fixed hourly rates over the daily and seasonal cycles in consumption. The utilities believed that it was in their highest priority self-interest to lobby for and to negotiate an increase in these average retail rates to meet the "revenue requirements" of previous capital investments that were "stranded" (that is, were believed to be unable to recover their costs under competition). This preoccupation with the past, and with irrelevant average revenue/cost thinking by regulators and regulated alike, ill prepared both the utilities and the state for the consequences of having no contingent dynamic mechanisms for prioritizing the end-use consumption of power.

As expected by many observers, and already experienced by liberalization in other regions and countries, the traditional volatility in the marginal cost of generated electricity was immediately translated into volatile intraday and interseasonal wholesale prices. What was not expected was that a combination of low rainfall (reducing Pacific Northwest hydroelectric output), growth in demand, unseasonably hot weather, generators down on normal maintenance schedules, and so on, caused the temporary normal daily peaking of prices to be greatly accentuated, and to be much more lasting

---

[2] These early "natural occurring" pricing experiments enjoyed different longevities in their day. Today the rise of the Internet has allowed such experimentation to become continuous and at a low transactions cost.

than had occurred earlier in the Midwest and South. Events of small probability happen at about the expected frequency, and since there are many such events, the unexpected is not that unlikely. When supplies are short and demand is unresponsive to price increases, this invites gaming of price determination in the spot market, and gaming of ill-advised bureaucratic incentives/rules governing access to constrained lines and retail sale credits. These rules allowed suppliers to fake congestion relief, "buy" at retail, and sell at the much higher wholesale prices – a deregulatory right that should have been used to empower every end-use customer who desired to reduce consumption in response to a high price and thereby "profit" from the savings.

Constructivist planning failed to deregulate retail distribution. It also failed to make it possible for competing retail energy suppliers to experiment with programs allowing consumers to save money by enabling their lower-priority uses of power to be interrupted in times of supply stress or in any circumstances that cause a price increase. Interruptible deliveries are a direct substitute for both energy supply and energy reserves, and are an essential means of ensuring adequate capacity and reserves that cover all the various supply contingencies faced by the industry. They constitute the means by which end-use consumers can in effect buy low (consume less) at the fixed rate and sell high at the peak rate, and thus share some of the savings.

Because of the historical regulatory mandate that all demand must be served at a fixed price, the planning did not allow for the early introduction of demand-responsive retail prices and technologies to enable peak consumption to be reduced. Instead of mechanism design, fixed retail price "design" was implemented to generate average revenue that was supposed to cover average cost, and it failed. The regulatory thought process is as follows: The function of price is to provide revenue, and the function of revenue is to cover cost – estimate cost and revenue and set price accordingly. Everywhere, managers, customers, and regulators will tell you that this is "fair." But it is the antithesis of the market function of price: The current market price, and any seller's corresponding volume determined by demand, tells that seller the unit cost that she can afford to pay without losing money. If the price is already the best she can get and is below out-of-pocket unit cost, she may be in the wrong business. Regulators and those they regulate failed to understand this normal market principle and to apply it in the new deregulated regime. For neither management nor the regulators was it natural to think in terms of making a profit from selling less power. Yet that was precisely the route by which the California distributors could have avoided the loss of an estimated $15 billion: Every peak kilowatt-hour not sold at the average retail rate would have saved up to ten times that amount of energy cost. This continues to be the most significant characteristic of all power systems in California and elsewhere, and market designs that fail to confront this problem head-on continue to invite disaster.

Static substation technology – which had fostered for decades the fantasy that all loads can always be served – was protected from innovation by the legally franchised local wires monopoly. An entrant could not seek to win customers by offering discounts for switching from peak to off-peak consumption at the entrant's investment risk; nor could it install the required control devices on end-use appliances or simply a second watt-hour meter to monitor a night and weekend rate, or implement any of a number of other technologies between the substation and the outlet socket. This legacy – long entrenched and jealously sheltered by local franchised monopolies after deregulation – gave California dispatchers no alternative but to trap people in elevators and shut down high-end computer programming facilities at critical times of peak power shortage.

Many of us can remember when Bell Telephone would not let you install any telephone not made by Bell (Western Electric), or allow any but a Bell serviceperson access to your wires. This was a barrier to entry that fortunately was removed early in telephone deregulation. Unfortunately, in electricity, regulation has protected the right of the local distributor to bundle energy sales with the rental of the wires, blocking all competition among retail energy suppliers. Without such competition, there is no incentive to implement existing demand response technologies or invent new ones. Nor is there an incentive for competing suppliers to locate microgenerators and other small power sources between the neighborhood substation and end-use consumers. Regulation blocks entry by requiring the customers of such entrants to continue to pay their full share of the historical sunk cost of infrastructure wires that the entrant is bypassing. This is like forcing Henry Ford to pay the sunk cost of the carriage makers out of revenues from the sale of his Model T. This is not the way the world works, but it is the way that utility regulation has worked through franchised monopoly. It derives from the fundamental historical cost fallacy of regulation, going back to its inception and its political sale by the industry to state legislatures (Insul, 1915).

All power delivery systems are vulnerable to a combination of unfavorable events that will produce short supplies at peak demand. Constructivism alone, without competitive trial-and-error ecological experimentation with retail delivery technologies and consumer preferences, cannot design mechanisms that process all the distributed knowledge that individuals either possess or will discover, and that is relevant to finding an efficient mix of both demand and supply responsiveness. Designers do not have, and cannot assemble, all the needed information that has yet to be discovered; the best they can do is to try to create the conditions that will enable that discovery to occur.

It is because no one knows in advance what will work best that you have to open up retailing to the field experiment called "free entry and exit." In that trial-and-error discovery process, if an energy supply company tries a pricing program and it fails, it is the stockholders that suffer the business loss, as there is no regulatory mechanism to pass the cost of failed experiments through to the customers; if the program is successful, stockholders will earn the gains as a return on the capital they risked on the investment.

This discovery process requires the business of producing and marketing energy to be separated from the wires business. The Federal Energy Regulatory Commission (FERC) recognized this principle early and used its federal authority to separate wholesale energy production (generation) from high-voltage wires transmission. But FERC had no authority over the distribution of energy to retail customers over the local low-voltage wires network. Historically, the state regulatory regime endowed the distribution company with a franchised local monopoly of both the wires and the sale of energy. Far from having a separation of the wires from the provision of energy at the local distribution level, we had state legal protection of the monopoly right of a single company to tie its sale of energy to its rental of the wires. There are no cases (with the exception recently of Texas) of energy and wires separation at retail in the United States, although it is common in many foreign countries such as New Zealand, Australia, Chile, and the Nordic Pool. There is, however, an exception in the U.S. gas industry: The State of Georgia mandated the separation of the commodity, natural gas, from the local pipe distribution companies. The local gas pipe distributor is regulated, but retail customers can purchase their natural gas from any of some dozen retail gas supply companies.

Economic Systems Design[3]

What can we learn from experiments about how demand responsiveness could impact energy shortages as in the California crisis? Rassenti et al. (2002a) measure this impact by creating a market in which a modest

---

[3] I have never been entirely comfortable with this label because it is reminiscent of the idea that we can engineer best social arrangements, which the reader will see is emphatically not my interpretation of the concept as a trial-and-error process.

and practically achievable 16 percent of peak retail demand can be interrupted voluntarily at discount prices by competing energy providers to retail customers. In the experiments, demand cycles through four levels each "day" and is expressed in the wholesale market with two contrasting experimental treatments: (1) robot buyers who reveal all demand, whether interruptible or not, at the spot market clearing price; (2) four profit-motivated human buyers who are free to bid strategically in the wholesale market at contingent limit prices, implemented by their capacity to contractually interrupt load to many of their retail customers. In each case, bids to supply wholesale power are entered by five profit-motivated generation companies. Robot buyers followed the "must serve" rule characteristic of the local utilities: At each price, all load is dutifully satisfied. No attempt is made to interrupt demand strategically to discipline wholesale price spikes. In this passive demand treatment, prices average much above the benchmark competitive equilibrium and are very volatile, as suppliers seek to raise prices and withhold generation from the market in critical periods when they believe they can obtain a net increase in profit by such action. In the treatment with human wholesale buyers, prices approach the competitive equilibrium, and price volatility becomes minuscule. By empowering wholesale buyers, in addition to wholesale sellers, to bid strategically in their own interest, even though 84 percent of peak demand is "must serve," buyers are able to discipline sellers effectively and hold prices at competitive levels. Whenever sellers try to withhold supply at the margin, buyers are there with limit bids that trigger demand interruption at the margin, denying a higher price to these attempts at seller manipulation. Anything sellers can do to withhold quantities, buyers can neutralize.

The most widely agreed upon design failure in the California crisis was the rule preventing the distribution utilities from engaging in long-term contracts to supply power (Wilson, 2002, p. 1332). Beware this simplistic, myopic, popular explanation: It is a two-wrongs-make-a-right argument. Yes, of course, given that you were going to protect the monopoly power of the distribution utilities to tie the metering and sale of energy at a fixed regulated price to the rental of the wires, then one way to protect them temporarily from the consequent wholesale price volatility might be to encourage long-term contracts at a fixed average delivery cost that is compatible with their regulated retail price. Suppliers, however, can apply a little backward induction, and will want higher prices and/or shorter-term contracts if they anticipate shortages. Long contracts work to lower cost only to the extent that suppliers are surprised by subsequent high spot prices, but when it comes time to renegotiate expiring contracts, they will not replicate the error. The State of California discovered this when it intervened to sign long-term contracts – after the wholesale price run-up had publicized the price effects of shortages – and suppliers negotiated contracts at 6.5 cents per kilowatt hour, perhaps double the spot rates after the crisis subsided on its own. It was too late to obtain long-term contracts at the previously low available prices.

Arguments as to what went wrong in California have turned the design problem on its head; it was never feasible to deregulate the wholesale market alone, without attention to the need to deregulate the retail sector. To do otherwise is to count on enough surplus generation capacity to last through the transition until you get your act together to implement retail deregulation. High wholesale prices had already occurred in the Midwest well before the California debacle. You must (1) remove the legal power of the local wires monopoly to prevent competing energy suppliers from contracting with customers to discount off-peak energy, charge premiums for peak energy, and install the supporting control devices; (2) let this competition determine the dynamic price structure and investment required to implement it; (3) simultaneously, let financial instruments evolve to hedge whatever risk is left over as prices become less volatile. Financial instruments can hedge price volatility, not load and output volatility. Only demand-responsive interruptible loads can relieve supply stress and provide the demand-side (along with supply-side) reserves that reduce the risk of lost load. No one mind or collective can anticipate and plan the needed mix of technologies to enable the market to manage demand. Therefore, it is essential to remove all retail energy entry barriers and to allow firms to experiment through competition to discover and innovate efficient ways of organizing retail delivery systems. Claims that short-run retail demand is "notoriously inelastic" miss the structural point: How would you know how elastic demand might be if load-shedding technology is inflexible, price-insensitive, and impervious to entry? Retail energy competition and incentives to innovate have never been part of the structure.

This example illustrates the use of the laboratory in economic systems design. In these exercises, we can test-bed alternative market auction rules and the effect of transmission constraints on generator supply behavior (Backerman et al., 2000, 2001); vary the degree of market concentration or "power" in a nonconvex environment (Dentin et al., 2001); compare the effect of more or less strategic demand responsiveness (Rassenti et al.,

2002a); study network and multiple market effects also in a nonconvex environment (Olson et al., 2003); and test-bed markets to inform, but not finalize, market liberalization policy (Rassenti et al., 2002b). For a survey of many examples, see McCabe et al. (1991a); Smith (1997, pp. 118–20).

The two types of rational order are expressed in the experimental methodology developed for economic systems design. This branch of experimental economics uses the lab as a test bed to examine the performance of proposed new institutions, and modifies their rules and implementation features in the light of the test results. The proposed designs are initially constructivist, although most applications, such as the design of electricity markets or the auctions for spectrum licenses discussed in Chapter 6, are far too complicated for formal analysis (Rassenti et al., 2002a; Banks et al., 2003).

An early use of the laboratory to test-bed a proposed new market design was reported in Smith (1967). The question examined in this paper was whether the U.S. Treasury should change its sealed bid auction rules from the discriminatory procedure for filling the bids to uniform clearing price rules. Friedman (1960, pp. 63–5) had argued that the Treasury should adopt the uniform price rules, and this is the event that motivated the experimental study. In the winter of 1968–9, Henry Wallich (a well-known Treasury consultant) used these experiments to help persuade the Treasury to run some field experiments comparing the two auctions. He had long "agitated at Treasury" (as he put it) for the change. This new effort led to sixteen bond auctions by Treasury in the early 1970s comparing the two methods, and ultimately Treasury adopted the new procedures in the 1980s and 1990s, some fifty years after it had initiated the auctioning of Bills under the replaced rules. Many people were involved in finally making this change in policy possible. For a more complete report on that episode, see my memoir, now in manuscript (Smith, 2007).

When a design is modified in the light of test results, the modifications tested, modified again, retested, and so on, one is using the laboratory to effect an evolutionary adaptation as in the ecological concept of a rational order. If the final result is implemented in the field, it certainly undergoes further evolutionary change in the light of practice, and of operational forces not tested in the experiments because they were unknown or beyond laboratory technology at the time.[4] In fact, this evolutionary process is essential

---

[4] People often ask, What are the limits of laboratory investigation? I think any attempt to define such limits is very likely to be bridged by the subsequent ingenuity and creativity (the primary barriers at any one time) of some experimentalist. Twenty-five years ago, I

if institutions, as dynamic social tools, are to be adaptive and responsive to changing conditions. But how can such flexibility become part of their design? We do not know because no one can foresee what changes will be needed.

In Chapter 6, I will return to the subject of economic systems design and develop in some detail the FCC spectrum license auction design experience, and the role played by experiments in testing the FCC auction procedures that led to several combinatorial auction proposals and the Combo Clock mechanism, various forms of which have been or are being implemented in other industries in the field.

Constructivism as Rational Reconstruction of Emergent Order

As I have noted, David Hume (and Adam Smith) used reason to understand and interpret the intelligence captured in the emergent orders of economy, law, and society. That is, we can think of Hume as concerned with the use of constructivism to rationally reconstruct interpretations of the empirical observations from experience. Of course, the next step should always be to devise tests of the validity of that reconstruction – what the econometricians describe as explaining or predicting out-of-sample observations; what the experimentalists call replication or robustness with respect to variation relative to controls; and what the economic historians are doing when they look for new data that are thought to reflect developments that parallel or contradict previous investigations. The econometric exercise as it is commonly applied is severely limited by its inability to study that which is *not*, a condition that may be somewhat less restrictive for economic historians in their search for obscure data sets recording ephemeral institutions and arrangements that failed to survive down to the present. In laboratory experiments, the study of what is not, or might have been, is limited only by the imagination of the designer.[5]

The concept of that which is not is essential to all language and effective communication: The ostensive definition, "That is a pencil," has meaning only if we also know the set of things not a pencil. This is why Ludwig Wittgenstein says that it is correct for one to say, "I know what you are

---

could not have imagined being able to do the kinds of experiments that today have become routine in our laboratories. Experimentalists also include many who see no clear border separating the lab and the field.

[5] As noted in correspondence with me by A. Ortmann in 2005, "the lab frees us from having to do (only) the kind of thought experiments that Hume was restricted to."

thinking," but incoherent for one to say, "I know what I am thinking." I can feel either that I know or not know what you are thinking, but how can I not know what I am thinking? Language learning in chimps took a great if controversial leap forward when apes were taught words for things that they learned to recognize, and also were taught how to recognize things that were not the things that had been defined by ostensive reference. Then chimps began to use words to recognize novel new things; similar results are reported for birds and dogs.

Dog owners can tell you all kinds of amazing mental feats their pets can perform; recently, scientific studies are documenting them. For example, Bloom (2004) and Kaminski et al. (2004) discuss experiments with Rico, a nine-year-old border collie whose owners claim he knows the names of some two hundred objects in his huge collection of toys. Rico was tested by putting ten of his toys in a room isolated from his owners. The experimenters instructed Rico to fetch two randomly selected items at a time identified by name. In forty tests, Rico got thirty-seven correct. Moreover, the researchers then repeated the test, now putting seven of Rico's toys in the other room along with one he had never seen before. His owner then called out the unfamiliar name of the new toy. Rico correctly retrieved the new item in seven out of ten tries. Rico used language in a novel way because he could not only identify items in a set, but also items in the complement of a set of things he could identify.

Equally remarkable is the research applying social modeling theory (not classical operant conditioning) to learning to teach Alex the parrot abstract concepts of color – red, blue, and so on – as distinct from the objects that have the color (Pepperberg, 1999, pp. 20–9; Grandin and Johnson, 2005, pp. 248–51).

In our own natural spoken language acquisition as children, we learn the distinction between objects and concepts and their complements without formal instruction, but we are not aware of this "*A* versus not *A*" subtlety in language learning and its role in the brain's autonomic communication. Similarly, we are unaware of the socioeconomic forces that eliminate our best-laid plans because those forces find the plans to be unworkable. Therefore, we are not aware of the processes underlying that which has survived ecological tests of viability, and we are unaware that such selection processes are and must be part of science and social learning.

The preceding discussion of the unanticipated ecological emergence of the hub-and-spoke structure in airline competition presents an example of a more general clustering phenomenon arising from interactions between external and internal economies of scale and scope. Aircraft manufacturing in Wichita; the electronic, computer, and communications industry concentration in Silicon Valley; the biotech and pharmaceuticals industry complex in San Diego – all are examples of the interplay of different mixtures of complementarities, and economies of scale and scope. These phenomena elude predictability, but can yield to understanding by constructivist modeling after the fact, and the predictions of such models can be tested in the laboratory.

One such cluster has been widely replicated in all modern and contemporary urban communities: the retail shopping mall. The mall is a combinatorial package of competitive and complementary retail commodity and service establishments reflecting economies of scale and scope that enormously enhance competition by attracting both buyers and sellers to localities that allow efficient high-density mass contact. Moreover, the institution solves the problem of providing a mix of both private and local "public" goods in which firms share common walkways, parking, insurance, security, custodial, and other services while simultaneously greatly increasing competition – dozens of establishments vend clothing competitively in both specialty and general-service department stores. This is not a phenomenon that an economist could have explicated and predicted with the standard tools of economic theory. It emerged ecologically to solve an intellectually hard problem by the action of agents who were naïve in formal analysis. After the fact we are able to reconstruct formal models of the phenomenon as an abstract complex based on observation and conjectures about the underlying economic forces, and the evolutionary product of those features can be studied experimentally. For example, in a location choice framework, one could endow subjects with demand and product supply functions reflecting external and scope economies, and observe the emergent solution over time as the subjects discover by trial and error the profit benefits of clustering.

Arthur (1989) offers a nonstandard theoretical treatment of these and related issues; also see Shepard (1995, pp. 53–4) for a discussion of Arthur's great difficulty in getting his paper published, which illustrates not just my point concerning the original inaccessibility of the problem to economists but the professional resistance to the solution effort even after it is made. Although agglomeration economies are "standard" concepts in economics, it was not standard to use such concepts specifically to model a significant ecological phenomenon.

But the historical success of the shopping mall is under the new challenge of the dotcoms – Amazon, eBay, Overstock, and many smaller niche specializations that have been part of the invasion by new entrepreneurs and plagued by a high rate of bankruptcy. There are a few survivors, so far, that may be on their way to creating exceptional long-term value by defining new and more efficient retail supply chains to the final consumer for an unknown proportion of the existing retail product mix.

FOUR

Market Institutions and Performance

...the fact of our irremediable ignorance of most of the particular facts...
[underlying]... the processes of society is, however, the reason why most social
institutions have taken the form they actually have.

Hayek (1973, pp. 12–13)

Knowledge, Institutions, and Markets

Noncooperative or Cournot-Nash competitive equilibrium (CE) theory has
conventionally offered two specifications concerning the preconditions for
achieving a CE:

1. Agents require complete, or "perfect," information on the equations
   defining the CE, and also common knowledge; all must know that all
   know that all know... that they have this information. In this way, all
   agents have common expectations of a CE and their behavior must
   necessarily produce it.
2. Another tradition, popularly articulated in textbooks and showing,
   perhaps, more sensitivity for plausibility, has argued for the weaker
   requirement that agents need only be price takers in the market.

Hayek's comment on the assumption of complete knowledge clarifies the
methodological problem posed for economics:

Complete rationality of action in the Cartesian sense demands complete knowledge
of all the relevant facts... but the success of action in society depends on more particular facts than anyone can possibly know. And our whole civilization... must rest on
our *believing* much that we cannot *know* to be true in the Cartesian sense. ... There
exists a great temptation, as a first approximation, to begin with the assumption

that we know everything needed for full explanation and control... the argument then proceeds as if that ignorance did not matter... the fact of our irremediable ignorance of most of the particular facts... [underlying] ... the processes of society is, however, the reason why most social institutions have taken the form they actually have (Hayek, 1973, pp. 12–13).

Nash (1996) did not ignore the problem of how a group of agents might actually achieve a noncooperative equilibrium[1]:

In an unpublished section titled "Motivation and Interpretation" of his 1950 Ph.D. thesis he proposed two interpretations, with the aim of showing "how equilibrium points and solutions can be connected to observable phenomena" (Nash 1996, p. 32)... the first one was of a positive kind... the *mass action interpretation*... based upon an iterative adjustment process, in which boundedly rational players observed the strategies played by their opponents drawn randomly from a uniformly distributed population of players, and gradually learned to adjust their strategies to get higher payoffs. Nash suggested that the learning process would eventually converge to an equilibrium point, if it converged to anything at all, and remarked that in this interpretation it was unnecessary to assume that agents had full knowledge of the game structure or the ability to go through any complex reasoning process (Giocoli, 2003, pp. 24–5).

The alleged "requirement" of complete, common, or perfect information is vacuous: I know of no predictive theorem stating that when agents have such information their behavior produces a CE, and in the absence of such information their behavior fails to produce a CE. If such a theorem existed, it could help us to design the experiments that could test these dichotomous predictions. I suggest that the idea that agents need complete information is derived from introspective error: As theorists, we need complete information to calculate the CE. But this is not a theory of how information or its absence causes agent behavior to yield, or fail to yield, a CE. It is simply an unmotivated statement declaring, without evidence, that every agent is a constructivist in exactly the same sense as we are as theorists. And the claim that it is "as if" agents had complete information helps not a whit to understand the wellsprings of behavior. What is missing are models of the process whereby agents go from their initial circumstances and dispersed information, using the algorithms and public messages of the institution to update their status, and converge (or fail to converge) to the predicted equilibrium.

---

[1] Neither did Cournot, but each agent's decision based on the assumption that other agents would not change their decision in the next period was not a sustainable belief, as it was always revealed to be false and did not constitute an internally consistent process of equilibrium convergence.

The inherent difficulty in equilibrium modeling of the continuous (bid/ask) double auction (CDA) is revealed in the fact that so few have even been attempted. Wilson (1987), characteristically, has had the courage and competence to log progress. (Also see Wilson, 1993.) Friedman (1984) used an unconventional no-congestion assumption to finesse the Nash-Cournot analysis, concluding the attainment of efficiency and a final competitive clearing price. Wilson (1987) uses standard assumptions of what is common knowledge – number of buyers (sellers), each with inelastic demand (supply) for one unit, preferences linear in payoffs, no risk aversion or wealth effects, valuations jointly distributed, and agent capacity to "compute equilibrium strategies and select one equilibrium in a way that is common knowledge" (p. 411). This is an abstract as-if-all-agents-were-game-theorists constructivist model of a thought process that no game theorist would or does use when participating in a CDA. The model itself generates its own problems, such as degeneracy in the endgame when there is only one buyer and seller left who can feasibly trade – a game-theoretic problem that is not a problem for the subjects, who do not know this, and see imperfectly informed buyers and sellers still attempting to trade and thereby disciplining price. Extra marginal traders, not knowing that they are excluded from the equilibrium, provide opportunity cost endgame constraints on price. Agents need have no understanding of opportunity cost in order for their behavior to be shaped by it. Wilson (1987) implicitly recognizes such considerations and their implications in stating: "The crucial deficiencies, however, are inescapable consequences of the game-theoretic formulation" (p. 411). I believe we may be squarely up against the limitations – perhaps the dead-end ultimate consequences – of Cartesian constructivism. We have not a clue, any more than the so-called "naïve" subjects in experiments, how it is that our brains so effortlessly solve the equilibration problem in interacting with other brains though the CDA (and other) institutions. We model not the right world to capture this important experimental finding. I think new tools and ways of thinking outside the game-theoretic box are needed to better model these processes.

As a theory, the price-taking parable, (2) at the beginning of this chapter, is also a nonstarter: Who makes price if all agents take price as given? If it is the Walrasian auctioneer, why have such processes been found to be so inefficient (Bronfman et al., 1996)?

Hundreds of experiments in the past forty years (Smith, 1962, 1982a; Davis and Holt, 1993; Kagel and Roth, 1995; Plott, 1988, 2001) demonstrate that complete information is not necessary for a CE to form out of a self-ordering interaction between agent behavior and the rules of information

exchange and contract in a variety of different institutions, but most prominently in the continuous bid/ask double auction (CDA).

See Ketcham et al. (1984) for a comparison of CDA with the posted offer
(PO) retail pricing mechanism. CDA converges more rapidly and is somewhat more efficient than PO. So why does not CDA invade and displace
PO in exchange? One reason is the high cost of training every retail clerk
to be an effective negotiator for the retail firm. Institutions reflect the fine
structure of opportunity cost, and the loss of exchange efficiency in PO is
surely more than offset by the distributional productive efficiency of the
mass retailing innovation of the 1880s that led price policy to be centralized
in upper management. As I write, those policies are being modified on the
Internet, where prices can be adjusted to the opportunity cost characteristics of buyers, such as how many other Internet sites they have visited. (See
Deck and Wilson, 2006, who report that sellers who are informed as to how
many other sellers a buyer has already visited adjust prices over time so that
the average price strictly decreases with the number of other sites visited,
and is increased for buyers who have not visited any other sellers; but seller
total revenue is unchanged! Hence, individual prices reflect subject-specific
search opportunity costs.) Institutional changes in response to innovations
such as mass retailing are part of the emergence of an ecologically rational
equilibrium.

That complete information also may not be sufficient for a CE is suggested
(the samples are small) by comparisons showing that convergence is slowed
or fails under complete information in certain environments (Smith, 1976,
1980).

An important contribution by Gode and Sundar (1993; hereafter GS;
see Sundar, 2004, for an overview and for the references it contains) is to
demonstrate that an important component of the emergent order observed
in these isolated single-product market experiments derives from the institution, not merely the presumed rationality of the individuals. Efficiency
is necessarily a joint product of the rules of the institution and the behavior of agents. What Sundar and his coauthors have shown is that in the
double auction market for a single isolated commodity (we know not yet
how far it can be generalized), efficiency is high even for robot agents with
"zero" intelligence, each of whom chooses bids (asks) completely at random from all those that will not impose a loss on the agent. Thus, agents
who are *not* rational constructivist profit maximizers, and use no learning or updating algorithms, achieve most of the possible social gains from

trade using this institution. Does this example illustrate in a small way those "super-individual structures within which individuals found great opportunities... [and that] ... could take account of more factual circumstances than individuals could perceive, and in consequence... is in some respects superior to, or 'wiser' than, human reason... " (Hayek, 1988, pp. 75, 77)? We know only that the CDA has survived and is widely used throughout the world economy and that in laboratory experiments it is very robust in converging quickly to an efficient equilibrium predicted by the induced supply-and-demand conditions.

We do not know if the GS results generalize to multiple market settings as discussed in the next paragraph. Miller (2002), however, has shown that in a very elementary two-market environment – intertemporally separated markets for the same commodity – the GS results are qualified. Complex price dynamics, including "bubbles," appear, and there is loss of efficiency, although the loss is not substantial. On average the decline is apparently from around 94 to 88 percent. But whether or not the specific GS results are robust, the GS question is of prime importance: How much of the performance properties of markets are due to characteristics of the individual, and how much to those of institutions? Characteristics of both entities may be passed on to later generations by biological and cultural coevolution. What are the dynamics of these processes?

In multiple market trading in nonlinear interdependent demand environments, each individual's maximum willingness to pay for a unit of commodity A depends on the price of B, and vice versa, and in this more complex economy double auction markets also converge to the vector of CE prices and trading volumes. A two-commodity example appeared in Smith (1986), based on nonlinear demand (CES payoff function) and linear supply functions, one of many replications with both inexperienced and experienced subjects originally reported in Williams and Smith (1986); also see Williams et al. (2000). (These early results were further extended by Plott; see his report, 2001, and its references.)

In the two-commodity experiments, different numerical tables – dispersed among the undergraduate subjects – express the individual preference and cost information defining the general equilibrium solution of four nonlinear equations in two prices and two quantities. They buy and sell units of each of the two commodities in a series of trading periods. Prices and trading volume converge, after several trading periods, to the CE defined by the nonlinear equations. The subjects would not have a clue as to how to solve the equations mathematically. Hence, the experimenter applies the tools of constructivist reason to solve for the benchmark CE, but in repeat play this "solution" emerges from the spontaneous order created by the subjects trading under the rules of the double auction market institution. Numerous other experiments with many simultaneous interdependent markets show similar patterns of convergence (Plott, 1988, 2001). Clearly the double auction performance properties are robust to extensions to very complex environments.

Economists are frequently mystified that equilibrium theory works in these private information environments, or continue to pretend that nothing has changed, that equilibrium theory requires complete or perfect information and "sophisticated agents." Still others who believe passionately that "markets work" as conventionally taught have been said to claim that the experiments simply show that "water runs downhill." Consequently, the discoveries have yet to induce appropriate programs of general equilibrium theory modification that take account of the dynamic processes of convergence to Nash equilibrium.[2] In this sense it is as true today as it was in 1945:

> ... that there is something fundamentally wrong with an approach that habitually disregards ... the unavoidable imperfection of man's knowledge and the consequent need for a process by which knowledge is constantly communicated and acquired. Any approach ... with ... simultaneous equations, [that] ... starts from the assumption that people's *knowledge* corresponds to [all] the objective *facts* of the situation, systematically leaves out what is our main task to explain (Hayek, 1945).

Asymmetric information theory provided some important progress in solving the Hayek problem. This development brought a " ... search for things that I came to call signals, that would carry information persistently in equilibrium from sellers to buyers, or ... from those with more to those with less information ... " (Spence, 2002, p. 407). But that path-breaking progress was still confined to static equilibrium analysis, and followed neoclassical traditions in not explicitly examining convergence processes. Experimental studies, as we shall see, have filled that gap by examining many asymmetric information models.

---

[2] See, however, the important progress reported in Gode and Sundar (1997) who provide a model demonstrating how different aspects of the structure of the double auction rules affect allocation efficiency in single isolated markets. For experimental studies of the anatomy of the effect on convergence of dissecting the double auction rules – measurements with and without a bid-ask spread, with and without the bid/ask improvement rule, with and without an electronic queue for bids below and asks above the standing bids and asks, and with and without experienced subjects – see Smith and Williams (1982).

Also important in the early experimental discoveries was the extension in which traders carry over units purchased in one period for sale in another period, or in which the buyers do not trade directly with each other but through middlemen in spatially separated markets (Miller et al., 1977; Williams, 1979; Plott and Uhl, 1981). These conditions arise naturally in markets where buyers and sellers are temporally or geographically separated, and the gains from trade rely on the emergence of specialists who trade with the relevant parties when appropriate. In Miller et al. (1977) and Williams (1979), a subset of traders was allowed to buy speculatively in the odd periods for resale in the even periods in an economic environment with stationary supply and cyclical demand. These were the first asset market experiments since the intermediary traders held units temporarily as an asset before reselling them. The traders observed low prices in the first odd periods and higher prices in the subsequent even periods, and quickly learned to buy and hold for resale, causing the prices to converge toward the intertemporal competitive equilibrium level where the excess supply in the odd periods was equal to the excess demand in the even periods.

In the first asymmetric trading information experiments of Plott and Uhl (1981), the ultimate buyers in even periods (sellers in odd periods) are unable to observe directly the prices and trading behavior of the sellers (buyers). The middlemen buy from the sellers in one place and carry the asset over to resell to buyers in another place. They profit from arbitrage trades between producer sellers and consumer buyers by taking "advantage" of asymmetric information between the two separated groups. This "profit" is a return on the uncertainty they bear in deciding how much to buy for resale and whether they can resell at prices higher than those at which they buy. Over time, however, middlemen learn about the prices at which sellers are willing to sell, and also about the prices at which buyers are willing to buy. An open question was whether middlemen can persist in profiting from their asymmetric information or whether the cross-market competition will narrow or eliminate their profits. Specifically, in the experimental markets, four middlemen traders buy units from five sellers in the "buying season," then resell the inventory (no storage or carrying cost) of units purchased to five buyers in the "selling season." This two-stage process is then repeated for up to fifteen times. Only two experiments are reported, but the results are strong: Convergence over time to the competitive equilibrium is evident. The aggregate profits of middlemen decline to very low positive levels in one experiment, and to negative levels in the other experiment. I will return in Chapter 5 to a more extensive treatment of asymmetric information theory and experimental tests of the theory.

The Iowa Electronic Market

What evidence do we have that the laboratory efficiency properties of continuous double auction trading apply also in the field? One of the best sources of evidence, I believe, is found in the Iowa Electronic Market (IEM), developed by an innovative team of experimental economists at the University of Iowa (Forsythe et al., 1992, 1999). The IEM was begun with the 1988 presidential election and is now applied widely around the world. These markets are used to study the efficacy of futures, or "information markets," in aggregating widely dispersed information on the outcomes of political elections or any well-defined extralaboratory event, such as a change in the discount rate by the Federal Reserve or a war. The "laboratory" is the Internet. The "subjects" are all who log on and buy an initial portfolio of claims on the final event outcomes; they consist of whoever logs in, and are not any kind of representative or scientific sample, unlike a scientific pollster's sample, to which they are compared. The institution is the open-book continuous double auction.

In the IEM, traders make a market in shares representing parimutuel claims on the popular vote (or winner-take-all in an alternative contract treatment) outcome of an election, referendum, and so on. For example, the first IEM was on the outcome of the 1988 presidential election. Each person wanting to trade shares deposited a minimum sum, $35, with the IEM and received a trading account containing $10 cash for buying additional shares, and ten elemental portfolios at $2.50 each, consisting of one share of each of the candidates – George H. W. Bush, Michael Dukakis, Jesse Jackson, and "rest-of-field." Trading occurs continuously in an open-book bid-ask market for several months, and everyone knows that the market will be called (trading suspended) in November on election day, when the dividend paid on each share is equal to the candidate's fraction of the popular vote times $2.50. Hence if the final two candidates and all others receive popular vote shares (53.2 percent, 45.4 percent, 1.4 percent), these proportions (times $2.50) represent the payoff to a trader for each share held. Consequently, at any time, normalizing on $1, the price of a share (divided by $2.50) reflects the market expectation of that candidate's share of the total vote. A price, $0.43, means that the market predicts that the candidate will poll 43 percent of the vote. Other forms of contract that can be traded in some IEMs include the number of seats a party will win in the House and so on.

One IEM data set analyzed in 2000 included forty-nine markets, forty-one worldwide elections, and thirteen countries (Berg et al., 2000). Several results stand out. The closing market prices, produced by a nonrepresentative sample of traders, show lower average absolute forecasting error (1.5 percent) than the representative exit poll samples (1.9 percent). In the subset of sixteen national elections, the market outperforms the polls in nine of fifteen cases. In the course of several months preceding the election outcome, the market predictions are consistently much less volatile than the polls. Generally, larger and more active markets predict better than smaller, thinner markets. Surveys of the market traders show that their share holdings are biased in favor of the candidates they themselves prefer.

In view of this last result, why do markets ever outperform the polls? Forsythe et al. (1992) argue that it is their marginal trader hypothesis. Those who are active in price "setting" – that is, in entering limit bids or asks – are found to be less subject to this bias than those traders accepting (selling and buying "at market") the limit bids and asks. Polls record unmotivated, representative, average opinion. Markets record motivated *marginal opinion* that cannot be described as "representative." This analysis helps to provide a good mechanical, if not ultimate, understanding of how human interaction with the rules of a bid/ask CDA yield efficient predictions.

Other markets besides the IEM are known to have efficient information-aggregating properties. Thus 50 percent of the return variation in frozen concentrated orange juice futures prices are explained by local air temperature alone, although only 4.5 percent of winter days involve a threat of freezing temperatures. Those days account for two-thirds of the total winter return variance (Boudouku et al., 2004). Hence, this market focuses on predicting information that has economic consequences.

Parimutuel racetrack markets are an example where, significantly, the environment is much like the IEM: The settlements occur at a well-defined end state known to all agents, unlike stock market trading, where expectations float continuously with no clear value revelation endpoint. "The racetrack betting market is surprisingly efficient. Market odds are remarkably good estimates of winning probabilities. This implies that race track bettors have considerable expertise [sic], and that the markets should be taken seriously" (Thaler and Ziemba, 1988, p. 169). It is "surprising" primarily to those behavioral economists whose methodology is restricted to a search for deviations from the standard model. (Mullainathan and Thaler, 2001, p. 2). What is unusual here is that in racetracks they have recognized reportable evidence for efficient outcomes. For those who follow experimental economics, IEM, and similar controlled-environment market studies, efficiency is not only commonplace (but not universal), it cannot be attributed to agents with "considerable expertise." The agents are mostly

naïve in the professional economics sense, although they get repeat interaction experience, which, from the evidence, clearly gives them enough
expertise. But, as in the IEM and experimental markets, racetrack markets
are not perfect: There are inefficiencies in the "place" and "show" options
and the favorite-long-shot bias, with the latter more pronounced in the last
two races of the day. Various hypotheses have been offered to explain these
inefficiencies, but more remarkable is the claim by Thaler and Ziemba that
computer programs have been written to arbitrage (yielding returns of
some 11 percent per bet) the place, show, and long-shot inefficiencies.[3]

The IEM has other advantages besides performing at least as well as the
average poll in prediction (lower average absolute error) and providing less
volatile predictions: The market provides a continuous evaluation of the
likelihood of the event's occurrence, does not require representative samples,
and is relatively inexpensive to operate.

We have seen that the markets that emerge in experimental environments
are ecologically rational in the sense that they economize on information,
understanding, the number of agents, and consciously rational individual
mental effort and action. Can they also economize on the need for external
intervention to protect particular interests from strategic manipulation by
others if all are empowered by the trading institution to act in their individual
interests?

Strategy Proof-ness: Theory and Behavior

Preferences are private and unobservable, and institutions have to rely on the
messages reported by agents, not their true preferences. This follows from
the fact that no one mind has all the dispersed information known in the
aggregate by all those in the market. It is therefore possible in principle for an
agent to affect prices and outcomes in a market by strategically misreporting
preferences. This prospect has motivated the literature seeking strategy-
proof mechanisms: "[A]n allocation mechanism is strategy-proof if every
agent's utility-maximizing choice of what preferences to report depends
only on his own preferences and not on his expectations concerning the
preferences that other agents will report" (Satterthwaite, 1987, p. 519). This
requires each agent to have a dominant strategy to report true preferences,

---

[3] It is asserted that good profits have been accumulated on these programs – apparently (so
far) without neutralizing the arbitrage opportunities. Let the good times roll! If it is indeed
true that irrationality is such a big white hole, profits may run indefinitely, but I would not
bet on it.

and has led to impossibility theorems establishing the nonexistence of such a mechanism under certain conditions.

In view of such negative theoretical results and the narrow conditions under which solutions have been investigated, it is important to ask what people actually do in experimental environments in which the experimenter induces preferences *privately* on individual subjects. We know what is impossible, but what is *possible* in more open-ended systems than are modeled by theory? Is it possible that when all are free to choose from a large space of strategies, ecologically rational strategies emerge that will immunize against strategic manipulation? Given that information is inherently dispersed, has society evolved institutions in which forms of behavior arise that result in practical if not universal solutions to the problem of strategy proofness?

The double auction is a well-known example yielding CE in a wide range of economic environments, including small numbers. Are there other examples, and if there are, what are the strategic behavioral mechanisms that people adopt to achieve strategy proof-ness?

One example is the sealed bid-offer auction. In each contract period, the submitted bids are ordered from highest to lowest, the offers (asks) from lowest to highest, with the intersection (cross) determining the uniform clearing price and volume exchanged (McCabe et al., 1993; also see Cason and Friedman, 1993; Friedman, 1993; Wilson, 1993). Smith et al. (1982) report comparative studies of different versions of the sealed bid-offer mechanism with the continuous double auction.

Consider sealed bid-offer experiments with stationary supply and demand. Initially both buyers and sellers greatly underreveal their true individual willingness to buy or sell. In the first period, with inexperienced traders, volume is very low (10–15 percent of optimal), the market is inefficient, and all agents can see that at the initial clearing price they are leaving money on the table. In repeat play, they increase their value (cost) revelation, but mostly of units near the last period's clearing price. As volume increases and the clearing price closes in on the CE, the realized inverse demand and supply become very flat near the true clearing price, with many tied or nearly tied bids and asks that exceed the capacity of any single buyer or seller. At this steady state, and given this behavior, if anyone withholds purchases or sales, she is denied an allocation as other more competitively traded units substitute for hers. This results in a "behavioral strategy proof equilibrium," which is not due to a conscious deliberate strategy but reflects participant intrinsic unwillingness to bid much above (ask much below) the emergent expected price, which is a good enough strategy when you know nothing about the economic environment or its stability beyond recent experience with it. Such is the power of motivated, privately informed agents in trial-and-error repeat interaction who work out their own adaptation.

These experimental results make it plain that the theoretical condition for a strategy-proof equilibrium – that each agent have a dominant strategy to reveal true willingness to pay or willingness to accept for all units, and not just units near the margin – is much too strong. The preceding description from two-sided auctions, however, also shows that there is a social cost to the achievement of a strategy-proof equilibrium: Two-sided sealed bid auctions converge more slowly to the competitive equilibrium than continuous double auctions; upon converging, they may not be quite as efficient if agents occasionally attempt manipulation, and are then disciplined by reduced profit before returning to the full exchange volume.

A second example is the uniform price double auction (UPDA), a real-time continuous feedback mechanism clearing all trades at a single price in each trading period. This is a "designer market" invented by experimentalists who investigated whether they could combine the continuous information feedback advantages of the continuous double auction with the uniform price advantages (zero within-period volatility) of the sealed bid-offer auction. As we have seen, with the sealed bid-offer auction, several repeat interactions are required to reach optimality, with many trades being lost in the process. Can we accelerate the price discovery process by continuously feeding back information on the tentative state of the market, and allowing bids (asks) to be continuously updated in real time within each period?

This institution is made possible by high-speed computer and communication technology. It comes in several flavors or variations on the rules. In all versions at each time, $t \leq T =$ time market is "called" (closed), the tentative clearing price, p_t, is displayed and each agent knows privately the acceptance state of all her bids (asks). This allows bids and asks to be adjusted in real time. (See the chapter by McCabe et al., 1993, pp. 311–16, for a report of forty-nine UPDA experiments comparing these different versions with the continuous double auction.) UPDA exhibits even more underrevelation of demand and supply than the two-sided sealed auction previously discussed, but efficiency tends to be much higher, especially in the first periods, and, in one form (endogenous close, open-book, the "other side" rule with conditional time priority) exceeds slightly the efficiency of the continuous double auction for the particular environment studied.

Experiments using UPDA in this randomly fluctuating supply-and-demand environment routinely exhibit efficiencies of 95 to 100 percent,

Table 4.1. *Summary data experiment UP 43*

Trading period	Equilibrium price[a]	Observed price	Observed volume[b]	Efficiency (%)	% Surplus revealed
1	295	300	16	91	22
2	405	400	18	100	7
3	545	540	18	100	14
4	460	448	18	92	14
5	360	350	18	100	9
6	500	500	18	98	12
7	260	250	17	96	26
8	565	553	15	92	28
9	300	300	18	100	28
10	610	610	18	100	33
11	365	350	15	85	88
12	550	558	15	88	55
13	450	450	18	100	31
14	410	410	18	100	5
15	485	484	19	89	39
Mean, all periods			17.3	95	27
Standard deviation, all periods			1.3	5	21

Notes:
[a] Listed price is the midpoint of the range ($\pm$ 10) of the equilibrium set of prices, all periods.
[b] Equilibrium volume in all periods is 18 units.

sometimes with as little as 5 to 10 percent of the available surplus revealed. This is shown in Table 4.1 for summary data from UPDA experiment UP 43.

Most agents enter bids (asks) equal to or near the clearing price, as the latter are continuously displayed in real time. It is of course true, hypothetically, that if all agents reveal their true demand or supply with the exception of one intramarginal buyer or seller, then that agent can manipulate the price to his or her advantage. But this parable is irrelevant. The relevant question is: What behavior is manifest when *every agent* has the potential for manipulating the price? Without knowledge or understanding of the whole, and without design or intention, the participants use the rules at their disposal to achieve three properties observed by the experimenter: (1) high efficiency, (2) maximum individual profit given the behavior of all other agents, and (3) protection from manipulation by their protagonists. This ecologically rational observation illustrates the perceptive insight of Hayek (1988): "Rules alone can unite an extended order.... Neither all ends pursued, nor all means used, are known or need be known to anybody, in order for them to be taken account of within a spontaneous order. Such an order forms of itself ... " (pp. 19–20).

Did Gresham Have a Law?

There are many examples of experimental markets in which Cournot-Nash equilibrium theory predicts the observed behavior. Do we have contrary examples in which we observe efficient market outcomes that are not supported by equilibrium theory? Yes. Gresham's Law: Bad money drives out good. This "law," while sometimes claimed to be an observed phenomenon in countries all over the world, is not a Cournot-Nash equilibrium.

Hayek (1967) notes that Gresham's Law is not due to Gresham nor is it a "law" in the theoretical sense, and " ... as a mere empirical rule is practically worthless" (p. 318). In the 1920s, when people started using dollars and other hard currencies in substitution for the depreciating mark, the claim emerged that Gresham's Law was wrong – that it was the other way around, with good money driving out bad.

If currencies A and B are both available, A having an intrinsic worth, whereas B is worthless fiat money, then the theory predicts that A will drive out B. This is because each agent is assumed to believe other agents are rational, and will accept only A in exchange. Each agent will therefore avoid getting stuck with the inferior B by accepting only A, which becomes the dominant circulating medium of exchange, while B is "horded." This argument involves a constructivist model of how the agent thinks about the problem. How well does its prediction organize the data?

Experiments have confirmed that if both types of money are initially available, subjects use only the superior currency (an interest-bearing consol) as a medium of exchange. Hence, a constructively rational model of behavior corresponds to the observed ecology of behavior, and the two concepts of rationality coincide. But in treatments in which subjects first experience a history of using fiat money – it being the only medium of exchange available and therefore the only way to achieve gains from real product exchange – and then the consol is introduced, subjects continue trading with the fiat money, hording the interest-bearing consol (Camera et al., 2001). This is entirely rational if each agent trusts that others will accept the fiat money in exchange, and this trust generates a belief that is reinforced by experience. Under these conditions, the two concepts of rationality fail to coincide. Think of Gresham's Law as a trust-belief equilibrium in which theory alone is unable to *predict* (as distinct from ex post rationalization) when it might occur. As proved by Ledyard (1986), any Bayes-Nash equilibrium can be rationalized by some prior set of beliefs. This is a form of the well-known Duhem-Quine problem discussed in Chapter 13: If a prediction fails, is this due to failure of the research hypothesis or to failure of an auxiliary hypothesis? But the problem here is not only that posterior predictions are conditional on prior beliefs; they are conditional on all the simplifying assumptions of the probability framework, such as the strong postulate of complete knowledge of all of the elementary states of nature.

Complementing these results, another experimental study shows that when fiat money is the only currency, it will be used even under the condition that it is abandoned and replaced with new fiat money issued at the end of a finite horizon. In this study, the real economy is found to suffer some loss in efficiency relative to the use of "backed" (commodity) money, but the economy does not collapse even in short horizon treatments. Collapse in real-sector efficiency is observed only when a "government" sector prints fiat money to purchase real goods from the private sector. Moreover, additional experimental tests show that the collapse cannot be due to the resulting inflation, but to interference with the relative real-price discovery of markets when some agents are able to crowd out private real purchases with printing-press money. This is demonstrated by comparison experiments in which there are no government agents, but fiat money is inflated each period by the average rate that is observed in those experiments with government agents present (Deck et al., 2006).

Market Power and the Efficacy of Markets

In the previous chapter, I have discussed the market-power effects of cost structure and the ownership of generators, and how demand-side bidding in wholesale electricity markets can neutralize the market power of generation companies.

Many experimental studies have examined the effects of other potential forms of market power, including a large firm competing with several smaller firms, the potential for contracting practices to be anticompetitive, allegations of anticompetitive behavior in gasoline markets, predatory pricing behavior, and contestability and natural monopoly theory. Brief summaries of these examples are provided in this section.

Equilibrium with a Dominant Firm?
The dominant firm model assumes that one large firm and a competitive fringe of several small firms supply an industry. Each of the competitive firms is assumed to choose its profit maximizing output given the price charged by the dominant firm. The net demand of the dominant firm is obtained at each price it might charge by subtracting the supply of the competitive

fringe, Scf(p), from the total demand, Dt(p). Hence the dominant firm demand function net of fringe supply is $Ddf(p) = Dt(p) - Scf(p)$. Given the dominant firm's marginal costs, the price is chosen to maximize the profit from supplying the net demand it faces.

Notice that the preceding is not an equilibrium model. The predicted market price and dominant firm output depend on arbitrary simplifying assumptions about behavior – in particular, that only the large firm is aware or perceives that it faces a downward-sloping demand. All other firms are presumed to behave as price-takers, reminiscent of the theoretical myths that a CE is driven by "price-taking" behavior. One implication of this assumption is that if the fringe supply costs are rotated using the equilibrium as a pivot so as to leave the dominant firm's equilibrium price and output unchanged, this will not affect anyone's behavior.

In view of the discovery that the traditional assumptions used by economists to characterize market behavior are incorrect, there is little reason to suppose, without empirical demonstration, that the dominant firm model is likely to have good predictive power. Similar conclusions have emerged from less arbitrary equilibrium modeling of oligopoly where firms are choosing supply functions rather than price-quantity pairs, although these models also make assumptions about behavior that may fail to be accurate predictors of behavior (Klemperer and Meyer, 1989).

Rassenti and Wilson (2004) study behavior in a dominant firm environment using each of two separate institutional market rules applied to all firms, dominant and fringe. The first is the *posted-offer* treatment, common in retail markets. As Rassenti and Wilson implement it, each firm posts a price and a maximum quantity that it is willing to supply at that price, and the buyers purchase successively from the lowest price sellers first. They then are able to observe whether members of the competitive fringe behave as price-takers.

Second, they study the *sealed offer auction,* in which the firms all submit sealed-offer schedules that, when aggregated, are crossed with the actual demand to determine a uniform market clearing price. This institution is directly relevant to deregulated electricity markets, which currently implement this institution; variants of the dominant firm model have been applied to that context, that is, the model is not tested but simply applied to the situation on the assumption that the model correctly represents the situation (Borenstein and Bushnell, 1999).

Subjects were informed that the buyer was computer-programmed to buy from the lowest posted prices first, in ascending order, and to choose to purchase quantities that reveal true demand. In keeping with the information assumptions of the model, all sellers were given complete information on demand, the dominant firm was provided the fringe supply schedule, and this was known to all sellers.

The results indicate that the dominant firm model of price leadership poorly organizes behavior in both the posted-offer and sealed-offer auction markets. Even in those cases where they cannot reject the null hypothesis that the transaction prices are equal to the predicted dominant firm price, it is because the standard errors of the estimate are quite large, and it is hard to reject any point prediction hypothesis. The residual demand construct of the theory appears to be the factor most responsible for the poor performance of the model. In the sealed-offer market, the results parallel those of Rassenti et al. (2003) in electric power environments where the offer curves become pretty competitive when the incentives to reveal the cost of marginal units are altered. This is manifest when Rassenti and Wilson (2004) change the elasticity of the fringe supply. In the low-elasticity treatment, the fringe and the dominant firm reveal correspondingly less than when the elasticity is high, yielding higher prices in the latter case. The dominant firm does not base his offer curve on the residual demand implied by the model. The theoretical construct of the residual demand also fails in posted-offer pricing. The fringe suppliers do not behave as price-takers and the dominant firm does not price as if facing the model's assumed residual demand.

The Ethyl Case and Antitrust Policy

Tetraethyl lead is an antiknock compound that was used for decades as an additive to blend gasoline in the gasoline industry, but is now prohibited as an environmental air pollutant. In 1979, the Federal Trade Commission (FTC) filed a complaint against the manufacturers of tetraethyl lead alleging that four common contracting practices in the industry were anticompetitive: (1) Customers were assured of a thirty-day advance notice of price increases; (2) prices were quoted as delivered prices at the same level regardless of differing transportation cost to the delivery point; contracts guaranteed that (3) the seller will sell to no other at a price less than that quoted to the buyer, and (4) the seller will match any lower price in the market or release the buyer from the contract.

The product was sold by four firms, two of about equal size, with 70 percent of sales, and two of about equal size that satisfied the remaining 30 percent. Sixty percent of sales were to eight large buyers, and the rest to a large number of small companies.

Advance notice of price increases, item (1), and no secret price discounts, items (3) and (4) in the FTC complaint, were the key elements studied experimentally as treatments by Grether and Plott (1984). The hypothesis was that these factors enabled de facto collusion to occur by contractual arrangements nominally justified on the basis of competition and good customer relations. The "reasonable" justification of the thirty-day advance notice to customers by public announcement was to facilitate customers by preparing them for cost increases. The FTC's strategic interpretation was that it allowed a seller to signal an increase, without committing to an increase if it was not matched by the other three sellers. Since all sellers know that the leader can rescind the announced increase if other sellers do not also raise prices before the deadline, this contracting practice compromises an independent competitive response in which other suppliers undersell the price leader to increase market share. For example, if an industry price increase raises the expected profit for each firm at the higher price, it is in their common interest to match it. Decreases in price will also be matched in the unlikely case that industry demand is increased enough for this to be profitable.

Condition (3), offering to sell to no other buyer at a lower price, can be interpreted as a guarantee of "fair," nondiscriminatory treatment of all buyers. Notice, however, that this practice also provides contractual proof that the firm does not violate and cannot be in violation of Robinson-Patman Act strictures on price discrimination, under which price differences are permissible only if they are warranted by cost differentials. Might it be true that these Robinson-Patman strictures are the motivating origin of these contractual practices?

What are the strategic implications of condition (3)? All firms using this clause are credibly precommitting not to reduce their prices to anyone. Why? Very simply, it makes the cost of any such move high by requiring the price to be lowered to *all* customers. Consequently, the indirect incentive effect of the agreement is to lower the price to no one, unless it happens to be optimal to lower it to everyone.

Finally, (4), standing ready to match any lower price offered by another seller, has the feel of an attractive gift to a prospective buyer, but the overall market effect is to remove all incentive for a competing seller to make such an offer to the buyer in the first place. Without this provision in the standard agreement, a competing seller may have his offer accepted and therefore has motivation to make the offer.

Taken together, these conditions provide a potential gridlock form of market pricing in which each is contractually bound not to price-discriminate. This may look "desirable" from the perspective of the Robinson-Patman Act, but the system effect may compromise the core of the competitive process. As observed in countless experiments, the competitive process involves "price discrimination" in the out-of-equilibrium dynamics of market adjustment, as first one firm, then another, undercuts existing prices to some customers. This is what happens in unrestrained competition. My interpretation of the contracting practices (3) and (4) is that they may be anticompetitive but also that the laws against price discrimination – which serve to justify, if not cause, these contractual practices – may be unintentionally anticompetitive. The strictures against price discrimination are about unjustified price differences in *equilibrium*, but charging different prices to different buyers is a normal part of the dynamics of competition and is an integral part of the problem of discovering equilibrium prices. Equilibrium discovery and temporary disequilibrium price discrimination are not separable components of market behavior.

Grether and Plott (1984) ask whether this anticompetitive hypothesis is supported, and their experiments show that it is. Trading was organized as a negotiated contract market with subjects located in different rooms and communicating by telephone. Under the treatment conditions, the practices listed in (1), (3), and (4), were in force. Using induced supply-and-demand conditions, trading was typically started under the control conditions in which contracting was unconstrained, then repeated for several periods. This allowed prices to stabilize. Then the contracting provisions were introduced into the trading rules and continued for many more periods. The effect of advance notice of price increases combined with no secret price discounting was to raise the mean contract price.

Grether and Plott (1984) report that the price increases when these contracting features are present are statistically significant relative to unconstrained negotiated pricing. Moreover, the conjunction of all the practices was essential to this result. Experiments that used only advanced notice of price increases, or used only the disallowance of secret price discounting, did not lead to a significant increase in prices. The samples, however, are small.

Finally, although the conjunction of these practices raised prices significantly above competitive levels in the statistical sense, the Grether and Plott data clearly reject the joint profit maximizing model. Prices, though higher, were still significantly (both statistically and economically) below the monopoly level. Hence, these practices do not allow their users to substitute monopoly for competitive prices.

It is unfortunate that the Grether-Plott experiment has never been replicated with a larger sample focusing on one-at-a-time variation of the practices along with conjunction treatments. The economic environment might also be reconsidered. Thus demand should probably be modeled to reflect the fact that the lead additive is a substitute for further refining at higher cost, an element that asymmetrically limits the ability of the firms to raise prices. Getting this right would require some research in the older literature on gasoline blending. On the supply side, the four manufacturers almost certainly had decreasing unit cost to different capacities – two large and two small. An unfettered market would have been brutally competitive because the most profitable unit is the last unit sold (e.g., Coursey et al., 1984b). In that environment, would the practices have been *more* effective, or would they fail to be *as* effective? What are firms supposed to do in declining marginal cost environments where the traditional competitive model does not apply?

Guarantees of no price discrimination can be interpreted as consistent with Robinson-Patman provisions, and may even have been encouraged by them. Both the practices and the antitrust law provisions are potentially antithetical to market competition for the reason that the path to a competitive equilibrium is strewn with instances of charging different buyers different prices. To interfere with that process either by law or by contract has potential to compromise the dynamic core of the competitive process. I suspect that the culprit here is textbook static equilibrium theory, which has been used so often to justify antitrust law. One of the important contributions of experimental economics has been to enable us to better understand the sequential dynamic process whereby markets achieve competitive equilibrium states.

The hypothesis suggested by these considerations is that the static constructivist model of equilibrium price discrimination leads to policy strictures contrary to an ecologically rational discovery process of equilibrium price formation. That dynamic process naturally generates different prices for different traders as part of convergence to a competitive equilibrium outcome. If antidiscrimination price policies are not to interfere with that process, they somehow must distinguish between the dynamics and statics of market pricing. This is a large order given that we do not have well-articulated dynamic constructivist models whose veracity can be tested. This is a subject ripe for experimental inquiry. In this perspective, the Robinson-Patman Act may commit a type-two error in which inappropriate strictures aid and abet the inducement of practices consistent with those strictures, but which are antithetical to the competition that antitrust policy is trying to encourage. A type-two antitrust error is one in which an unnecessary policy fix deals with an incorrectly specified problem and the policy itself induces less competitive behavior. It is comparable to iatrogenic treatments (physician-induced illness) in the practice of medicine.

Gasoline Market Behavior and Competition Policy

Increases in the pump price of gasoline commonly generate complaints to the FTC and to other government authorities that something should be done to curb the market power of the oil companies. Two industry practices that have come under attack are zone pricing and vertical integration:

- In *"zone pricing,"* as it is called, some retail stations located in less competitive locations charge a higher price than in locations with more station rivals. Moreover, the wholesale suppliers to these stations charge a higher price for delivering gasoline to the more remote retail stations charging higher prices than to those stations charging lower prices. Some policy makers have inferred that this means that the higher retail prices are *caused* by the higher wholesale prices, and have proposed legislation that would ban zone pricing by wholesalers, requiring them to charge a uniform price to all the retail stations they serve in any particular region. The intention is to force the greater price competition in the city center onto the remote stations, so that the latter prices would fall to the lower levels at the city center. The "logic" appears to be simple and clear: If the law of one price in a competitive market fails, then you take action designed to impose it as a regulatory constraint to increase customer welfare to the more desirable competitive level. That the existence of a competitive market implies the emergence of one price (A implies B) does not mean that the imposition of a single price will yield a competitive market (that B implies A).

- *Vertical integration* of the wholesale and retail operations by companies selling different gasoline brands has been common. The structure and practices of the industry are to pipe a common pool of indistinguishable gasoline from refiners to storage tanks owned by the various major oil companies and located near the metropolitan demand areas. Here the stored gasoline is branded by mixing it with different advertised additives that differentiate the various branded products for marketing. Tank trucks owned by each company transport their brand from their storage tanks to their own retail stations. This vertical integration has been thought to be undesirable, and some states (Maryland was the first) have passed laws forcing *divorcement* of their retail stations from the refinery/wholesale operations. In such cases, the divorced stations are operated independently by lessee-dealers who set their own retail prices, while the company charges them a wholesale price. If the stations are company-operated, then the company sets the retail price and there is no separate wholesale market or price. Divorcement was intended to lower consumer prices and reallocate profits from producers to consumers.

Deck and Wilson (2004, 2007) compare the effect of a law prohibiting zone pricing by wholesalers with a baseline treatment allowing unfettered competition and with stations divorced from the refiners. The latter is further compared to a treatment in which there is no such divorcement. They use the following spatially separated gasoline service environment:

A seven-by-seven city block grid on each subject's monitor screen has four stations in the city center, each separately branded A, B, C, and D. Four remote stations are located in each of the four corners of the city, one owned by each of the brands A, B, C, and D. Buyers are robots randomly distributed across the city grid. Each brand is preferred by 20 percent of the buyers, who will pay up to 11 percent more than a common unit value for gasoline by all buyers. That leaves 20 percent of the buyers who have no brand preference and will pay no premium. Each robot buyer incurs a cost for each block driven to one of the station intersections. The further she drives, the lower the maximum willingness to pay for a tank of gasoline. Robot buyers know the price at each station and drive to the one with the lowest net cost for a tank of gasoline.

Prices are set by four retailers and four wholesalers who choose initial posted prices however they please and can adjust prices at any time. They see each other's prices, see a buyer emerge every period (1.7 seconds), and drive to the lowest-cost station net of driving cost. Each experimental session lasts 1,200 periods.

The authors conduct four independent experiments (four distinct groups of subjects) under each of the following conditions:

- *Baseline zone pricing with divorcement,* in which retail and wholesale operations are independent (divorced) and wholesalers can charge different prices to their central city and remote branded lessee-stations. The question here is whether lower city center and higher remote station prices would emerge naturally out of this spatial competition. It did, and the different station prices tended to reflect their differing location costs to the consumers.
- *Regulation enforcement of a uniform price,* that is, the law prohibits zone pricing by wholesalers. The question here is whether this rule causes the higher prices charged by the remote stations to fall to the levels prevailing in the city center to yield one uniform competitive price, using the baseline experiments as a benchmark. This did not happen. The opposite occurred. The city center prices rose while the remote

Table 4.2. *Division of surplus among consumers, service stations, and refiners*

Sector	Zone pricing, divorcement (%)	Uniform pricing, divorcement (%)	Company-owned stations (%)
Consumers	56	44	74
Stations	9	26	26
Refiners	35	30	

Notes: Based on results reported in Deck and Wilson (2004, 2007).

station prices stayed essentially at the same levels that were attained in the baseline treatment, although both time series showed random variation around mean central tendencies. Hence, the equilibrating process is one in which the pattern of price discovery at the retail level defines the spatial willingness to pay for wholesale gasoline, which in turn determines the pattern of nonuniform wholesale prices.

- *Company-owned stations, no divorcement.* Here there are no separate wholesalers, and each company freely sets the retail prices at each of its two stations. The question here is whether consumers would suffer under this treatment relative to the baseline requiring divorcement. They did not suffer. Consumers were better off when company-owned stations competed directly.

Table 4.2 lists the average percentage of the available gains from exchange – "profits," or surplus – captured by the consumer, retail station, and refiner/wholesale sectors of the industry, under the zone pricing, uniform price law, and company-owned treatments. Consumers do best (74 percent of total surplus) when stations are company-owned, worst (44 percent) under the uniform pricing law (and divorcement), and in between these extremes (56 percent) with zonal pricing and divorcement. Zone pricing favors the refiner/wholesalers, while the station owners do best with divorcement and a uniformly enforced wholesale price. The effect of the two regulations combined – station divorcement and the prohibition of zone pricing – is to inflict the greatest harm on consumers and the greatest benefit on station owners.

Predatory Pricing

Isaac and Smith (1985) report eleven experiments, aspects of whose design were motivated by the conditions claimed to give rise to predatory pricing in the antitrust literature. These conditions included (1) a large "predator" incumbent firm and a small "prey" entrant firm, (2) economies of scale on the part of the large firm, who (3) also had a "deep pocket" relative to the small firm and (4) sunk entry costs that discourage entry by the smaller prey. The experimental design gave the predatory pricing literature what Isaac and Smith and others thought was a "best shot" to predict behavior in their experiments. The idea was to introduce *all* the various features discussed by one or more of the authors in this theoretical literature. But nothing approaching the behavior reasonably thought to define and separate predatory from competitive behavior – specifically pricing below marginal cost – was observed in any of the experiments.

The Isaac and Smith research program also included a planned second phase designed to examine proposed antitrust remedies to prevent predatory pricing in industries thought to be prone to predation. Unfortunately, after searching for and failing to find predatory pricing, this implied that the second part of the author's original research plan had had its thunder stolen. They proceeded anyway, reinterpreting the second series of experiments as a test for the existence of type-two antitrust error, a version of which was previously discussed in connection with the ethyl case. Thus, will introducing antipredatory pricing rules, although perhaps unnecessary for that purpose, have unintended anticompetitive effects?

Predatory pricing theory is based on the proposition that, if the conditions postulated in the preceding section apply, then an incumbent firm may respond to a new entrant by lowering price below marginal cost, drive the new entrant out of the industry, and then raise prices to monopoly levels. If the threat to lower price is credible, thereafter no new entrant will enter. The remedy based on this theory is twofold: (1) If an incumbent firm lowers price in response to entry, then the firm is required to maintain the lower price for a specified period of time after the entrant exits (Baumol, 1979); (2) whenever the smaller firm enters the market, the incumbent firm is also subjected to an output expansion limit (Williamson, 1977). The two rules were of course each intended to nullify the alleged predatory acts of lowering prices, expanding output, then subsequently raising price and restricting output to monopoly levels after the new entrant exists. Implicitly, the Baumol model assumed that the remedy would do more good than harm. No harm was anticipated in this one-issue-at-a-time modeling exercise.

Two conclusions followed from the Isaac and Smith study:

- "Based on the results of 11 predatory pricing experiments, our principal conclusion is that, so far, the phenomenon has eluded our search" (Isaac and Smith, 1985, p. 776).

- "The antitrust regulation imposed on a market that might be thought to be susceptible to predatory pricing caused the market to perform less competitively and less efficiently than in the absence of any regulations against predation" (ibid., p. 778). Thus efficiency was over 12 percent lower when the restrictions on price and output were in effect than in their absence, and this difference was statistically significant. Although only one of the seven experiments applying the antipredation rules yielded the monopoly price, on average these experiments showed higher prices and lower efficiency than the eleven "predation" experiments that did not apply these rules. Hence, a type-two antitrust error is identified: An unnecessary remedy had the effect of facilitating collusion between the incumbent and the entrant. For other studies of predation, see Harrison (1988) and Jung et al. (1994); also see Gomez et al. (2008), who report that they were unable to replicate Harrison (1988).

But there is a more serious and fundamental problem with the predatory theory literature: its failure to take account of, or even acknowledge the impact of, predatory pricing on the market value of a prey's capital goods market. Entry can occur only with investment, whereas exit implies divestiture. Hence, if the capital stock of an existing prey is specialized, any purchaser will rationally pay the exiting firm a price for it that fully discounts the predatory proclivities of the incumbents' pricing policies. This allows the acquiring firm to enter at a capital cost that has already taken account of the expectation of predation. Profitability is thus ensured for the new entrant under continued predation, whereas supernormal profits are obtained if the incumbent abandons the practice. Consequently, a predator simply creates long-run competitive problems for herself by initiating predatory pricing.

If the capital stock is not specialized (for example, trucks, automobiles, or other multipurpose capital goods), then there is a large competitive resale market for the capital, and the exiting prey recovers his invested financial capital with minimal loss. A new entrant can also enter at relatively low sunk cost, and predation may simply invite a costly succession of hit-and-run entrants. Consequently, under these conditions the incumbent is forced to charge a price that limits entry, as in the theory of contestable markets. If this analysis is correct, it suggests that the appropriate antitrust policy is to do nothing about predatory pricing complaints.[4]

---

[4] Predatory pricing theory shares with many theories limitations inherent in partial equilibrium analysis based on backward induction. How far ahead is the consequential endpoint

There is, however, another view of "predatory pricing": It exists in fact, but it is irrational. But should antitrust policy be directed at protecting firms from alleged "irrational" behavior by competitors?

Entry deterrence games, such as Jung et al. (1994), following the spirit of Camerer and Weigelt (1988), finesse these considerations by using fixed strategic form matrix payoffs. In such experiments, the cost/value structure faced by firms is independent of capital investment, and therefore entry deterrence behavior does not have an opportunity to be discounted into the market for specialized capital to effect the relevant costs and taken into account by the protagonists. But the scientific contribution in these experiments is to elementary sequential reputation theory, not predation or "trust," as in the Camerer-Weigelt story; they consciously oversimplify both the economics of predation and of "trust." These labels should not be confused with their true purpose, and their more restricted area of contribution, which is to test game-theoretic models of reputation formation in repeat play of normal form matrix games.

Entry Cost and Competition: Contestable Markets Theory

Both the process and the consequences of airline deregulation brought a new awareness of the limitations of the neoclassical competitive model to economic theorists. The basic learning was this: The existing stock of aircraft could be reallocated with planes shifted from one route to another at relatively low cost. With low entry and exit costs, passenger rates on any route were highly competitive (the redundant new word was "contestable"), even if the numbers serving those routes were very small. "Small" includes one. Yes, a route served by only one airline might well charge little above a competitive price; otherwise, the airline would soon find itself facing other airlines operating on the same route. Hence if the route passenger demand were such that it was a "natural monopoly," with only one firm needed to serve the route at lowest cost, this implied not a monopoly price, but a price that would limit the entry of a second competitor. There is no way that any single airline thinking it had a monopoly on a particular route could charge more than the opportunity cost of a second entrant. Such a price is a competitive equilibrium price: The market clears with one firm scheduling,

---

where backward induction should begin? Ultimately this is an empirical and behavioral question and not a problem in economic theory alone. I will return to this issue in Chapter 6, where there is ambiguity in interpreting the implications of common value English auctions.

say, one flight a day to serve all those willing to pay at least the clearing price, with no firm finding it profitable to schedule a second flight.

I have already described the unanticipated emergence of the hub-and-spoke route network structure in the late 1970s and early 1980s, after deregulation. In addition, apropos to serving individual point-to-point routes, what economists were discovering in the airline industry was an ecologically rational equilibrium in which competitive prices were being observed on routes that had one or very few competitors. This contradicted the usual constructivist models of monopoly and market power. Airline managers were unintentionally finding sustainable equilibrium states point to point that were not previously described, predicted, or rationalized by the theory. The market was more sophisticated than the theory, and this motivated theory modification. The models were clearly inadequate and needed to be reexamined to bring constructivist rationality in line with ecological rationality, and this is what happened.

The observation that there exists only one firm in a market routinely led – and too casually still leads – to the informal inference that it must be a "monopoly," and that this state cannot constitute a competitive equilibrium. This is wrong; the error arises from the confusion created by neoclassical theory and its focus on the false claim that any competitive equilibrium requires large numbers of buyers and sellers actively serving any particular market rather than a focus on the opportunity costs incurred by those who are potential suppliers to the market, but do not find it in their interest to be active because of inadequate demand. This is another example of the significance of knowing what is *not* (the unobserved opportunity cost that explains why a second supplier is not active) if you are to understand what *is* (the fact that there is only one active supplier).

For a simple illustration, consider an airline route connecting a small city with a hub city. Let there be enough passengers per day to support fully one Boeing 737 flight per day with each passenger willing to pay at least the (marginal demand) value, $V1$, for a ticket. (I ignore important network effects in which what a customer is willing to pay, and what the airline charges, depends on his total trip and therefore includes any flight that he will connect with in the hub city.) I assume that the seat demand is elastic in the range up to the plane's capacity, which is implied by the universal tendency of airlines to overbook flights – otherwise they could profit by underbooking and charging a higher price. Similarly, let $V2 < V1$ be the marginal demand value for a potential second flight per day that would serve the lower priority (value) customers unwilling to pay $V1$. On the supply side, let the lowest-cost carrier able to serve that route have a flight

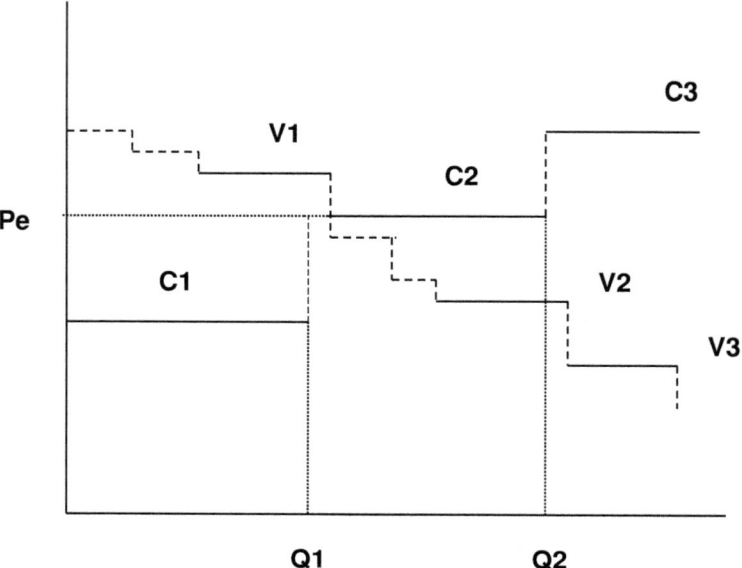

Figure 4.1. Competitive Equilibrium with One Supplier.

supply cost of C1, where V1 > C1, and let C2 ≥ C1 be the opportunity supply cost of an independent second potential supplier on that route. Figure 4.1 provides supply-and-demand representations of this market. Shown dashed are values for blocks of passengers willing to pay more than V1 (the ruling marginal value), and also some blocks willing to pay less than V1.

The unit of supply (demand) is one Boeing 737 (capacity Q1); a third unit value, V3, and opportunity supply cost, C3, are also shown, but both are well distanced from being economically active. The market-clearing equilibrium price in this market is C1 < Pe = Min (C2, V1).

The equilibrium is either the opportunity cost of the second potential seller or of the marginal buyer, whichever is smaller. Observe the following:

- If V1 > C2, the carrier charges Pe = C2, and net unit revenue from the flight is C2 – C1. Also note that if C1 is not much below C2 in Figure 4.1, then most of the surplus is obtained by the passengers, not the flight. Hence, if all airlines have comparable costs for the same equipment, profits are nil, as the surplus is competed away to the customers, even though one or a very few serve particular routes.
- If C2 > V1, Pe = V1, and net unit revenue is V1 – C1, this is the "monopoly" case in which an entrant's opportunity cost is above the

marginal buyer whose value determines the competitive price and rent of the only supplier. (Note, however, that if the seat demand is inelastic, the supplier will benefit by raising the price above V1, underbooking the flight capacity, and flying partially loaded deliberately; in this case, C2 may still limit the posted price for the flight and the rent of the supplier; also, we have some modest, if constrained, monopoly power that is a natural part of the competitive process. But given the strong observed tendency to overbook, this cannot be the prevailing situation.) Moreover, while this case may very well apply to some routes, it is not pervasive since the airlines persistently fail to make an overall profit.

- If V2 increases (due to growing willingness-to-pay demand), so that V1 > V2 > C2 < C3, then C2 < Pe ≤ Min (C3, V2).

Hence, the second carrier will enter, schedule a flight, and Pe = C3 or V2, whichever is smaller.

These cases are simply illustrations of a supply-and-demand market-clearing equilibrium under differing opportunity cost (and value) conditions: The active seller is always constrained to post a price sensitive to these conditions.

If we have a commodity or service produced only in integer units (or integer batches of units, such as airplane loads as in Figure 4.1), then for an equilibrium volume of Q units, the algorithm for the competitive price is *any* Pe in the set defined by

$$\text{Max}\,(C(Q), V(Q + 1)) \le Pe \le \text{Min}\,(C(Q + 1), V(Q)),$$

where $Q \ge 0$, making it transparent that there is nothing special about the fact that Q might happen to be zero or unity in a particular supply-and-demand equilibrium. If $Q = 0$, we have the autarky case with no positive difference between the highest demand value and lowest supply cost unit (V1 < C1): An efficient competitive equilibrium can yield zero output and consumption. Figure 4.2 illustrates one of the four cases defining the set of equilibrium clearing prices, Pe, for equilibrium volume Q.

In the standard theoretical case in which a firm produces multiple units, a single firm has incentive to withhold enough of its production to support the monopoly price, provided that its demand is inelastic and the price does not attract entry. Unregulated natural monopoly theory is therefore a story about one firm being sufficient to satisfy an inelastic demand, and its lowest cost potential competitor having an opportunity cost high enough to discourage entry at the monopoly price. Otherwise the equilibrium price is the traditional competitive price plus the premium necessary to limit

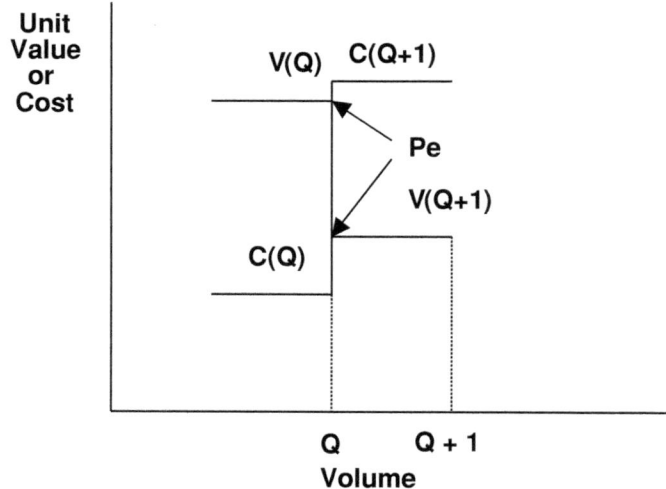

Figure 4.2. Competitive Equilibrium (Pe, Q) with $C(Q) < V(Q + 1) < Pe < V(Q) < C(Q + 1)$.

the entry of a second firm. As the (lowest) opportunity cost of a competitor approaches that of the incumbent supplier, the standard neoclassical competitive price prevails, with only one firm serving the market, but this polar case is not necessary to see the disciplinary function of entry.[5] If the incumbent's demand is elastic, no competitor with higher cost will have any motivation to enter.

My view is that the neoclassical concept of "large numbers" competition was and is fatally flawed by assuming away the opportunity cost of potential entrants (both buyers and sellers) waiting in the wings – these opportunity costs should be part of the definition of equilibrium efficiency in which firms forgo a smaller expected return in an alternative activity to enter an industry currently having only one incumbent. Pricing on a firm's marginal *internal* cost – as if it were an island isolated from the rest of the economy – is not what dynamic competition is all about. That competition is about entry capital allocation across a menu of activities. Every new product innovation by a new

---

[5] Artificial increases in entry cost result from the success of state lobbying efforts to regulate and/or raise cost of entry for businesses and professional services – beauty operators, dental technicians (whose practice by law must be combined with that of dentistry), trade licensing, new local automobile franchises, and restrictions on automobile sales over the Internet. Such regulations are usually justified in terms of quality, but this motive is confounded by the seller's desire to limit entry and maintain a higher price. On regulating automobile dealers, see Singleton (2000).

start-up firm, or an existing multiproduct firm, initiates a "monopoly," but pricing by that monopoly is constrained by substitute products (dotcom retailers must compete with the existing supply chain and price to divert sales), and hence by the opportunity cost of buyers and the opportunity cost of potential entrants. The appropriate concept of efficiency requires price to approximate *external* marginal opportunity cost. The monopoly price rules and output is restricted only in the event that the price is lower than this external opportunity cost.

Contestable market theory had already been implicit in the earlier concept of a "limit entry price," but that concept had been too restrictive to enable economists to see its full implications. The basic idea stretched back to John Bates Clark, who suggested that prices could be disciplined by potential competition – new firms that might enter help to discourage monopoly prices. But contestable market theory was a consciousness-expanding breakthrough. It explains today the "anomaly" that Northwest Airlines dominates the Minneapolis airport hub, but that this does not make NWA a monopoly and a profitable investment; TWA dominated the St Louis hub, but lost money after 1989 and was bought out by American Airlines, which dominates Fort Worth–Dallas.

Experimental economists had already had many of their traditional economic biases favoring the standard competitive model shaken to the core by what we had observed in the laboratory from 1956 to 1980. Consequently, to experimentalists, the rash of papers on contestability in markets around 1980 was like hearing an updated sermon to the choir (see Baumol et al., 1982, and the references therein). The main elements of contestability theory were eminently reasonable to this audience. But more important, the new theory provided specific environments and hypotheses that would enable experimental learning to be tested and extended in new ways that would not have been possible without this new body of principles. Conceptually, the contestability literature greatly expanded a horizon that had already grown rapidly, but it was always possible that there was an unanticipated wrinkle in the theory that would create problems. It therefore needed to be tested.

Coursey et al. (1984a, 1984b) provided the first explicit experimental tests of contestability and natural monopoly theory. The baseline consisted of four monopoly experiments in which no entry was possible. The monopolist had a declining marginal cost curve to his capacity, and demand at the capacity output was such that the demand price of the last production unit was slightly above the break-even "Ramsey price" (minimum average unit cost of that unit). The seller did not know his demand curve and had to grope for the monopoly price by trial and error. By the eighteenth period (the last), the average price charged was close to the monopoly price.

In the contestable natural monopoly market, two firms, with identical marginal and average unit costs to that in the monopoly experiments, vie for a market that can be fully supplied by either firm; that is, demand is the same as in the monopoly experiments. The firm with the lowest price in any period gets the entire demand at that price. If the two firms tie, they split the market 50-50 and each incurs a loss because they each fail to capture the full economies of scale implied by decreasing marginal cost. There is zero entry opportunity cost, and therefore the environment enabled the phenomenon of "hit and run entry," hypothesized by contestable market theory, to be manifest in the experiment. (Positive entry cost is introduced in Coursey, et al., 1984a.)

In every period, the mean ruling (lowest) posted price across the six contestable monopolies is below the corresponding mean price posted by the four baseline monopolies. Moreover, although the mean posted monopoly price begins above and falls below the monopoly equilibrium price, it recovers and approaches the optimal price from below. The mean ruling contestable monopoly price falls steadily across all periods and ends just above the Ramsey price, or minimum average unit cost.

Clearly, contestable market theory is supported by the dynamic trends observed in these test comparisons. For further results on contestable market experiments, see Harrison and McKee (1985b).

Plott et al. (1994) report three experiments extending these early results to a more general environment in which sellers operate in a double auction market A, where they are assured of an equilibrium rent. Market A provides the opportunity cost of entry into a posted-offer market B, in which they may choose a plant scale but without exposure to a major out-of-pocket loss if they expect volume to be inadequate to support their choice of plant scale. This was implemented by allowing those who were choosing to enter market B to see each other's tentative precontract asking prices before committing to the chosen plant scale. In this manner, an effective form of "hit and run entry" was made operational. Particularly important here was the provision of the alternative double auction market A, whose profit provided the natural means of implementing a foregone (opportunity) cost of entry into the contested market – the external marginal opportunity cost as indicated previously.

In this environment, the authors observed (1) a single seller entering market B, and (2) prices closer to the predictions of contestability theory than those of monopoly. This demonstrated that the Isaac-Smith results are robust to several important changes in the economic environment.

In this chapter, I have provided brief reviews of a large number of experimental variations on the many replications of standard supply-and-demand markets: multiple interdependent markets, the Iowa Electronic Market (now studied generally as information markets), the attainment of strategy proof equilibria, money and Gresham's alleged "law," and many examples in which constructivist models of market power are tested against their ability to predict ecologically rational market outcomes. The standard model of competition is deficient, and I have indicated the role that entry cost, buyer and seller opportunity costs, and other market constraints and opportunities play in promoting contestability and efficiency.

FIVE

Asymmetric Information and Equilibrium without Process

An exact mathematical theory means nothing unless we recognize an inexact non-mathematical knowledge on which it bears and a person whose judgment upholds this bearing.

Polanyi (1969, p. 195)

Rationality in Asymmetric Information Markets

In 2001, on the one-hundredth anniversary of the Nobel Foundation, the Swedish Academy recognized one of the more influential developments within the economics profession since the prize in economics had been established: the burgeoning field of information economics. Because of the large literature that has emerged on models of asymmetric information, no one in economics could doubt that the award and its three recipients were highly deserving of this landmark recognition of that influence. Here is how two of the recipients saw the new developments they helped to pioneer (the third, Michael Spence, is discussed later in this chapter):

In the late 1960s there was a shift in the job description of economic theorists. Prior to that time microeconomic theory was mainly concerned with analyzing the purely competitive, general equilibrium model.... In some markets, asymmetric information is fairly easily soluble by repeat sale and by reputations. In other markets, such as insurance markets, credit markets, and the market for labor, asymmetric information ... results in serious market breakdowns ... the elderly have a hard time getting health insurance; small businesses are likely to be credit-rationed; and minorities are likely to experience statistical discrimination in the labor market... (Akerlof, 2002, pp. 365, 368).

For more than a hundred years formal modeling in economics has focused on models in which information was perfect. One of the main results of our research was to show ... that even a small amount of information imperfection could have a profound effect on the nature of the equilibrium.... If [labor] markets seemed to function so badly some of the time, certainly they must be malperforming in more subtle ways much of the time. The economics of information bolstered the latter view (Stiglitz, 2002, p. 460)

These two quotations reflect the emphasis of George Akerlof and Joseph Stiglitz on their claims that the findings of information economics have enabled theory to escape its reliance on perfect information, and that this has very negative consequences for the role of key markets – labor, health care, insurance, finance – in economic betterment, requiring a variety of interventionist public policies. Their contributions, however, primarily concern abstract models of asymmetric information and the consequences for static equilibrium in such markets, not to modeling the interventionist policies.

Regrettably, the breadth of these claims exceeds what can be supported by more soberly and cautiously expressed thoughts, and by the empirical results from many experiments. This is the rhetoric of science and political policy, not the substance of these pioneers' scientific contribution, and serves to distract from the underlying importance and deserved merit of these contributions to understanding the role of information in the static equilibrium modeling of markets.

Consider insurance markets for health and life. An ideal effective market must be able to distinguish the service requirements of different customers, and fit the price of the service to the higher costs of delivering more segmented service. Advanced age (or a history of bad health) invites more maladies, making it transparent that with or without perfect information on customer differences in service delivery requirements, almost all older (or any illness-plagued) customers form a class that must be charged a higher price for the greater services needed to insure life or health if that service is to be self-sustaining in a market. More perfect individual information simply enables a better means of distributing the higher prices among particular individuals. If they are not charged appropriately higher prices, then other lower-risk patients are being required, arbitrarily (and "unfairly"), to pay for their higher-risk cohorts and have the incentive to switch to firms that sweep them into a lower-cost category, allowing all subgroups and their supply firms to reduce cross-subsidization. Some of the high-risk individuals may be able and willing to pay the higher price, others will not, just as some may be able and willing to buy a Mercedes, others a Toyota, while some will walk or take the Metro.

If words are to have meaning, this is not an instance of market breakdown, only a breakdown in the traditional flawed *theory* of markets. It is an illustration of a market struggling to do the best it can with real people to fit the service or product to an individual customer's willingness to pay, and does not try to serve those who cannot pay, arbitrarily shifting inappropriate costs into their price. Nor is it a market "failure" if the costs of distinguishing and monitoring different customers' risks, prior to insuring them, are too great to be worth the benefit. Customers are perforce grouped into nonhomogeneous classes, and charged a common price in each class – a plainly economic adaptation to the realities of functioning in a world in which it is too costly to tune each product to the fine structure of tastes and requiring demand characteristics to be lumped together. It is no different than when the market for garden shovels produces two products, short- and long-handled, although a mathematical physiological model of customer demand might "prove" that a continuously variable array of shovel lengths fits better the continuously variable height of customers. As noted by Stigler (1967), "There is no 'imperfection' in a market possessing incomplete knowledge if it would not be remunerative to acquire (produce) complete knowledge: information costs are the costs of transportation from ignorance to omniscience, and seldom can a trader afford to take the entire trip" (p. 291). The challenge is to model these realities, test the model's implications in controlled environments, and ask how it can improve the interpretation of whatever adaptations to these realities might be reflected in observed markets.

Of course, none of this says anything about the social desirability of some form of nonmarket support of health care services for the aged, infirm, and poor. In these quotations, Akerlof and Stiglitz confuse their analysis of market performance (success or failure) with their personal heartfelt concerns for the less fortunate, a topic on which neither they nor anyone else has any unique claim to expertise. They are not examining alternative nonmarket programs for the needy, how such programs might be efficiently designed, or whether, as currently constituted, these alternatives also suffer from failure to be feasible, let alone optimal. If those programs replace the individual's with the state's or the employer's payment for service, the system still faces the problem of asymmetric information. (As I write, General Motors, United Airlines, and other companies are confronting bankruptcy over accumulated incentive – incompatible employee benefit packages that can no longer be sustained out of customer sales.) How is that problem to be solved? And how is the problem of excess consumption of services, induced by third-party payment systems, private or public, to be solved?

The fundamental problem with state and employment-based programs to solve the problem of extending medical care to all risk classes is as follows: A (the physician, hospital, or other medical service supplier) recommends to B (the patient) what he or she should buy from A, and C (the insurance company or the government) reimburses A for the services. This is an incentive nightmare, and it explains why the price of medical services persistently rises faster than almost all other economic products and services. Education is another service whose price has risen "out of control," and for the same reason: A (educational institutions) define what B (students) should buy from A, and C (government or private donors) either pays for all of it, as in elementary or secondary education, or subsidizes below-cost tuition rates and/or provides low-interest education loans to B, as in university education. These are examples in which consumer sovereignty is compromised by lack of direct experience and knowledge, and the supplier, who harbors an inherent conflict of interest, is considered best capable of deciding what the consumer should buy.

There is no equilibrium in this A-B-C statement of the problem.

The same asymmetric information considerations have applied to the capital loan markets going back to the Old Testament and Middle Ages' usury laws that prohibited or limited interest charges for loans: "If you lend money to *any* of My people *who are* poor among you, you shall not be like a money lender to him; you shall not charge him interest" (Exodus, 22:25). A market in which lenders can measure and monitor each borrower's risk, reliability, and productivity will have no difficulty charging higher prices for the greater costs incurred, and will also deny loans to those for whom the risk yields an insufficient opportunity cost return at all positive levels of lending. Again, pooling customers into imperfect classes – auto financing, consumer financing, credit cards, and so on – may be the best that any practical market can do in attempting to attenuate the problem of adverse selection and specialization in the information requirements of diverse classes. If such be impossible and no sustainable market emerges, or if firms weed out the higher-risk cases based on their own experience and then ration them, this is most certainly *not* an example of market failure; rather, it is the way dynamic, well-functioning markets adapt their institutional structure to the opportunity cost peculiarities and the realities of each industry. Is it market failure that today oil does not come from coal, although a thriving market for coal oil in the nineteenth century before petroleum was discovered in 1859, quickly bankrupting the coal oil industry? The opposite is true: The nonexistence of certain markets that are not self-sustainable, or their failure to serve all potential customers, or that otherwise encounter incentive compatibility problems that fail to align prices with individual costs and benefits, is evidence of market success, not failure.

In fact, those borrowers in loan markets – who proclaim high moral intentions, who are overoptimistic about their projects yielding outcomes enabling them to pay the highest interest rates, and who supply inadequate equity protection – may present the greatest risks to the lender and be the ones who take the money and run. This is just everyday common sense, and was not lost to astute early observers such as Adam Smith; but the Stiglitz twist is, ".... [Adam] Smith, in anticipating later discussions of adverse selection, wrote that as firms raise interest rates the best borrowers drop out of the market." Stiglitz goes on to quote Smith saying that if the interest rate be high, "'... the greater part of the money would be lent to prodigals and profectors, who alone would be willing to give this higher interest. Sober people, who will give for the use of money no more than a part of what they are likely to make by the use of it, would not venture into the competition'" (Stiglitz, 2001, p. 475, text and footnote 8). Adam Smith was not, I suggest, "anticipating later" developments. He was laying an innovative foundation, as he did in so many other areas, for others to build on, but it was ignored by the neoclassical perfect information theorists. Consequently, others have stepped in to invent a better wheel to explain capital rationing based on the analysis of default risk. In short, a lender should know his borrower, limit his risk, perhaps require him to put up an equity cushion and vest himself in its success, but certainly should not lend to all comers all the funds each wants at a common interest rate. To do otherwise would constitute the market failure. Being a careful observer, Smith was pointing out how the lending industry works differently than other markets. The neoclassical synthesis lost all this wisdom with its predominant focus on solving the problem of how prices in the standard ideal commodity and service markets are determined by the conjunction of supply cost and demand value at the margin. For example, that synthesis corrected the important flaws in Smith's paradox of the prices of diamonds and water, and the inconsistencies in the labor theory of value. This new development articulated partial and general static equilibrium theory, but then made up the story about "perfect information." What Stiglitz did was to rediscover, with formal modeling, some of the wisdom in Adam Smith. But there are deeper lessons to be learned from better modeling of these markets than the assertion that the perfect competition model was deficient.

Markets as they necessarily must be constituted are not a magic solution to every social problem. There is certain terrain to which viable markets have not emerged naturally, and this must be counted as a resounding success for the economy, not a failure. If markets were extended to the loan and health care industries without modification for their special information characteristics, this would be the market failure. As for the widely shared concerns for the less fortunate, no one was more eloquent than Smith in discussing the psychology of human sympathy. It does not follow from any of these comments, however, that some of the recalcitrant cases would not yield to contemporary economic design efforts and testing.

It is further claimed that all these considerations of imperfect information and alleged market failure in turn open the way for a role by government, although that role is not formally and systematically modeled in ways that make it plain which alternative might be best, and under what circumstances: markets under imperfect asymmetric information, or the government, which must somehow deal with the same imperfections of information. The government cannot solve the problem just because it has access to the taxpayer's deep pockets, and without systematic analytical/empirical comparisons, we do not know whether government solutions exist and are better than imperfect private voluntary solutions. So far, in health care, the foreign government solutions that have been attempted have yielded good examples of the failure to deal with excess demand and sickness promoting waiting lines for treatment because marketlike incentives are not permitted to assist in solving the rationing problem.

What we have are serious problems of incentive failure that may not find good solutions, but the welfare stakes are high, and I think it is premature to give up on the incentive and institutional design issues.

While information economics was changing the job description of economic theorists, public choice economics was changing the presumptive faith of economists that where markets fail there is an ipso facto case for government intervention. Incentive as well as information failures also plagued the world of the traditional idealization of the dedicated public servant (Buchanan and Tullock, 1962). Hence, asymmetric information theory, while alleging market failure, does not provide us with theorems showing the superiority of government intervention based on equilibrium models of both asymmetric information and the actions of public decision makers whose incentives may not align with those of an ideal government competent and informed enough to operate in the "public interest," however defined.

What seems remarkable about these two essays is that the authors take seriously the flawed rhetorical claims made about the equilibrium theory that they were taught in graduate school, and then presumed its relevance to the performance of any and all markets. Similarly, based on the hypothesis that "now we have it right," they seem to suggest that the new models at long last constitute the truth of market performance. But is this rhetoric of the new asymmetric information theory any more believable than the rhetoric of the earlier general equilibrium theorists, who truly believed the naïve proposition that they had discovered the "institution-free" core of economics and, without a line of proof, asserted that their models required complete perfect information?

When these neoclassical equilibrium models are tested in the laboratory, it is found that complete perfect information is not necessary for achieving the competitive outcomes they predict. Hence, there is full agreement between experimental market economists and these asymmetric information theorists in their critique of the relevance of the claimed perfect information requirements of neoclassical market models. But there is little basis for agreement concerning the claims made for the new asymmetric information equilibrium models. Both the old and the new models are about equilibrium outcomes; neither of them is about dynamic processes of convergence, or failure to converge, to the predicted equilibrium.

Yet it was in the prescription of the information requirements where the old models failed in the laboratory: What failed were the information conditions specified by the models, not the markets whose equilibrium they defined. Neoclassical markets performed very efficiently with strictly private information, provided that they were implemented by extant observed market institutions. The models correctly predicted the competitive outcomes but erred grossly in "predicting" the conditional need for complete information. They also failed to predict that in certain environments with complete information the market performed worse than when information was private, suggesting that perfect information was also not sufficient for equilibrium to be attained.

Models of markets with asymmetric information require similar testing exercises. Many have been tested in the laboratory; a few of these cases and the issues raised will be examined in this chapter. I have already referred to the first asymmetric information arbitrage experiments by Miller et al., 1977; Williams, 1979; and Plott and Uhl, 1981.

The primary justification of the new asymmetric information models cannot be that they explain the nonexistence or failures of certain markets. Adam Smith had no difficulty using a little scientific common sense, based on observation, to explain the difficulty that plagues the loan market. A far better justification and correct defense of the new models is in the prospect that more precise models of information allow sharper tests, going much beyond "explanation," and perhaps even laying the basis for better market or other institutional designs.

The third recipient of the 2001 award, Michael Spence, writes in a far different and more cautiously positive mode than indicated by Akerlof and

Stiglitz. In effect, he came not to judge, blame, or find fault, but to better characterize certain issues not previously modeled explicitly by the science:

> ... my puzzlement about several aspects of the consequences of incomplete information in job markets, pretty much launched me on a search for things that I came to call signals, that would carry information persistently in equilibrium from sellers to buyers, or ... from those with more to those with less information. ... Of course ... one needs to know ... who has an incentive to establish a reputation through repeated plays of the game. ... While it is perhaps interesting to note that the signaling effect causes overinvestment in the signal by the standard of a world in which there is perfect information, that isn't the world that we live in. Thus it may be more interesting to examine the equilibrium outcomes in comparison to various second-best outcomes that acknowledge that there is an informational gap and asymmetry that cannot be simply removed by a wave of the pen (Spence, 2002, pp. 407–8, 423).

Spence's last point goes to the crux of what it means to find fault with systems whose characteristics necessarily differ from what one might imagine would exist in a world of pencil and paper. In particular, it means that if we have a market failure, and look to a government to correct the failure, that government must not somehow have miraculous access to a waving pen. Otherwise, we are in danger of doing well-intentioned harm, as illustrated by the many flawed and failed policies in welfare programs, health programs, foreign economic aid, and poverty reduction. It may turn out that none of the models are very good relative to an imaginary world of perfect information, but they may point the way to the design of better management systems for aligning individual incentives with improved outcomes. This dichotomy, between the ideal and that which is empirically achievable by any mechanism, is at the heart of what was wrong with traditional conceptions and interpretations of neoclassical market equilibrium and applies with equal force to the interpretations of the new contributions. Nothing much has changed except to give us much more thoughtful and precise static equilibrium models that are more explicit about who has what dispersed information. Theorists make important contributions that need to be clearly distinguished from any rhetorical leaps to untested economic policies or general claims about market failure.

The Neoclassical Synthesis

I want to attempt a brief sketch of the development of the fundamental ideas as I see it that underlie equilibrium theory, and how things got off track and have tended to continue to be stuck in that off-track rut after the marginal revolution of the 1870s. I am not a scholar of the history of economic ideas, and only record here my perceptions as a practicing economist and experimentalist who with several colleagues has worked alongside practitioners on many market design applications to industry.

The fundamental two-part theorem of Adam Smith is that wealth is derived from specialization – the "division of labor" – which in turn is "determined by the extent of the market." Thus, we have

$$(\text{exchange}) \rightarrow (\text{specialization}) \rightarrow (\text{wealth creation}).$$

This division of labour . . . is not originally the effect of any human wisdom, which foresees and intends that general opulence to which it gives occasion. It is the necessary, though very slow and gradual, consequence of a certain propensity in human nature which has in view no such extensive utility; the propensity to truck, barter, and exchange one thing for another. . . . As it is the power of exchanging that gives occasion to the division of labour, so the extent of this division must always be limited . . . by the extent of the market (Smith 1776/1909, pp. 25, 31).

Hence, the wellsprings of wealth formation were not part of a deliberately conceived plan of the stumbling and fumbling human enterprise. It was part of what today we would call an emergent, spontaneous, self-ordering system, of which there are a great many examples in nature, from snowflakes to human systems. These emergent human processes may of course be in fits and starts that are not themselves orderly. Nor is there anything here supporting the following popular misconception: "The argument of Adam Smith, the founder of modern economics, [was] that free markets led to efficient outcomes . . . " (Stiglitz, 2002, p. 472). The latter was the neoclassical interpretation recast and updated in the form of equilibrium theory. I do not interpret Smith as claiming that the markets emerging from his "very slow and gradual" process were efficient or the best of all possible worlds. In fact, much of the analytical spirit of information economics can be interpreted as a return to the thoughtful framework articulated by Adam Smith, as in the previous quote on loan markets.

The proposition that equilibrium in a competitive market requires all agents to have complete information appears to derive originally from W. S. Jevons (1871; 1888):

A market, then, is theoretically perfect only when all traders have perfect knowledge of the conditions of supply and demand, and the consequent ratio of exchange [price] . . . (p. IV.18).

A page earlier he tells us his belief about the actual achievement of this state:

The theoretical conception of a perfect market is more or less completely carried out in practice. It is the work of brokers in any extensive market to organize exchange, so that every purchase shall be made with the most thorough acquaintance with the conditions of the trade. Each broker strives to gain the best knowledge of the conditions of supply and demand, and the earliest intimation of any change (ibid., p. IV.17).

Plainly this is a convenient belief asserted by Jevons, not a theorem; neither is it an empirical demonstration. It stands as a poorly articulated fallacy of constructivism that has done much damage – real and intellectual – to the way economists think about the world. This is because the assertion has been elevated to the status of a Folk Theorem. Moreover, the resulting frame of thought continued to infect the foundations of microeconomic theory after theorists – as sooner or later they were bound to do – had turned to the task of attempting a formal analysis of information and equilibrium. Jevons' false Folk Theorem – together with the universal acceptance by economists that (1) welfare economics depended crucially on the validity of static competitive regimes, and that (2) competitive equilibrium outcomes did indeed require perfect or complete information – set the stage for a large industry of revisionist theoretical attacks on both the competitive model and how that model was undermined once you relaxed its claimed dependence on complete information.

A false proposition spawned a host of new equilibrium theorems sharing the same faults that characterized the original claims. Why?

Equilibrium theory avoided confronting the problem of process, and those heroic research programs attempting to confront the problem mostly failed because, I hypothesize, the problem is just plainly beyond the competence of the standard tool kit and the full optimization methodology of game theory. (Again, refer to Wilson, 1987, p. 411, on problems of modeling the enormously successful double auction institution.) Subsequent attempts to model information, and in particular its fundamental and essential asymmetry, recognized by the Swedish Academy in 2001, perpetuated this style of failing to confront process, although the information content of the equilibrium itself was now certainly and laudably more precisely and accurately articulated in some important cases. Consequently, the new "paradigm" failed to confront the flaws in the old paradigm. Neither did it confront the essentially accidental and unintended discovery by experimental economists that the competitive equilibrium is attainable as an ecological equilibrium with strictly private information, provided that you fudge a bit by using the rules of one of humankind's several market institutions.

But if you begin with the discovery in controlled experiments that complete information is not necessary and probably not sufficient for achieving efficient competitive outcomes, what are the implications of the numerous theoretical results showing that equilibrium either does not exist or is inefficient when there is asymmetric information? Well, you are back to square one. There is no reason to believe in the truth value of any of these cases short of an extensive program of testing them. And without any satisfactory means of describing equilibrating processes, the results of those tests are likely to yield more enigmatic surprises.

If human agents with private and therefore inherently universal asymmetric information can converge to equilibrium outcomes by means that theorists can neither model nor predict, why would you believe that the market failure theorems derived from asymmetric information theory apply to those agents? We are indeed back to "go," armed, I believe, with models that are still incomplete in dealing with equilibrating processes, as distinct from static equilibrium (symmetric or asymmetric) specifications. We need some reappraisal that is not plainly in the interest of anyone trained in the SSSM and its new extensions that deal in great depth with revised models of information and equilibrium, but leave untouched the treatment of the ecological processes through which economic agents evolve, or fail to evolve, from their initial circumstances and knowledge to the equilibrium prediction derived from one of the new asymmetric information models. The fundamental error has been in the universal tendency to think about the world in terms of equilibrium models and the consequent welfare theorems that are devoid of process, rather than a world composed of humans with the demonstrated capacity to achieve efficient outcomes in a wide class of neoclassical environments and who may be able to surmount the negative implications of some of the new asymmetric information models.

Hayek and the Hurwicz Program

Beginning seventy years ago, Hayek (1937, 1945) challenged the meaningfulness of the statement of the conditions necessary for equilibrium, and recognized that the statement was in a sense a tautology. Thus, in the spirit of Coase (1960), but predating him by twenty-three years, Hayek argued:

The statement that, if people know everything, they are in equilibrium is true simply because that is how we define equilibrium. The assumption of a perfect market in this sense is just another way of saying that the equilibrium exists but does not get us any nearer an explanation of when and how such a state will come about.... [I]f we want to make the assertion that, under certain conditions, people will approach that state, we must explain by what process they acquire the necessary knowledge.... these ... subsidiary hypotheses ... that people learn from experience, and about how they acquire knowledge ... constitute the empirical content of our propositions about what happens in the real world (Hayek, 1937, reprinted 1948, p. 46).

In his better-known elaboration eight years later, he expanded:

Fundamentally, in a system in which the knowledge of the relevant facts is dispersed among many people, prices can act to co-ordinate the separate actions of different people.... The problem is precisely how to extend the span of our utilization of resources beyond the span of the control of any one mind; and, therefore, how to dispense with the need of conscious control and how to provide inducements which make the individuals do the desirable things without anyone having to tell them what to do (Hayek, 1945, reprinted 1948, pp. 85, 88).

Hayek's perceptive critique was almost completely ignored by prominent mainstream theorists, but did not escape the close attention of Leonid Hurwicz. Hurwicz clearly grasped the importance of the idea that private information was inherently and fundamentally asymmetric and dispersed; that this dispersed information was summarized in equilibrium prices; and that the pricing system itself was a decentralized information system whose signals enabled distributed agents to act appropriately without anyone having to issue orders telling them what to do. These hypotheses by Hayek, however, were neither proved nor empirically demonstrated. There were no fundamental "Hayekian" theorems that explicated his critical insights.

Hurwicz' work brought to the Hayek problem a powerful intellect whose contributions to equilibrium theory and stability analysis were widely respected. But in all those contributions, Hurwicz was persistently bothered by the failure of that analysis to come to grips with the Hayek problem as stated in the preceding quotations – he frequently cited Hayek (1945) when he was discussing this issue. No one was more qualified than Hurwicz to tackle such an important and challenging problem. Moreover, he inspired many brilliant students and colleagues to work on these very difficult process problems. Hurwicz sought a rigorous foundation for a Hayekian concept of "informational efficiency," which would show how equilibrium could grow out of the exchange of information – messages – with the resulting prices being both the carriers of information and the result of that message exchange. A large probing literature grew from that program, limited as it was to a few mathematical economic specialists working to crack the Hayek problem. Here is a small sample: Reiter, 1959/1981; Hurwicz, 1960, 1973; Hurwicz et al., 1975.

This program failed, however, in the strict sense that it was unable to produce a general fundamental theorem on the informational efficiency of decentralized processes, in the tradition of the Hayekian conception of the pricing system, and also in the general equilibrium tradition of *full optimality*: a theorem comparable to but deeper than the standard welfare theorems. I think the program came far closer to following the Hayek conception than the full optimality tradition, which came to be dominated by Nash theory and game theory. The Hurwicz program was enormously productive, and insightful, but unfortunately it was swept aside by the rush to solve "easier" problems, more amenable to the familiar methods of the day, and the continuing commitment of mainstream theorists to restrict sufficiently the problem formulation to make tractable full (Nash) equilibrium solutions. The Hurwicz program deliberately gave priority to formulations that preserved an information processing perspective, on the obvious grounds that this was the essential idea that had been avoided in traditional theory, and was at the core of the Hayek critique. Hence, Hayek's ideas did not fail because no one attempted a rigorous treatment that yielded many interesting results; what failed was the attempt to link process with equilibrium and efficiency. Essentially the "failure" merely indicates how difficult the problem was that Hayek had posed.

Reiter's (1959/1981) remarkable contribution was also closer to the experimental tradition and its discoveries than was generally appreciated, and was not subject to the Hayek critique, nor to the critique of economic theory that arose from the experimental market discoveries. This is conveyed in a small measure in the following summary:

The model presented in this paper deals with the dynamic behavior of a market not in equilibrium. It is a characteristic feature of markets that are out of equilibrium that opportunities to trade on different terms exist simultaneously. One explanation of how a uniform price comes to prevail in a market is in terms of arbitrage. In broad terms, the model studied here formalizes a dynamic arbitraging process and explores the extent to which that process can bring about results of the kind usually envisaged for it.

The literature on markets that are out of equilibrium contains models which deal with search behavior and models in which a market "process" is formalized as a game. The model studied in this paper differs from these in many details but mainly in two respects. First, the behavior of agents is modeled from a "bounded rationality" point of view rather than, say, in terms of optimal searching. Second, the main focus is on the dynamic process and in its long-run tendency rather than on a solution concept such as Nash equilibrium or market clearing. In these respects, my viewpoint of 1959 is retained here.

A dynamic market process cannot satisfactorily be based on static excess demand behavior of economic agents (traders, in a pure exchange setting) because it is unsatisfactory to assume that agents will maximize utility, treating the budget constraints as if it [*sic*] were a certainty in a situation in which it is necessarily uncertain. Away from the static equilibrium, no agent can be assured of his opportunities.... While it is possible that behavior in equilibrium should be in some sense a limit of disequilibrium behavior, we take the view that behavior of agents should, in some sense, be quantitatively the same throughout the process.

The role of information and its effect on the appropriate behavior of agents then becomes important. We try to take account of two aspects of the role of information. First, the institutional structure of the market determines the information agents get from the market process, the "structural" aspect. Second, the restricted capacity of economic agents to handle information restricts behavior, the "bounded rationality" aspect..." (ibid., p. 3).

We are left with Hayek's critique, his statement of the problem solved by decentralized pricing, and the experimental evidence supporting Hayekian efficiency results in a wide variety of environments and institutions, but the theory showing how this works eludes articulation by means of the economist's standard tool kit.

Some of the dispersed information characteristics posed by Hayek are addressed formally by the work in asymmetric information theory. In particular, the case where private information includes willingness to pay (or willingness to accept) and also a feature of the product that is different for the buyer than for the seller. Thus, if it's a used car of a given make, year, and model, the seller has a supply limit price and knows that the car is or is not mechanically flawed. The buyer has a demand limit price conditional on the state of the car's quality, known to the seller but not the buyer. Since quality information is known only to the sellers, it is impossible to separate cars into the two "lemons versus nonlemons" information classes and to apply two market-clearing prices. The market does the best it can with this reality, and blaming it on imperfect information is like cursing the wind because it has the temerity to blow. There are still gains from trade, and so the market discounts all the used cars with a price that reflects the risk that some unidentifiable individual units are mechanically flawed. If this means that new car manufacturers are hurt by lower car resale prices, they have an incentive to innovate transferable warranties if the benefits of such exceed their cost. By thus absorbing the cost of mechanically flawed units, they may leave less money on the table and squeeze out further gains from trade – a prospect that if true is discoverable by ecologically rational market processes (Hayek, 1978/1984). If so, the new market for warranties supports the specialization that enables the creation of some new wealth. This is desirable, as Pareto improving. If all these things are true, we observe an imperfect adjustment to a market with imperfect information, a corresponding betterment, and hardly have cause to panic over allegations of "market failure" because information is not perfect – an irrelevant and uninteresting comparison.

Experimental Markets with Asymmetric Information

Many important competitive issues are raised when a subset of agents has asymmetric advance information on product quality and value characteristics. Constructivist equilibrium analysis shows that such conditions can generate market failure in the sense that the allocations are less than fully efficient. Some of these problems, however, arise because the analysis is inadequate in examining both sides of the market and the full implications of the information content of prices. Experiments have established that constructivist inefficiency may be alleviated by one of several ecologically rational response mechanisms: competition among sellers for reputations, quality (brand) signaling, product warranties, and the aggregation of private asymmetric information into public price patterns that self-correct the alleged problems. See, for example, Plott and Wilde (1982), Plott and Sundar (1982, 1988), and Miller and Plott (1985).

Consequently, much of the experimental evidence – while supporting the idea that asymmetry of information has potential effects like those in the new standard models – reveals that there are countervailing responses that mitigate the proposition that asymmetric information necessarily implies market failure. Models showing that there can be failure relative to perfect information regimes are not the same as models showing that observed markets have failed in the environments to which they have become adapted.

Special cases where markets (or their models) perform less than ideally do not add up to a general market failure theorem. In such "policy" analyses, several things may be omitted from the formal analysis:

- If someone has asymmetric information yielding an abnormal return, what prevents self-correcting entry by competitors to exploit the advantage? Even if the entry is inefficient, what is the alternative? The market is inefficient relative to what? (Recall Spence's point about second-best adaptations.)
- Do agents with asymmetric information take actions that reveal their position of advantage, allowing others (buyers or sellers) to correct for it?
- What is revealed in the analysis of the costs and benefits of intervention to correct the efficiency loss from asymmetric information? If it costs more to correct the inefficiency than it is worth, then it is not inefficient,

except relative to the imagined world of perfect information inherited by successive generations of theorists.

- Are there extramarket sources of information about reputations, even where there is no repeat trade that serves a correctional purpose?

Markets for Quality

A pioneering early contribution to the study of markets with asymmetric information on product quality is that of Plott and Wilde (1982). They model a market such as arises between doctors and patients, or automobile customers and repairmen. Buyers typically lack the private information and technical competence to choose among the products available (surgery versus prescription treatment; a new part for their car versus a less expensive repair or adjustment of the affected item). Buyers depend on the recommendation (diagnosis) provided by the vendor, whose short-run incentive may be to misrepresent what is best for the customer to enable a higher profit.

In the Plott and Wilde design, a seller's diagnosis is based on uncertain information (or "clues") as to the true state of the buyer's needs, and if a seller biases his diagnosis in his self-interest and other sellers do not, then his sales may be adversely affected. A principle finding from two experiments was that competition among the sellers tended to result in a common diagnosis pattern – a reputation-based, approximately "truth-revealing" pattern – that is supported by a reluctance of buyers to purchase from sellers who deviated from the recommended norm observed to emerge in the market among the sellers. They strongly reject the hypothesis that sellers will systematically deceive buyers. The qualified conclusion is that "sellers who wish to make sales *at all* (of either the low-profit item or the high-profit item) must give advice similar to other sellers, and in the absence of collusion the best strategy is the 'truth' as seen by the seller" (Plott and Wilde, 1982, p. 97).

The fact that there was no search cost incurred by the buyers was brought into question by Pitchik and Schotter (1984), who allowed consumers to engage in costly sequential search for expert advice. This raises the opportunity cost of buyers, and alters a feature not present in the model. Even so, they report a high level of truth-revealing diagnosis. Behavior is more robust than the models.

The power and influence of asymmetric information theory has been to establish that economic theory has long failed explicitly to come to grips with the role of information in equilibrium analysis. What the new theory made clear is that there was a problem to be solved if markets were to be efficient. But observed markets were exposed to that problem long before it was modeled in economics. The new models, which focused on the deficiencies of the old symmetric information tradition, did not address the question whether the long economic exposure to the problem might have created an endogenous demand for countervailing responses in the social and institutional framework within which observed markets operate.

One can argue that some markets, like the physician and auto repair markets, seem not to show much search behavior, and therefore that the preceding experiments do not really rule out an equilibrium with deceptive diagnosis. But costly individual search is only one response to money being left on the table, and in any case needs to be explicitly modeled in making efficiency evaluations. Are there other extramarket response modes? Various organizations have in fact emerged to provide low-cost complaint procedures – the local Chamber of Commerce, the Better Business Bureau, or the automobile manufacturer's hotline that can serve to keep local dealer repair operations honest. Moreover, gossip, in which neighbors voluntarily report to each other their experience with various physicians or repairmen, provides a potentially effective means by which consumers pool their information and seek out the most reliable service providers, especially where repeat trade is nonexistent and the stakes are large. Local gossip rapidly transmits local experiential information on who are the best eye surgeons, the most reliable automobile repairmen, and, while they last, who is not trustworthy. To my knowledge, "cheap talk" among buyers as a laboratory treatment variable in these particular environments has not been explored. Recall Hayek's perceptive suggestion, quoted earlier, that socioeconomic institutions evolve the forms that they do because information is not complete (Hayek, 1973, pp. 12–13).[1]

In the spirit of Spence's work on signaling equilibrium, Miller and Plott (1985) design an asymmetric information experiment in which sellers can choose costly signals that are observable in the market. If a high-quality product can be signaled at lower cost, a separating equilibrium can exist that enables buyers to determine quality from the signals. Products were either "regulars" or higher-quality "supers." The induced value to a buyer consisted of a part that depended on true product value and a part depending on the seller's signal. Prior to purchase, buyers could not observe product quality, but they could observe the seller's signal. Double auction bids and asks specified both a price and a quality signal. In the experiments, signaling

---

[1] For example, I recently had cataract surgery – you usually do that just once – in Tucson, where one can easily find out who are the best surgeons for this specialty. We learn from each other's "experiments" through gossip. Indeed, gossip may have had evolutionary fitness value because of its ability to pool information.

often started out being excessive – more than needed for separation – so that excessive costs were incurred, but then signaling of "supers" would decay dynamically to just enough to yield an efficient separation. Efficient signaling occurred less frequently in the results when the difference in signaling cost between supers and regulars was reduced, suggesting that the achievement of efficient outcomes was not worth the subjective effort cost, a component omitted in standard theory. A similar regulars versus supers product quality environment was studied by Lynch et al. (1986) using a number of different treatments. When sellers were subject to a truth-in-advertising-supers policy rule, the market was highly efficient. When sellers were free to misrepresent their product, and the identities of the sellers were also suppressed, this combination resulted in inefficient "lemons market" outcomes, as predicted by asymmetric information theory. If, however, sellers' identities were observable, this led to a seller reputation effect that yielded a higher proportion of efficient supers being traded, although many inefficient outcomes were still observed. (Note that if sellers try to avoid revealing their identity, this can yield great returns to a seller who reveals himself.) Finally, a strong shift toward efficient outcomes was apparent with a treatment that gave sellers the opportunity to offer a warranty under which the seller would compensate the buyer if the product was not as advertised. Hence, with money being left on the table, there are powerful incentives to alter past practices and profit from the improvements in efficiency.

Labor Markets and Efficiency Wages

Competitive wage theory implies that wages depend on workers' ability, firms bidding for worker services, markets clearing at a wage that balances supply and demand, and a lack of involuntary unemployment. Discrepancies between this prediction and empirical observations have given rise to efficiency wage theories that postulate that firms pay supracompetitive wages. One approach models the labor contract as a partial gift exchange, or the fair wage–effort hypothesis. Employers offer a premium above the market wage, and in return workers offer above-minimum levels of effort. The idea is similar to "gratuities," or tips, in which restaurant and taxi customers voluntarily pay (after delivery, however, not before) a bonus above the billed rate in return for above-minimum service, except that a key extreme assumption driving the results in efficiency wage theory is that employers cannot observe, measure, and take into account employee effort as part of the employment contract. Employers depend completely upon the voluntary good offices of the employee. Note that this model leaves out gossip in which employees know and observe each other's effort and the circulating information may be leaked to the employer. Nor does the model allow the employer, who cannot observe "effort," to compare different output levels of different employees and *infer* differential unobserved effort/ability levels – by their works (fruits) shall ye know them.

This is a good environment for experimental study because you can use reward monies to induce an explicit value on the product of effort, and a cost on supplying that effort, with employees choosing effort levels that cannot be observed *before or after* accepting the wage offers of employers. This allows an equilibrium wage, effort, and surplus to be defined based on strict self-interest (dominance), and compared with the most efficient (maximum net value) outcomes, based on the "perfect information" known to the experimenter but not to the subjects.

Fehr et al. (1993) report good support for the efficiency wage model: Subject employers offer wages that average above the equilibrium level and propose high levels of costly effort to their matched employee; employee subjects respond with above-equilibrium levels of effort at the accepted contract wage. Moreover, these cooperative outcomes are observed to persist over time in repeat play with random rematches of employers and employees in an environment with a fixed number of each. Note that random rematching operates like a temporary employment pool. The observed outcomes are much above the narrowly self-interested equilibrium, but are also well below the efficient outcome based on "perfect information." Thus, wages are neither fully efficient nor at the competitive level, and efficiency wage theory leaves plenty of surplus on the table so efficiency wages do not prevent a "market failure" problem.

Fehr et al. (1993) did not, however, investigate the robustness of their results to alternative parameterizations, and Charness et al. (2004) report data showing that these earlier results were not robust to procedural and subjects effects. Rigdon (2002) notes two potential problems[2] with the Fehr et al. (1993) experiments; the first deals with an important methodological consideration, the second with motivation or payoff adequacy:

- In the Fehr et al. (1993) experiments, the employer and employee dyads are anonymously matched. This implements the condition that employee effort levels are inherently unobservable – the key element in efficiency wage theory. The experimenter, however, knows the circumstances of each pairing: the contract offers, acceptances, proposed and

---

[2] On similar points, see Shaked (2005), Charness et al. (2004), and Trivers (2004, pp. 964–5) on the rhetoric in Fehr and Schmidt (1999).

actual effort levels, and the earning payoffs of each individual by subject name. Someone is therefore "watching" and the subjects know that this is the case, thereby violating the key assumption of unobservability in the model.

- The marginal cost of high levels of employee effort is unrealistically low; it costs the individual only 0.32 Swiss Francs given up to supply the average observed level of effort.

Rigdon (2002) exactly replicates this wage contract environment, but conducts the experiments as "double blind," a procedure whereby no one, including the experimenter, can ever know what decisions and payoffs were made by any individual subject. Thus, the individual effort levels really are "unobservable." Moreover, the payoffs were economically more significant: $10 (employee) and $25 (employer) at the asymmetric competitive equilibrium, and $25 and $40 respectively at the efficient cooperative contract under full information. She reports experiments showing that the Fehr et al. (1993) results are *not* robust with respect to these changes in the economic environment: Across sixteen repeat transactions with random rematching of contracting subjects, the observations converge to inefficient competitive equilibrium levels of effort, and the average wage declines from an initial level of over $20 to $12.10 as the free riding by employees gets discounted into the wage. Aggregating across all periods, workers shirk in providing the promised effort by choosing to provide less effort in 79 percent of the contracts, and in some 68 percent of the contracts choose the minimum equilibrium effort available in the choice set.

Yet to be established in this development is how the two changes in the experimental environment separately impact the substantial change in the observed results. Which of the two have the largest effect on market inefficiency? Other two-person experimental game environments have established that the "double blind" condition has strong effects on strategic two-person interactions, but also that this condition interacts with other conditions such as payoffs and variations on the procedural context. (See Chapters 10 through 12.)

Efficiency wage theory has evident flaws to anyone experienced with an employer, and this has led to some institutional adaptations ignored by the theory:

- The boss cannot observe effort, but can usually observe that some employees are more productive than others, even if their effort cannot be monitored, and keep the more productive at the earliest opportunity. This is why employers want to be free to select their most productive workers for overtime, and labor unions have favored work rules preventing such actions.

- Employees observe each other, and some may try to get ahead by telling on the others. Gossip, an integral part of human sociality, can serve to expand that which is observable and compromise asymmetric information.
- The explosion of the part-time-help industry has enabled employers and employees to observe each other *before* a hiring contract is locked in. This enables the employer to preselect employees who can be trusted in advance before committing to a permanent contract and the large start-up cost/benefit package.

As noted by Hayek, institutions do indeed take adaptive forms that deal with the fact that information is not complete. The part-time-employment agency is a new market-enabled specialty that facilitates trial-and-error selection to help solve the shirking problem and the high cost of hiring a permanent employee.

SIX

FCC Spectrum Auctions and Combinatorial Designs

Theory and Experiment

The best-laid schemes o' mice an' men Gang aft agley An' lea'e us nought but grief an' pain For promis'd joy.

Robert Burns

Introduction

In this chapter, I discuss a case study in the constructivist design of a seller's auction mechanism, the use of both laboratory experiments and the field to test-bed those designs to determine their ecological fitness, and their modification in the light of that experience. In the 1990s, a series of Federal Communications Commission (FCC) auctions led to numerous changes in the auction rules with many instructive parallels to the earlier laboratory learning, but at far higher cost. Moreover, the FCC design contained flaws arising from several interdependent assumptions that from the beginning were never critically examined and studied empirically: a common value environment, revenue maximization as the criterion for auction design that in turn implied the English auction format, and the public revelation of complete information on who was bidding how much for what in real time.

Following a series of spectrum auctions, the U.S. Congress mandated an independent evaluation of the Simultaneous Multiple Round (SMR) auction mechanism that had been developed by the FCC to award licenses to bidders. This evaluation was required to include an experimental study that would increase understanding of the problems and the complexities of the SMR mechanism. The study was also to examine alternatives to the SMR that might better facilitate the acquisition of efficient combinations of the elementary licenses where some of those elements are complements.

To clarify the issues involved, I will provide a brief review of the theoretical issues and some previous and recent experimental findings that bear most directly on the conceptual and behavioral foundation of the FCC design problem.

The Combo Clock auction, discussed later in this chapter, evolved from earlier experimental research, followed by tests of the SMR, and is another example of how a proposed solution to a complex design problem can benefit from the discipline of test bedding in the laboratory before applying the solution in the field. Does the design guide behavior that produces ecologically rational (efficient) allocations in the laboratory? If not, can the test environment be used to inform design modifications that improve allocations? In this development, the next step is to test the mechanism using industry and other professionals, as has been done in industries such as electric power and emissions trading in the Los Angeles Basin. (The recent Virginia nitrogen dioxide emission rights (NOX) auctions used an English Clock but not a Combo Clock; see Porter et al., 2005.)

Auctions: Modeling Institutions

The subject matter in this case also affords a convenient opportunity to review some of the experimental literature on the simpler naturally occurring auctions, and to evaluate the extent to which the constructivist models of historical auctions predict subject performance in the environments and institutions that have been modeled. (See Smith, 1991, for several papers on the results and methodology of tests of the standard auctions.) This is important to the theme of this book because auction theory has had some notable predictive successes as well as failures in laboratory tests. In particular, as we shall see later in this chapter, the failure of the strong efficiency property of the English auction to generalize to multiple units led to experiments designed to explore alternative implementations. This laboratory exercise led to the discovery that the problem could be solved by using a clock to announce successive prices in the auction process. Hence, experiments in the 1980s paved the way for the discovery of a better institutional form for multiple-unit English auctions, and this was subsequently generalized for combinatorial auctions, originally invented and studied using the sealed bid format (see Cramton et al., 2006).

I believe it is accurate to say that auction theory provided the first formal analysis of the property right rules governing the dynamics of an exchange institution that predicted its equilibrium outcomes conditional on postulated properties of the economic environment and of bidder preferences toward uncertainty. Auction theory provided a constructive partial equilibrium response to the Hayek critique of equilibrium theory, although it did not address the information role of prices and the discovery process for economic organization in general equilibrium. It was a baby step, but it showed clearly how institutions (rules) and dispersed information could predict messages, prices, and outcomes that were precisely testable in experimental environments. In the field, testability was imprecise in that it was conditional on confounding assumptions about behavior and the distribution of values, and, for tractability in making precise predictions, testability required risk-neutral agents or special functional forms for utility in some auction models.

Thus, the various simple auction forms for the sale of single units are sharply delineated by the rules governing the information that is or is not made public during the auction, and the way in which bids are processed to determine outcomes:

- In the open English progressive auction for a single unit of a good, bidders progressively raise the standing previous bid until no new bids are forthcoming. The auctioneer then declares the item sold to the last and final bidder at her bid price.
- In the Dutch auction, an auctioneer or clock is used to announce successive decreasing bid prices. The item is declared sold to the first buyer who shouts "mine," or, in clock auctions, presses his accept button to stop the clock at a particular price. The buyer pays the last standing "stop" price.
- In sealed bid auctions, each bidder submits a private secret bid to buy. In the "first-price" auction, the winner is the person who submitted the highest or first bid in descending order and pays the amount bid.
- In the "second-price" auction, the highest bidder also wins but pays a price equal to the bid of the second-highest bidder. For more descriptive detail, see the classic treatment by Cassady (1967); for the definitive treatment of the theory, see Vickrey (1961); and for the first experimental tests see Part IV of Smith (1991).

Economics of English Auctions

Consider the theory of English progressive auctions under alternative assumptions about auction values. First, individual valuations for an auctioned item are private and independent, or second, the valuations are common (or affiliated) across all individuals.

The so-called "English" auction was imported to the British islands by the Romans, providing a nice illustration of both the antiquity and the durability of the most widely available auction institutional form. Its antiquity is directly indicated by word origins. Thus, the word *auction* is derived from the Latin root *auctio*, meaning "an increasing" (procedure). More revealing, the Oxford English Dictionary notes that an old English word, fallen into disuse, is *subhasta*, which is still used in Spain to refer to an auction or auction house. This word is compounded from the Latin words *sub hasta* meaning (a sale) "under the spear," which refers to the practice of the auctioneers who followed the Roman army to auction the spoils of war so that the soldiers would not have to be paid in kind with goods not easily transported.

Independent Private Values

In this environment, there are both advantages and disadvantages to the progressive auction when it is compared with sealed bid (first or second price) or Dutch auctions (Cassady, 1967).

The principal advantage of the progressive auction is the transparency of the optimal bidding strategy. The auction requires minimal bidder sophistication for choosing a noncollusive strategy, and bidders have no incentive to invest in acquiring information as to other bidders' values or strategies. A bidder who has estimated her value (defined as maximum willingness to pay) for the item merely follows two simple (dominant strategy) rules: (1) after the first bid has been announced, and at all times until the auction ends, if the standing bid is below her value, the bid should be raised by the minimum bid increment such that the new bid is not greater than the bidder's value; (2) never raise your own bid, that is, "jump" bid. This transparency and strategic simplicity apparently accounts for the fact that the vast majority of auctions are of the English form both on and off the Internet (ibid., 1967; Reiley, 2000). The implications of (1) and (2) guarantee theoretically that the bidding will stop on each item auctioned as soon as the bidder with the highest private value (v_1) bids $b^* \geq v_2$ (second-highest value), given that the minimum bid increment is sufficiently small.[1] No one will rationally invest in estimating the values or bidding strategies of other bidders because these will be revealed at no cost to everyone by the open bidding procedure.

---

[1] It is possible that the minimum bid increment will eliminate all remaining bidders simultaneously, in which case the item is randomly awarded to one of the remaining bidders.

Note, however, that rationally the two-part strategy is not quite the same as stating that a bidder's dominant strategy is to bid actively until the price reaches the value of the object to him. The latter is correct for all but the highest-value bidder, who, following rule (2), *discovers* that he need bid only high enough to displace the second-highest valuation bidder, and rationally, will no longer remain active. When bidders do not announce bids from the floor, as in the English Clock auction, the clock ticks up until all but one bidder drops out; then stops, determining the winning price. Historically, English Clocks were introduced and tried, but failed to be adopted in the field, although Dutch Clock auctions are over a century old (Cassady, 1967, p. 196). This fact informs us of the event that a constructivist plan to sell English Clock mechanisms was not ecologically fit at the time it was tried. Since the 1980s, experimentalists have regularly used the English Clock auction when they wanted to implement an incentive procedure that will induce participants to reveal their true willingness to pay for a gamble or any object.[2]

There are, however, prominent disadvantages of the progressive auction: If bidders are risk-averse, then revenue will be higher in the first-price sealed bid auction, which the seller will prefer; (2) the procedure has high transactions cost in the sense that bidders must be present at the auction, tending to it in real time; and (3) it greatly facilitates the opportunity for collusive arrangements, or "rings" among the bidders, a fact that accounts for much of the colorful history of auctions. (See Cassady, 1967, pp. 177–92, on rings, and pp. 212–18 for other forms of collusion routinely observed in the history of English auctions.)

Concerning (2), some auctions reduce transaction cost by defining off-floor bidding procedures; stamp (and some fish) auctions permit "book bids" by buyers not present at the auction, and Sotheby's, for example, allows buyers to monitor the auction by telephone and submit oral telephone bids to the auctioneer for paintings and collectibles. Book bids led practicing auction houses to discover that the second-price sealed-bid auction is isomorphic to the English auction – ecological rationality preceded Vickrey's formal analysis of the second-price sealed bid auction. Hence the rule that if there is one book bid, the auctioneer calls for a bid from

---

[2] But see Berg et al. (2005), showing that the English Clock auction determines cash values for gambles that are statistically different than those measured by other mechanisms. This work, when it first appeared in prepublication form, provided the first well-documented results showing that the standard theory of risk-averse behavior (expected payoff maximization) was not independent of context.

the floor. Given a bid from the floor, the auctioneer advances the bid for the account of the book bidder by the standard increment, Δ. If this bid is further advanced from the floor, the auctioneer again advances the bid by Δ for the book bidder, and so on, until either the book bidder is eliminated by a higher bid from the floor or bidding from the floor ceases, in which case the item is knocked down to the book bidder at a price equal to the last standing floor bid plus Δ. If there are two (or more) book bids, then the auctioneer starts the bidding at the second-highest book bid plus Δ. If there are no counterbids from the floor, the award is to the highest book bid at Δ over the second book bid.

As for (3), the great transparency of the English auction, if you know who is bidding and how much, greatly facilitates the monitoring of collusive arrangements in real time. The English auction provides free real-time information that enables ring members to identify those who are not living up to their advance collusive agreement. It also allows ring members to determine who among their competitors outside the ring should be included in their ring at subsequent auctions. (This last consideration applies also to sealed-bid auctions where the seller announces all bids after the auction – an unnecessary procedure, as bidders need only know what the award price is to determine the success of their bid and to verify that their bid was correctly processed.) The English auction also enables collusion in the complete absence of prior agreements or communication at auction time: Associates who know each other find it natural not to raise each other's bid. (Outer-continental-shelf petroleum lease auctions use sealed bids that the government publicizes after each auction, thereby providing costless information services to any ring and facilitating its success; see Cox et al., 1983.)

The essence of the English auction is that the successive bids are known to all bidders, not that bidders are publicly identified. For this reason, auction houses have long permitted bidders, by prior prearrangement with the auctioneer, to use signals to transmit bids. Such practices are common among auction houses and have two important consequences: (1) They make collusion more difficult, and (2) if bidder valuations are not independent, secrecy in bidding allows asymmetric information between bidders to remain private. By bidding covertly, bidders who are known or believed by others to be better informed are protected from revealing any asymmetric information. This is a natural development in the evolution of property rights – private investment in information and expertise is protected by this rule. The effect of (2) tends to lower the selling price, whereas the effect of (1) is to raise it.

These observations are directly relevant to the various FCC auction mechanisms as they have evolved, since the original designers postulated, or believed, that asymmetric information was empirically so important that bidder identities were made public to allow all to update the common value (see the following section). Left out of the analysis was that it greatly facilitated gaming (strategic manipulation attempts) and collusion, something that auction houses appear to have discovered long ago, as indicated by rules that facilitate bidder anonymity; this significant real-world learning was ignored in the FCC auctions, and was painfully rediscovered at high cost. The subsequent attempt to control such behavior has led to more and more complex rules and constraints on bidding. (See McAfee and McMillan, 1996; Milgrom, 2000.) It is not evident that this sequential process, of rule modification in step t in response to designer conjectures about the interpretation of bidder behaviors in auction t-1, is convergent. The problem here comes from a second source of asymmetric information that is not part of the model of auction design: what the auction designers know about the bidders, their values, and value-independence. Theory is developed conditional on assumptions about bidders, and is relevant to optimal design only to the extent that those assumptions are valid.

All of the preceding considerations apply to offers of multiple units in English auctions under either of the two simultaneous auction procedures suggested by Vickrey (1961, 1976). The case in which multiple units are offered, discussed below, has led quite prominently to the phenomenon of "jump bidding," observed and then studied in laboratory experiments by McCabe et al. (1988), and prompting them to evaluate English Clock auctions as a means of controlling for the behavioral urge to "jump bid." Any claim that such observations were a consequence of "low stakes" was rendered false by its subsequent emergence in the high-stakes FCC auctions.

Common Values

In a path-breaking theoretical contribution, Milgrom and Weber (1982) introduced the concept of the common value auction, as distinct from the independent private values auction. But the path broken, leading to important theoretical insights, turned out to be littered with potholes in practical applications not sensitive to the negative aspect of design criteria that focused only on the common value issue. The paradigm here was said to be a feature of what was called the "mineral rights" problem: Bidders for rights to explore for oil, gas, and minerals on land or off-shore tracts each estimate (by seismic readings in the oil and gas industry, for example) the quantity of the recoverable mineral, if any, and base their bids on their own estimates.

Thus, Milgrom and Weber (1982) assume that "To a first approximation, the values of these mineral rights to the various bidders can be regarded as equal, but bidders may have differing estimates of the common value" (p. 1093). Such approximations are critical to the development of theorems, but anathema to design applications if their limitations are then ignored in testing and evaluating the design. Again, constructivist rationality is good at producing variations, but not at selection. As we have seen, this is why most ideas fail and why such failures are essential to selection and to the emergence of ecologically rational outcomes.

Suppose the value of an auctioned item to individual i is of the linear convex form

$$\tilde{V}_i = \alpha \tilde{P}_i + (1 - \alpha)\tilde{v}, \quad 1 \geq \alpha \geq 0 \tag{1}$$

where $\tilde{P}_i$ is the item's uncertain, strictly private, value to i, and $\tilde{v}$ is the item's uncertain common value to all bidders. The distribution of $\tilde{v} = \tilde{v}(s_1, \ldots, s_N)$ is assumed to depend upon information (signals), s_i, available to each i. As indicated in the previous quotation, Milgrom and Weber (1982) assume, as a first approximation, that α, a characteristic of the commodity and its economic environment, is approximately zero; it is this assumption that leads to their fundamental theorem on common value actions. Although this is a useful abstract exercise in developing the theory of auctions, and in providing an understanding of possible issues in auction design, the theorem is about a polar case, and we are lacking in objective, testable criteria for determining α. There are no systematic procedures for identifying α from the observed characteristics of an industry. That α is indeed greater than zero in applications to mineral extraction industries is consistent with the observation that oil and mineral companies buy and sell proven reserves. Thus an oil company whose marketing proficiency outstrips its supply capability may value a petroleum reserve more highly than the selling company. Indeed, empirically it may be that α is closer to unity than to zero in the petroleum sector. We don't know.

For mixed value environments of the form (1), when $0 < \alpha < 1$, we have decisively negative theoretical results: A bidding equilibrium does not exist in auctions defined in this mixed environment. The difficulties are foundational. "With multidimensional types (as represented by $\tilde{P}_i$ and $\tilde{v}$ in (1)), inference from prices may be non-monotone and complicated. It may not be possible to order the bids of agents with multidimensional information and incentives in a way that is consistent with what they can infer from prices and their resulting incentives" (Jackson, 2005, p. 2; also see

Klemperer, 1998, which examines a host of hazards consequent to the occurrence of a small amount of common value-ness in English auctions where bidders are identified and information about them is common knowledge). An important implication of this is that it cannot be claimed that, although we have no equilibrium theory for the FCC auctions, we can still apply game-theoretic "intuition," developed from complete information examples, to articulate useful guidelines about auction design (McAfee and McMillan, 1996, pp. 172–3). Another implication is that confidence in the postulated advantages of the open English format is undermined by the dissemination of too much information – more information need not imply "better" if it facilitates collusion.

The principal (risk-neutral) theoretical result that follows from the hypothesis that individual valuations are common can be stated as follows: Prices are higher in the English auction than in the first-price (or Dutch auction, which is strategically equivalent to the first-price) and the second-price sealed-bid auctions. Prices are higher because each bidder's private information on the common value is valuable information to other bidders and revealed in the open bidding process. This potential pooling of private information to produce a higher revenue outcome, and the implicit unexamined assumption that higher revenue is socially desirable, is the essential feature that drives the judgment that English auction procedures dominate the alternatives. That judgment, however, is based on the assumption that α is zero, and formal theory alone cannot say why this should be so. Somewhere along the way over the last twenty-odd years, the preceding provisional and tentative simplifying stricture, "To a first approximation, the values of these mineral rights to the various bidders can be regarded as equal...," became a central presumed "truth" that drove the FCC and other auction designs.

We note several possible or potential contraindications that provide a basis for substantial doubt and caution concerning the postulated superiority of the open English auction, conditional on values being common.

- If information not already publicly known is costly to acquire and all bidders are both rational and sophisticated, then bidders are not well motivated to invest in acquiring information if they will subsequently be compelled to reveal it in open bidding. It appears that all but the high bidder, given that the investment is a sunk cost, will reveal the value implications of their information. If so, then the decision problem devolves to assessing the probability of being the high bidder, then computing how much investment in information is justified by weighing cost against that expected benefit. One would think that "sophisticated bidders," using consulting firms hiring economic/game theorists and skilled in applying the principle of backward induction, will anticipate this future state of the world and refrain from investing in the acquisition of information that is not private to their own special circumstances. For all nonprivate information, each should let others invest and free-ride on the information they reveal at auction. But then no one rationally invests. We lack a positive theorem that addresses these needed extensions of the theory.

- This line of reasoning suggests that the common value information that will be revealed in the bids will be that which is dispersed initially among the bidders as a consequence of technological, institutional, and legal activities that had been undertaken for other business purposes, and are then discovered to be complementary to the commodity (or right) that is proposed for auction. But investment in such activities is justified independently and prior to the sale of the commodity.
- Ironically, it follows that buyer rings or consortia are not prima facie to be judged undesirable in common value auctions with costly information. Such consortia, if they pool the information of their members, provide an incentive to invest in information acquisition by reducing the free-rider problem. Rings can improve efficiency if there are subsets of the bidders whose commonality is distinct from that of others; that is, private. To the extent that a common value $\tilde{v}_K$ for bidder subset K is a technological or marketing reality, then allocations can be improved by allowing consortia to each internalize these effects through a joint venture and then bid competitively against each other. Each consortium has a common internalized value distinct from the others, and optimality is restored by competition among the K consortia that have alternative uses for the resource.
- Given the hypothesis that commodity "X" has predominantly common value components, one can evaluate alternative auction procedures, deriving theorems, as in Milgrom and Weber (1982), showing the implication of common valued-ness for auction performance. What the theory does not tell you is the answer to the following: Given commodity "X" and a description of its operating and technological characteristics, what is the value of α, $0 \leq \alpha \leq 1$, in equation 1? In particular, is α small enough to justify the choice of English over other auction forms, especially given the preceding implications for investment in information acquisition? Clearly there must be a trade-off

here. The fact that the English auction provides the greatest scope for collusion among sophisticated bidders, yielding lower prices and efficiency and requiring high transactions' (monitoring and participation) cost, must be weighed against the effect of the postulated size of α. Is α low enough for the revenue advantages of the English auction, under common values, to outweigh its disadvantages? No one knows the answer, and this information asymmetry creates hazards for auction design.

- In auction theory, where the value of the auctioned item to oneself or others is uncertain, this uncertainty is resolved at the end of the auction. This is the basis on which experimental studies have tested hypotheses derived from common value auction theory; that is, experimentalists can create these postulated conditions in the lab. But suppose, after the sale of commodity "X" at auction, the main uncertainty that is resolved is who won and who lost the rights awarded, with the winners still facing most of the uncertainty as to the value (common or private) of such rights. How should this reality affect the choice of auction format? Clearly, if most of the uncertainty will be resolved subsequently, based on "signals" purchased following the auction, this ex post auction information cannot be revealed for pooling purposes during the auction, and the case for English procedures based on common value is further compromised.

- We note, finally, that in a truly common value auction environment, efficiency is irrelevant – any allocation to any firm is efficient. The issue is exclusively one of maximizing revenue by incentivizing the public revelation of all information.

Later in this chapter, I return to some of these issues in the specific context of the FCC auctions.

Review of Relevant Experimental Results

This section provides a short summary of those well-established experimental findings that are directly relevant to the spectrum auctions. These findings cover (1) behavior in standard single object auctions; (2) behavior in common value auctions; (3) a winner's curse in private value English auctions for gambles; (4) jump bidding, impatience, inefficiency, and manipulation attempts in claimed "incentive-compatible" multiple-unit English auctions. In the next section, I discuss how English Clock auctions correct for this bad performance.

Single Object Auctions

Behavior in experimental private value English, Dutch, first- and second-price sealed-bid auctions was originally reported in Coppinger et al. (1980; reprinted in Smith, 1991; also see Kagel and Roth, 1995, Chapter 6, for a comprehensive survey). Oral English auction prices and efficiencies were found to vary insignificantly from the predictions of theory; that is, most awards were to the highest-value bidder at prices reasonably near to the second-highest value. A very few awards, somewhat above the second-highest value, could be attributable to minor jump bidding in the sense that bidders, free to select the amount by which they raised the standing bid, sometimes chose an increment larger than necessary. (Jump bidding is much more serious in multiple-unit English auctions "Jump Bidding and the Class of Badly Performing Multiple-Unit English Auctions.") A few awards were at prices well below the second-highest valuation (Coppinger et al., 1980, Chart 2). These may have been due to subjects who knew each other and refrained from raising each other's bid, which was indicated in one case. First-price sealed-bid auction prices were significantly higher than those for the second-price sealed-bid auction.

A more extensive comparison of 780 Dutch, first- and second-price sealed-bid auctions in an environment in which values are drawn from a rectangular distribution is reported by Cox et al. (1982; reprinted in Smith, 1991). Again, first-price auctions yield higher prices than second-price auctions. Also, Dutch prices are found to be significantly lower than those in first-price auctions. Efficiency was highest in second-price auctions (94 percent of the awards were Pareto-optimal), lower in first-price auctions (88 percent), and lowest in Dutch auctions (80 percent).

Common Value Auctions

Experiments finesse the ambiguity, indicated in the preceding section, in the mix of private and common values by creating a controlled valuation environment in which $\alpha = 0$ or 1. Initial experimental studies of first-price sealed-bid auctions with strict common values, $\alpha = 0$ (Kagel and Levin, 1986), identified serious problems: Subjects did not discount properly their private information except in small groups of three or four bidders, and suffered the "winner's curse." All bidders receive unbiased estimates of the value of the object, but the highest of several estimates is biased upward. The optimal bid must take into account this characteristic of the "order statistic." Hence, the Nash equilibrium bid of each bidder is below each bidder's private estimate. Inexperienced subjects tend to bid the value indicated in their "signals" or a larger value. Consequently, the highest signal bidder wins, but pays too much on average, and profits on average are negative.

However, this preliminary finding in Kagel and Levin (1986) did not hold up with experienced bidders. Cox and Smith (1992) found that with experienced bidders (who could elect to enter the auction, or eliminate themselves and receive a modest return each period), subjects tended to bid below their estimates of value and made positive profits. Subsequently, Garvin and Kagel (1994) confirmed these results in their study of "learning" (experience) in common value auctions. It is now well established that in most cases even unsophisticated bidders with training and experience have no significant difficulty overcoming the winner's curse in common value auctions; the few who do not subsequently eliminate themselves by self-selection (Dyer and Kagel, 1996, provide an interpretation of how the commercial construction industry corrects for the winner's curse).

Kagel and Levin (1986; see also Kagel and Roth, 1995, p. 548) report common value auction results comparing first price with English auctions. Average prices in the English auctions were lower than in the first-price sealed-bid auctions, and both were higher than the expected common value, yielding negative profits under both auction procedures. But losses were lower in the English auction. Consequently, the effect of the winner's curse is reduced but not eliminated in English auctions.

A "Winner's Curse" in Private Value English Auctions for Gambles?

Two studies (Berg et al., 2005; Chew and Nishimura, 1999) have reported results from English auctions of objects of uncertain value – gambles yielding alternative monetary prizes with known stated probabilities. The two sets of results are qualitatively very similar, although the auction formats, and commodity gambles, are quite different. Cox and Grether (1996) – incidental to their study of preference reversals – report similar findings. All three report English auction prices in excess of the expected value of the auctioned gamble, supporting the conclusion that uncertainty in these private value auctions leads to overbidding and losses.

Jump Bidding and the Class of Badly Performing Multiple-Unit English Auctions

Jump bidding in the form of bid increases higher than the minimum required increment and, equivalently, raising your own standing high bid ("upping yourself") occurred in all of the first three FCC auctions. This is a clear violation of one of the two dominant strategies of rational bidding rules in the theory of English auction bidding. It occurred most notably in the first two narrowband license offerings, but less in the third (wideband) auction, which might have been the result of "learning," the higher value of the widebands (these interpretations are suggested by McAfee and McMillan, 1996, pp. 168–9, but just as likely it could be due to the scarcity of bidders). Since the major bidders all used game theorists as consultants, what was there to "learn" about the most elementary principle in the constructivist model of "rational" bidding? Standard theory views such bidding behavior as transparently irrational. Such bidding, however, had been observed as early as 1988 in multiple-unit experimental auctions.

Vickrey (1961) proposed an extension of the single-object English auction for the multiple-unit case (each bidder desiring at most one of the m homogeneous units offered for sale) as follows:

In simultaneous auctioning the m items can be put up simultaneously, and each bidder permitted to raise his bid even when this does not make his bid the highest. When a point is reached such that no bidder wishes to raise his bid further the items are awarded to the m highest bids.... Bidders with the top m values then secure the article at a uniform price equal to the (m + 1)st value; the result is again Pareto-optimal (p. 24).

We note that this mechanism does not require or impose a uniform price. In the single-object English auction, the highest-value bidder, never raising the price by more than the minimum increment, will not need to bid higher, once her bid equals or just exceeds the second-highest value. Similarly, in the preceding quotation, Vickrey is stating that the m highest-value bidders will each rationally raise their bid only to the level of the (m + 1)st highest value (or just above), and as a consequence they will each be awarded an item at the approximately common bid price equal to the (m + 1)st highest value.

In their comparison of eight variations on English multiple-unit auctions, McCabe et al. (1991b; hereafter MRS)[3] refer to this version of the Vickrey procedure as the Simultaneous (SIM) Bid Auction, which is similar to the Japanese hand signal auction for selling fish in single lots, and the Japanese electronic auctions for fruit and vegetables (Cassady, 1967, pp. 63–6; 197–8).

Subsequently, Vickrey (1976) modified his earlier proposed simultaneous auction for n units – referred to as the Vickrey Matching (M) Auction by MRS – as follows:

---

[3] In McCabe et al. (1990), a ninth mechanism – the multiple-unit uniform-price Dutch auction – is modeled theoretically and tested.

A Pareto optimal procedure is available, however, if all the items are auctioned simultaneously, with up to n bids permitted at any given level, the rule being that once n bids have been made equal to the highest bid, any further bid must be higher than this. Within the "jitter" determined by the minimum acceptable bid increment, this assures optimal results . . . (p. 52).

The results are optimal because once n matching bids reach the $n + 1^{st}$ value, no one will raise the bid. In all of the experiments reported by MRS, four units are offered for sale to ten bidders, and the private resale values for those units are drawn independently, with replacement from a uniform distribution with support $(1, 224)$ (see ibid., p. 46, Table 1). These facts are common information to all subjects.

When MRS ran the "Pareto-optimal" Vickrey Matching Auction (and SIM), they "found that prices were higher than predicted, and in general, subjects were raising their bids by too large an amount. When one of the three highest value subjects raises the suggested price p^t by too much, all four units may not trade" (ibid., p. 54). Their results, showing the effect of such jump bidding under several treatment variations on Vickrey's proposals, are displayed in a series of figures along with those for the English Clock auction that emerges as uniformly the best procedure studied by MRS (see ibid., figures 1, 2, 4–8).

Finally, MRS (ibid.) implemented a backtracking procedure (called "Vickrey Backtracking"), allowing the price to fall, whenever price bids overshot the $n + 1^{st}$ value and there were fewer than four bidders in the contract. This "fix" greatly improved allocations, but there was gaming by some subjects who anticipated that the price would overshoot and then fall, and less cautiously jumped the bid by too much. The major problem with the procedure was that it took a much longer time relative to the other auctions, as backtracking was widely employed, and transactions (time) cost escalated. The same issues plagued the FCC auctions discussed later in this chapter.

These experiments illustrate the potential to initiate the following costly design cycle. You begin with a precise theoretically "optimal" auction procedure – both of Vickrey's proposals for multiple-unit English auctions seem transparently optimal. It was an "elementary" exercise in constructivism, but it was not ecologically fit. In implementation, the model encountered behavioral incentives or "strategic" problems not considered as part of the original theory and likely intractable from a theoretical point of view. You come up with a rule "fix" to provide a countervailing incentive. This creates a new problem requiring a new rule adjustment, and so on.

The English Clock Corrects Bad Performance

What MRS (ibid.) discovered was that the problem arose from a common feature in all the "fix" attempts: bidder control over price bids from the floor. Jump bidding seriously violated the elementary dominant strategy rule: Don't raise your own bid. Yet clearly, if the minimum bid increment is too small and people value their time, there will be a completely rational motivation to jump the bid. Alternatively, if you believe that jump bidding will signal determination to outbid, and encourage other bidders to drop out of the bidding, to you that is rational. But if you are correct in that belief, it means one of your competitors violates the rule that she should remain active and raise the bid anytime it is below her value. (Under common value bidding theory, such aggressive belief bidding is pooled as providing information of higher common value.) In any case, a critical piece of analysis was left out of the "rational" model, and it took real cash-motivated people to expose the error. Discovering the intransigence of jump bidding required many experiments.

This is an example of an incentive issue not readily anticipated by formal analysis that can surface naturally in experiments and make sense ex post. When you do pencil-and-paper theory, you just cannot intuit every assumption that will be relevant; it is part of the dirt of human frailty, and people like to sweep it under the rug. Logically, it makes sense that "impatience," if this be the right characterization, would disappear with high stakes, but it surfaced in the spectrum auctions where hundreds of millions of dollars could be at stake, and bid jumps in the millions occurred. (See Isaac et al., 2007, who extend MRS in offering a theory of jump bidding based on impatience and strategic considerations in the spectrum auctions.)

The bottom line, transparently evident in the MRS results, is that if you want to do English multiple-unit (incentive-compatible) auctions, the way to do them is to use the English Clock. In forty-four English Clock auctions, only one failed to award the item to a highest-value buyer. This method dominates all others in terms of efficiency (percent of maximum possible joint payoff that is realized by the participants). There can be no jump bidding because no one can bid a price. Thus, MRS concluded "that the English Clock is our best implementation and is likely to find acceptance in the field. This auction gives participants feedback during the auction, . . . produces consistent pricing and very high efficiency, [and] can accommodate programmed (or electronic) . . . bidding" (p. 70). Essentially, the procedure works well because it removes from bidders the right to announce bids from the floor – they can only indicate willingness to be in, or out, at the standing price, and once out they cannot reenter (in the MRS implementation). Bidding from the floor invites jump bidding, collusion, and longer auctions. This was discovered in the lab and subsequently confirmed by experience in the high-stakes FCC spectrum auctions in the field.

The English Clock, which failed to find acceptance forty to fifty years ago (Cassady, 1967, p. 196), is now in the process of finding many field applications. (See Banks et al., 2003, and particularly Milgrom, 2004; Cramton et al., 2006; Porter et al., 2005. But there is much entrenched resistance, which explains why it did not find ready acceptance earlier.

Combinatorial Auctions

The general concept of smart computer-assisted markets, and the specific version known as a "combinatorial auction," originated in the early 1980s (Rassenti, 1981; Rassenti et al., 1982; hereafter RSB). As noted by RSB:

To our knowledge, this study constitutes the first attempt to design a "smart" computer-assisted exchange institution.[4] In all the computer-assisted markets known to us in the field, as well as those studied in laboratory experiments, the computer passively records bids and contracts and routinely enforces the trading rules of the institution. The RSB mechanism has potential application to any market in which commodities are composed of combinations of elemental items (or characteristics). The distinguishing feature of our combinatorial auction is that it allows consumers to define the commodity by means of the bids tendered for alternative packages of elemental items. It eliminates the necessity for producers to anticipate, perhaps at substantial risk and cost, the commodity packages valued most highly in the market. Provided that bids are demand revealing, and that income effects can be ignored, the mechanism guarantees Pareto optimality in the commodity packages that will be "produced" and in the allocation of the elemental resources. The experimental results suggest that: (a) the procedures of the mechanism are operational, i.e., motivated individuals can execute the required task with a minimum of instruction and training; (b) the extent of demand under revelation by participants is not large, i.e., allocative efficiencies of 98–99% of the possible surplus seem to be achievable over time with experienced bidders. This occurred despite repeated early attempts by inexperienced subjects to manipulate the mechanism and to engage in speculative purchases (p. 672).

Subsequently, the idea of smart computer-assisted markets was applied in a proposal to deregulate the electric power industry in Arizona by separating

---

[4] Based on hindsight, an exception is in several market-like public good mechanisms that appeared in the period 1977 to 1984 using the Groves-Ledyard and other mechanisms. (See Smith, 1991, Part III, pp. 375–506; also Kagel and Roth, 1995, Chapter 2.)

the "wires business" from energy sales, and creating a smart market in the form of the Arizona Energy Exchange (Economic Science Laboratory Research Group, 1985); in a Federal Energy Regulatory Commission study of the application of linear programming algorithms to the processing of node-specific bids to buy delivered gas and offers to sell wellhead gas, and leg-specific offers of pipeline capacity by multiple rights holders (McCabe et al., 1989); in a two-sided combinatorial auction for trading pollution permits (Ishikida et al., 2001); and in a trucking logistic problem (Ledyard et al., 2002).

The RSB (1982) mechanism addressed three problems generic to the combinatorial features of the commodity space: (1) separating prices (Lagrange multipliers) in the optimization do not exist; (2) in view of (1), it must be determined which information should be reported to the bidders after each sealed-bid auction allocation (or round in the case of multiple-round auction mechanisms); and (3) the behavioral incentive properties of the resulting rules must be identified. An integer programming algorithm was devised that allocated integer elements $\{0, 1\}$ to packages that maximized reported surplus as contained in the bids submitted for the packages, subject to constraints on the supply of each elemental resource. Two pseudodual programs to this primal problem were used to define a set of *accepted packages*, A, and a set of *rejected packages*, R; also, a set of lower bound prices $\{w_i^*\}$ and a set of upper bound package prices $\{v_i^*\}$ were determined. Then, (a) if a package bid were greater than the sum of its component values in the set $\{v_i^*\}$, it was in A, and, except in rare marginal cases, the bidder paid less for the package than her bid, providing good incentives not to under reveal true value; (b) if it were less than the sum of its component prices in $\{w_i^*\}$, it was in R; (c) all bids between A and R were in a region where acceptance or rejection was critically dependent upon the integer constraints on the allocation of the elemental resources. Thus, each bidder knew that in a subsequent auction (or round if an iterated procedure were used) whether a best reply would certainly be accepted, certainly be rejected, or depended on the integral "fitness" of the bid.[5] A bulletin-board after-market was used to allow subjects to adjust further for elements needed to fill out packages after each primary auction using the combinatorial auction (CA) mechanism. This was compared with independent auctions (IAs) in which each resource was

---

[5] All the indicated value information is private for each agent. The mechanism allows bids of the form, "Accept no more than p of the following q packages," or, "Don't spend more than $M; accept package A only if B is accepted." Any logical constraints linear in the (0, 1) variables are acceptable. The experiments did not utilize these bid options.

auctioned simultaneously in a sealed-bid uniform-price auction, followed by the bulletin-board after-market.

This computational and feedback reporting process in RSB yielded efficiencies for experienced bidders no lower than 97.8 percent in a "difficult" combinatorial environment, and no lower than 83.2 percent in an "easy" combinatorial environment. Efficiency in the difficult environment with experienced subjects weakly dominated efficiency in the easy environment. In both environments, efficiency in the combo auction mechanism tended to be higher than the independent auction mechanism.

The lower efficiencies indicated in some of the experiments were a consequence of subject attempts to acquire units strategically in the primary auction for resale in the after-market. The independent auction invited these manipulative attempts much more prominently in both the easy and difficult combinatorial environments. With experienced subjects, the combo mechanism invited such attempts slightly more in the easy than the difficult environment. This small difference was conjectured to be due to increased transparency when combinatorial complexity is minimal. In more complex environments, it is especially risky to fail to obtain the desired packages: Bidders are motivated to better reveal true package value and thereby rely on computer assistance to put together valuable packages. This was the primary learning experience that subjects went through in using the RSB combo mechanism. The fact that it performed better (relative to the independent auctions) in the more complex combinatorial environment is consistent with the findings reported in Ledyard et al. (1997), in which single-item commodity awards were less efficient than package-item awards. I suggest that the more transparent environments simply invite manipulation attempts more readily than complex environments where, with private information, it is more difficult to capture benefits that outweigh the cost risk of manipulation. The combinatorial auction performed so well in the primary auction that there was little room left for improvement in the after-market.

Tests of SMR and a Proposed Alternative

In response to an FCC call for a new combinatorial auction design, Charles River and Associates, Inc. (1998a, 1998b, 1998c), proposed and Banks et al. (2003; hereafter BOPRS) tested a hybrid auction system that combines multiround and continuous bidding periods. It also implements various activity rules that are like those emerging from the original SMR auction along with some new variable and more complicated activity rules.

The BOPRS test auction environment consisted of one unit of each of ten items for sale. Six bidders each had different values for various packages or subsets of the ten items. A strategic impediment to efficient outcomes in any auction mechanism for package bidding is the possibility that value interdependence will create the following threshold problem: Suppose that each of two bidders is bidding on a separate item, but a third bidder is bidding on a package that contains both items. If each knows the values to all three, then the two bidders must implicitly coordinate through their bidding to ascertain what price each will pay so that the sum of both bids exceeds the package bid. If the sum of the values for the two bidders exceeds the package value to the third bidder, and their coordination fails, then the award will be inefficient. Even if real bidders in the world and in the experiment *have only private information*, the value environment should include the possibility that there are threshold effects even if unknown to the bidders. This is because of the large number of experiments – for example, the electric power experiments discussed in Chapter 3 – in which market participants with a strategic advantage find ways to use it effectively simply by trial-and-error learning in repetitive environments. However, RSB found that the complexity of their combo mechanism caused manipulation attempts to fail, and these effects were self-correcting with experience. This, of course, is one of the objectives of mechanism test bedding: to find efficiency-promoting rules that discipline manipulation attempts. In effect, we look for ecologically fit rules that are used to modify constructivist designs before trying to implement them in practice.

The threshold problem motivated BOPRS to use variations on the following two conditions for generating challenging distributions of bidder values to test the efficacy of auction designs:

- Condition 1: The Joint Value Factor. This measures the relative difference between the value of the optimal allocation (V^*),[6] constructed to include several smaller bidder packages, and the next highest value allocation (V),[7] constructed to include a single bidder's package covering the optimal set of smaller packages. Define the Joint Value Factor as the ratio $J = (V/V^*)$. The larger the value of J, the greater is the need for the smaller-package bidders to coordinate their bids in order to

---

[6] V^* is the value of the problem max $\Sigma v_j \cdot x_j$ subject to $x_j \in \{0,1\}$ and $\Sigma x_j \cdot y_j \leq Y$, where v_j is the value for package j, y_j is the vector of elements of package j, and Y is the vector of total supply.

[7] V is the value of the max $\{v_{1Y}, \ldots, v_{nY}\}$, where v_{iY} is participant i's value for the entire supply Y.

"defeat" the bidder with the highest package bid. This coordination is made difficult by the incentive bidders having to free-ride off the bid increases of other bidders. But note that as J approaches 1, the relative loss in efficiency from not having the big bidder win approaches 0. J directly measures the opportunity cost of an auction, and thus fails to solve the threshold problem.

- Condition 2: The Own Effect. This occurs when b is the large package bidder who demands V, but b is also one of the small package bidders included in V^*. To achieve the optimal allocation, b must forgo his large package to be included in the optimal allocation of smaller winning packages. Efficiency may be hurt because b may not collaborate in his role as a small package bidder, for two reasons: First, he may feel he is in a stronger negotiating position, since he owns the large package, and may demand more of the surplus than other small package bidders; or second, he may think that displacing himself, even if it is apparently profitable, may create unpredictable dynamics, as in the RSB experiments.

The preceding reasoning stems from the perspective of complete information, and the best way to control for all such sources of inefficiency is to maintain privacy in the auction design and procedures. But it is still an open question as to the ability of agents to achieve efficient outcomes in nonclassical environments as they do in classical environments. (See van Boening and Wilcox, 1996, and Durham et al., 2004, for problematic private information environments using double auction and posted-offer trading rules that likely require combinatorial or other mechanisms to achieve efficient outcomes.)

Tables 6.1 and 6.2 show the optimal and second-best allocations to six bidders for the various experiments for awarding ten licenses (A through J), with different values to different bidders. Student participants, bidding anonymously through a local area network of computers, earned significant cash payments for their bidding successes.[8]

In Environment 1 (Table 6.1), bidder 6 has a value for the super-package and is either included in the optimal allocation or not depending on whether he or bidder 4 is assigned the package (F, G). The Join is either high, 0.81 (350/430), or low, 0.70 (301/430), depending on which of the listed values are assigned to bidder 6 for the super-package.

---

[8] Complete value parameters (not shown in Tables 6.1 and 6.2) for these experiments can be viewed at http://ices3.gmu.edu/FCC_Parameters.

Table 6.1. *Environment 1, optimal and second-best assignments*

				Optimal allocation of 10 licenses							
A	B	C	D	E	F	G	H	I	J	$ Value	Bidder ID
◆	◆		◆							100	1
		◆					◆			80	2
			◆			◆				80	3
					◆	◆				120	4 or 6
									◆	50	5
			2nd-best allocation of 10 licenses								
◆	◆	◆	◆	◆	◆	◆	◆	◆	◆	350 or 301	6

Table 6.2. *Environment 2, optimal and second-best assignments*

				Optimal allocation of 10 licenses to six bidders							
A	B	C	D	E	F	G	H	I	J	$ Value	Bidder ID
◆										17.37	4
	◆									36.27	5
		◆	◆	◆						88.59	1
					◆					24.00	2
						◆				30.00	4
							◆			36.00	3
								◆		48.00	1
									◆	54.00	5 or 6
			2nd-best allocations of 10 licenses								
					◆	◆	◆	◆	◆	180 or 153	5
◆	◆	◆	◆	◆						114	1

In Environment 2 (Table 6.2), the licenses were divided into two separable groups of five with separate join factors and own effects within each group and with no bidders valuing packages containing licenses in both groups. In the first group, 2a, one bidder had value for the entire license group from A to E, and also value for a package that was part of the optimal allocation. In the second group, 2b, of licenses F through J, each bidder had values for single licenses except bidder 5, who had value for the entire group of licenses F through J.

Table 6.3 lists the efficiencies achieved for thirty-one experiments comparing the FCC's SMR with the proposed CRA auction mechanism (the

Table 6.3. *Combo Clock, CC versus CRA versus SMR*

Case	J Factor	Own Effect	Auction	% Efficiency % Allocation efficiency
			CC	100,100,100
1	.81	Yes	CRA	78, 79, 78
			SMR	59
			CC	100,100,100
1	.81	No	CRA	97, 79
			SMR	63
			CC	100,100,100,100
1	.70	Yes	CRA	100, 100
			SMR	70
			CC	100,100,99,100,99,100
2a	.80	Yes	CRA	99, 99, 99, 95, 94, 95, 95
			SMR	100, 99, 95, 95
			CC	100,100,100
2b	.94	Yes	CRA	91, 94, 94
			SMR	100
			CC	100,100,100
2b	.94	No	CRA	95, 95
			SMR	100
			CC	100,100,100
2b	.80	Yes	CRA	100, 91
			SMR	100

results listed for Combo Clock (CC) will be discussed in the next section). As the join factor increases, efficiency falls. The own effect also reduces efficiency. Even though efficiencies are low for the CRA auction in Case 1, they are higher than for the SMR auction. In Case 2b, in which bidders have values only for single-license packages, except for one bidder who has a value for all the licenses, the SMR auction outperforms the CRA auction.

The FCC Auction Design Process

The FCC conducted several high-profile and high-priced auctions for communications' spectra during the 1990s. The SMR emerged after much early discussion, debate, and repeated modification in the light of experience with previous auctions. The FCC did not originally opt for a combinatorial form of auction, for three stated reasons:

Computational uncertainty: On any round during a sequential combinatorial auction, the selection of the round's set of winning bids and what bid levels would be necessary for competition to displace them typically requires the solution of integer programming (IP) problems. These problems are notorious for being computationally burdensome; they are technically described as "NP complete" or hard problems: There is no guarantee that an algorithm will be able to find the solution for such a problem in a "reasonable" amount of time, if the number of bidders and items becomes large.

We indicate later in this chapter how this problem is easily solved, for the cases considered, by the Combo Clock, in which optimization calculations are required only once after all rounds have been completed. Moreover, with computational assistance, unsophisticated bidders have no difficulty achieving very efficient outcomes, and the mechanism reveals no information enabling sophisticated bidders to manipulate the mechanism to their advantage.

Bidding complexity: Combinatorial auctions would be burdensome and difficult for participants and the auctioneer for at least two reasons. First, a bidder might want to place bids on any number of packages, and selecting any subset may be strategically awkward and provide the auctioneer with incomplete information. Second, the bidder has the computational burden of determining how much to bid to be successful.

In the auction design literature, this is sometimes referred to as the $2N - 1$ bogeyman since this is the magnitude of the many combinations of items a bidder must consider. McAfee and McMillan (1996) argue that combinatorial calculation is impractical because for values of N likely to be encountered, there are more alternatives than there are particles in the universe. But we all routinely face similarly complex problems, such as the problem of shopping in supermarkets. Suppose you have $100 in your pocket and you are standing at the entrance to the supermarket. The preceding argument suggests that you ought to realize despondently that you may never come out, as you must first consider every possible way you might fill your shopping cart. I think it unlikely that those concerned about the computational burden in FCC auction design avoid shopping in supermarkets, but their brains are untroubled by this supermarket shopping problem. The brain is good at reducing these problems to manageable proportions – based on one's past discovery of own preferences – by routines that make no attempt to call on the mind's conscious modeling resources to evaluate preferences over everything in the available set. Similarly, bidders for components of the spectrum will be specialists in certain uses that will limit the number of alternatives that are worth considering. This is a good example of the difference between constructivist and ecological rationality: Few complete information constructive models of optimal behavior will likely be ecologically fit.

Financial exposure: Bidders in an SMR auction who had super-additive values for a particular package of licenses were subject to financial exposure if they missed acquiring at least one of those licenses. The FCC auctions implemented a withdrawal rule to "reduce" the potential financial exposure of the bidders: that is, bidders had the right to withdraw bids on particular licenses subject to the fact that they were obligated to pay the difference between their own and the final winning bid if it was less. This and other SMR rules led to results in various FCC auctions that revealed some interesting, perverse strategies. In particular, individuals would withdraw and then bid just below their withdrawn bid to signal a willingness to not compete. To manage eligibility, bidders would bid on items they did not value in order to maintain activity without showing their hand on items for which they were interested in bidding; this was called "parking" (see Porter, 1999). This unintended effect of a rule designed to solve another problem prompted several auction rule adjustments. (Recall that the Baumol rule discussed in Chapter 4 for controlling predatory pricing had the unintended effect of facilitating collusion that did not otherwise occur.) Two notable rule changes implemented were restrictions on jump bidding – the allowable increment when raising your bid – and restrictions on the number of withdrawals allowed, a change precipitated by concerns that bidders were using such withdrawals to "game" the auction, as opposed to using them to eliminate financial risk. This illustrates the anatomy of rule revision to control gaming incentives, which induces new gaming incentives, new rule changes, and so on, as in MRS. But in combinatorial auctions, it is easy to have bidders submit their financial constraints in advance, let the computational assistance apply them, and avoid using the rules of the auction mechanism to control a purely mechanical problem.

As I see it, these problems are a direct consequence of the initial failure to use the laboratory to test-bed the proposed rules and all changes in the rules before modifying them at great strategic, transactional, bankruptcy, and opportunity costs in the field. Casual empiricism dismissing these means of testing because "it's not the real world" fails to confront the demonstrated capacity of experiments to uncover at low cost the many "mistakes" in auction design that also characterize field applications. In fact, parallelism (see Smith, 1976, 1982a) in the learning from field and laboratory environments has been evident from the first studies of Treasury auctions to designs for electricity liberalization. (See the discussion in Chapter 3 on economic systems design.) After the laboratory exercises have eliminated all the elementary errors made in the FCC auction designs, there is time enough to see what further modification issues need to be addressed because of any unique implementation problems that arise in the field.

Based on these problems with the SMR design and spurred by the FCC-sponsored debate, several new designs have emerged.

Auction Design for Complex Environments

It is quite natural for auction participants to wish to reveal as little as possible concerning their interest and maximum value for particular items or packages in an auction that simultaneously sells multiple items. In general, auction systems that provide feedback and allow bidders to revise their bids seem to produce more efficient outcomes. This feedback feature is the cornerstone of the recent SMR spectrum auctions by the FCC and many others worldwide.

An iterative combinatorial auction could allow bidders to explore the bid space without having to place bids for all possible items. Among the iterative designs, prior to the much simpler Combo Clock, several alternatives had been examined:

- *Continuous Auction* (Banks et al., 1989). A timer is started and bids can be submitted in real time. The best initial bid combination that fits within the logistic constraints of the auction is posted. New bids can be tendered at any time and can either be placed on a standby list to be "combined" with others or used immediately to replace tentatively winning bids directly. The standby list is a place where participants can enter nonwinning bids to signal to others a willingness to combine to outbid a large package bid. The auction ends if no trades occur during a prespecified time interval.
- *Multiround* (Charles River and Associates, Inc., 1998a,b,c; Kwasnica et al., 2003). A round begins and sealed bids are submitted. An integer program (IP) is solved to find the highest value combination of winning bids. The winners are posted and a new round is started. New sealed bids can be submitted and the IP is run again to see whether a new solution, with some new winners, is found. When there are no new winners or no new bids, the auction is over.

The most obvious approach to facilitating solutions to the threshold problem is to use a Vickrey auction in which it is theoretically in the interest of the participant to bid her true values – each bidder's assigned cost is determined by the bids of others, not her bid. In particular, Isaac and James (2000b) use the standard Vickrey auction that requires the running of an IP for each winner to determine an item's price. The RSB auction summarized previously eliminates the need to run individual IPs for each winner and instead uses pseudocompetitive prices by running a specialized linear program to get prices for each item.

Banks et al. (1989) implemented a computationally intensive iterative process where in each round Vickrey prices are computed for each item and are delivered to each bidder along with a set of bids that are best replies to the other bidders. Following the pseudocompetitive pricing of RSB, Kwasnica et al. (2003) use a procedure to calculate prices as feedback information to bidders. They find that provision of this price information allows bidders to identify competitive bids more easily and, in so doing, generate a more efficient assignment of items.

Although more efficient, these new designs still require solving multiple IP problems across rounds and thus are still subject to computational complexity and scaling-up concerns. To solve this problem, Rothkopf et al. (1998) would simply limit the type and number of bids that can be submitted to achieve computational feasibility, which has an arbitrary and unknown effect in reduced efficiency and could easily be dominated by other mechanisms. Banks et al. (1989) would eliminate the computational burden of the auctioneer through decentralization, requiring the bidders to execute any computations to determine what to bid on and how much to bid.

The Combo Clock: Simple Solutions for Complex Auctions

This section describes a transparent, efficient, and practical combinatorial buyers' auction for multiple distinct items with perhaps multiple units of each available (Porter et al., 2003; hereafter, PRRS). The design and its experimental testing were motivated by the preceding critique of the FCC auction formats. Many of the design principles ultimately implemented by PRRS are easily extendible to more complex two-sided[9] negotiations with technical and capacity[10] constraints. As noted previously, this mechanism grew out of laboratory experience with market failure when the bidders

---

[9] A buyers' auction becomes two-sided when sellers submit inelastic offers (in effect, at zero asking prices) of their available units for sale, then participate as buyers who are free to buy back any portion of their own offerings as a means of ensuring minimum acceptable exchange prices.

[10] For example, conservation of flow in network distribution systems or storage and channel capacity limits can be applied as allocation constraints in the auction proposed.

announced multiple-unit English auction bids, and was used to modify the combinatorial auction principles used originally by RSB and subsequently developed by others. A sequential version of the mechanism (without the simultaneous combo feature) was adapted recently for live application to the Virginia State NOX auctions (Porter et al., 2005).

During February through October of 1999, while testing versions of the FCC's SMR and the CRA proposal, PRRS had the chance to observe carefully hundreds of auction participants' behavior and the outcomes that their strategies generated. PRRS took seriously the computational burden of generating round-by-round prices even in moderately sized combinatorial auctions; they also wanted to avoid imposing activity constraints that interfere with transparency and raise transactions and strategizing cost, as had been done in the FCC auctions. The result was a simple new auction design: the Combinatorial Clock (CC) auction. The objective was efficiency in achieving all gains from exchange, task simplicity for the bidders, efficacy in handling complexity in the allocation problem, and computational feasibility.

As described previously, MRS demonstrated that even simple ascending English auctions can become inefficient because of overstated ("jump") bids motivated by impatience or the use of signals in the hope of forestalling competition. In experiments with the SMR, financial exposure aggravates this behavior. The MRS solution is a simple upward-ticking clock controlled by the auctioneer to remove active bidder price control. The CC, following this lesson, created a unique price for each item that is controlled by that item's clock.[11] The price ticks upward only if there is excess demand for an item.

The clocks, one for each item, are started at low prices. Each round, bidders are given a fixed amount of time to nominate which packages or individual items they would like to purchase at the current clock prices. A simple algorithm counts up the demand for each item by each bidder, making sure not to double-count licenses that appear in multiple bids. The item demands are then aggregated across participants. For items that have more than one bidder and more units demanded than are available, the clock price is raised. A new round is started and new bids are requested. For example, suppose that on some round, bidder #1 in Table 6.1 (showing optimal and second-best assignments only) has nominated the package (A, B, D) and individual items E, F, and G (individual values are not shown).

---

[11] If there is more than one unit of an item available, then the clock price for that item applies equally to all the units available.

On the next round, each price advances, and the price of E is now above bidder #1's value for E, so he nominates (A, B, D) and F and G.

The CC is transparent because each bidder is free to indicate the contents of any package whose value is above its cost at the given vector of clock prices. Each bidder's previous bids for packages at previous prices are remembered and eligible for consideration by the auctioneer unless they are deliberately withdrawn. The fact that a clock has stopped and there is only one active bidder does not as yet guarantee that bidder the award.

The auction continues as long as there is excess demand for one of the items being offered. If the auction reaches the point where there is exactly one bid left for each unit of each item available, then the auction ends and the standing bidders are awarded the items at the current clock prices – no computation is required!

But it can happen that after a particular clock price increases, the demand for that item becomes less than is available. In the case where there is excess supply for at least one item, and demand exactly equals supply for every other item available, then the auctioneer must compute the solution to an IP to find the allocation of items that would maximize surplus. The auctioneer includes all bids at current and previous clock prices that have not been withdrawn. The solution will tend to use old bids to allocate units for which there is excess supply at the current clock prices. If the resulting solution does not seek to displace those holding the standing winning bids on items where supply equals demand, the auction is over and bidders pay the prices bid. If the solution does seek to displace at least one bidder holding the standing winning bid on an item where supply equals demand, then that item is now considered to have excess demand, the clock price is ticked upward, and the auction continues.

Therefore, the final allocation is one in which all standing bids win items at the final clock prices, and items with excess supply at the final clock prices are awarded at previous clock prices. The auction process that generates this allocation is simply a greedy algorithm to discover pseudodual upper bound prices[12]: The lowest prices (final clock prices) at which everyone who submitted a bid is definitely declared a winner. In complex environments, these prices are often not unique, and this allows the auction mechanism some flexibility in achieving an efficient outcome.[13]

[12] See Rassenti et al. (1982) for a more thorough description of pseudodual price computation.

[13] For example, if a winning bid is for a package containing one unit of item C and one of item D, then the bidder isn't concerned with whether the winning prices are 40 and 60 or 60 and 40, respectively, as long as they eliminate her competition and have a total cost weakly less than she is willing to pay.

The CC auction is easy for the participants and places minimal computational requirements on the auctioneer. It trivially accommodates the sale of multiple units of multiple items. Bidders have complete freedom to move in and out of the auction bidding on any packages at will. It allows the bidder to impose logical constraints without increasing the computational burden during the auction. For example, a bidder may submit mutually exclusive bids and "if and only if" bids; the auction simply computes his demand for an item as the maximum number of units he could possibly win. The bidder is also free to blend current and previous clock prices in a current compound bid as long as part of her bid is at current clock prices.[14] The CC auction may be the most flexible known combinatorial auction. Moreover, strategic behavior is controlled by feeding back only that information bidders need to know (item prices) in order to avoid bidding more than their maximum willingness to pay. For this purpose, bidders do not need to know who is bidding, how many are bidding, and on which items or packages. Less information is better. Hence, in auction environments where certain items have few bidders – for example, timber and off-shore petroleum tracts – this fact may still elicit full-value bidding if every tract is potentially contestable, and bidders face much uncertainty about how active the bidding will be on any one item.

Table 6.3 shows the allocation efficiencies for all of the CRA, SMR, and CC auctions. The combinatorial clock auction was uniformly more efficient than both the other auction mechanisms across all the test environments. Moreover, twenty-three of the twenty-five CC auctions were 100 percent efficient and two were 99 percent efficient.

Implications for the Design of Spectrum Auctions

My discussion suggests several potential implications for the design of spectrum auctions and for other complex allocation problems. Auction design requires balancing a number of competing considerations, each one of which has an uncertain weight in the final specification of the mechanism to be used. Achieving that balance is a problem in trial-and-error selection among alternative constructively rational designs to find and choose an ecologically rational design; even if one has managed to come up with what is believed to be a sophisticated constructivist model of the process, it must be tested

[14] For example, a bid on a package containing items C and D at current clock prices in round t can be resubmitted as a package in round $t + 1$ by bidding on one item at the previous (*pre*) price as long as at least one item is bid at the current new price; for example, $\{C, D\}_t$ can become $\{preC, D\}_{t+1}$.

to see whether it is also ecologically fit because of the inherent uncertainty in conjectures as to which assumptions are relevant in abstract modeling. Consequently, the implications listed here should be viewed primarily as a basis for further theoretical, experimental, and field investigation, and for discussion among the growing number of scholars associated with applied mechanism design (Cramton et al., 2006).

Virtually all the complexity of the SMR bidding rules and SMR's antecedents derives from the assumption that spectrum licenses are predominantly characterized by common and/or affiliated values.

This assumption leads to the theoretical conclusion that revenue to the seller is greater if the identity of every bidder is made public in real time using the English auction format. The governing argument, originally, was based on a hypothetical example of the form, "If AT&T is bidding, its information is superior, and therefore what AT&T is willing to pay is an indicator of value to others." By full revelation of private identities, the dispersed information on common value is pooled, enabling the price to reflect fully all such information. This assumption, I believe, was based on remarkably casual empiricism and has been accepted as an axiom without careful examination or skeptical scientific challenge.

Even if the assumption has more or less validity, I believe the argument and theory based on it is flawed once you examine its implications beyond the given narrow analytical context in which it was introduced. The public information characteristics of the auction force revelation of value by the better-informed bidders, and thus undermine the bidders' original incentive to invest in acquiring information. To the extent that values have commonality, there will be underinvestment in information that is not firm-specific. The information revelation of the English procedure itself will therefore diminish the importance of common values in the environment, to the disadvantage of the procedure. In effect, the procedure undermines its own defining conditions.

The stated primary objective of the FCC auctions is efficiency – awarding licenses to those who value them most and can most effectively employ them. But if bidder values are in truth the same, an award to any bidder is efficient. Hence, the emphasis on common values has devolved into a preoccupation with maximizing FCC revenues and has preempted studies furthering the stated FCC efficiency objective.

There exists no litmus test for common value-ness; that is, we have no microanalytical procedure for going from the economic characteristics of the auctioned item to a statement about what proportion of a bidder's value is common to other bidders.

Nor is there a procedure for solving this problem from observations on price data. We have a theorem (Jackson, 2005), as noted in this chapter, showing that there is no equilibrium price for an item whose value is α percent private to bidder i, and $(1 - \alpha)$ percent common to all bidders. This is because there is no way to infer from the price alone what is the confounded mixture of common and private information that is reflected in that price.

All these considerations call for laboratory and field tests of alternatives to the SMR and its derivatives (those allowing combinatorial bidding, including the sealed bid format).

The examination of alternatives to the SMR auction, and its antecedents, is further indicated by the high transactions cost of participation, requiring advice from a wide range of consultants, not only to estimate maximum willingness to pay (value), which should be the primary focus of each bidder, but to deal with the strategic issues that arise in real time as the bidding proceeds, which is a deadweight loss.

The ideal incentive mechanism design should lead to a two-step procedure: (1) an estimation of the individual's maximum willingness-to-pay value of the auctioned item(s), followed by (2) the submission of this maximum value in the form of a bid, if needed, such action being a fair approximation to that which serves the interest of the bidder. Thus, in the RSB combinatorial auction summarized in this chapter, which was only approximately incentive-compatible, a difficult interdependent environment quickly yielded 98 to 99 percent efficiency with experienced bidders.

The key question for the SMR, and for combinatorial mechanisms based on open bidding, is the following: Does the benefit based on a judgment as to the importance of common value-ness outweigh the costs of higher complexity induced by the need to control manipulation? All the examples of strategic behavior – exposure, free rider, demand reduction, and so on – are based on postulated common information, known to the analyst and/or revealed in open English bidding. Consequently, the chain of causality is that common value implies English format (with bidders publicly identified) implies strategic behavior implies higher transaction cost design complexity to control that behavior. This chain needs fundamental examination.

The SMR auction has evolved over a sequence of field applications in which weaknesses and defects revealed in each application led to "fine-tuning," followed by the observation of further problems leading to new "fixes," and so on. Each fix, designed to limit strategic exploitation, tended also to generate complexity and its attendant higher transactions cost. This is reminiscent of the MRS series of experiments leading to a sequence of increasingly complicated modifications of English procedures until all such fine-tuning attempts were abandoned in favor of the elimination of price bidding – the feature that historically has been the defining characteristic of the English auction. Prices were quoted by an English Clock, and all bidders' anonymity is strictly protected.

In spectrum auctions, probably very little of the ultimate technological, antitrust, and demand uncertainty is resolved at the end of the auction. This feature is not formally part of the theory that led to the original conclusion that the English auction procedure should be used. But whatever might be the consequence of reexamining the basis for the SMR in the light of this feature, it is important also to challenge the assumption that these licenses constitute a single event allocation task to be privatized by a one-sided auction. It is more appropriate, I think, for the FCC to consider periodic two-sided auction exchange mechanisms in which misallocations caused by bidding errors or various FCC mechanism design faults in earlier auctions, together with subsequent resolution of earlier uncertainties about technology, demand, and antitrust issues, allow spectrum title holders to participate in the sale of incumbent rights in new auctions. This would facilitate repackaging of old as well as new rights in response to changing information and technology, whatever might be the auction format that is used.

If the open English bidding mechanism continues to be the primary procedure of choice, I believe all price bids should be supplied by the English Clock procedure as in PRRS, with bidders simply indicating whether they are active or no longer active, in expressing their demand for any licenses (or combination of licenses) in real time. This is straightforward for licenses that are independent, but combinatorial value bidding would require the support of optimization algorithms. Finally, the proposed and tested Combinatorial Clock auction demonstrates clearly that there exist computationally feasible procedures that are simple and transparent for the bidders to use.

The needs of the future are twofold. First, more laboratory tests by independent scholars are required, including explorations of alternative economic environments, with the objective of uncovering the Combo Clock's boundaries of validity – I believe that all mechanisms have limits to their robustness that can be determined only empirically, whether guided by theory or not. Second, tests are needed in the field in which users must be persuaded to see the merits of strict security that enables bidding to be driven primarily by anonymity. This latter need will be particularly difficult because the problem was not addressed up front – early designers were all inexperienced and hampered by asymmetry in their own information – and both users and designers have become accustomed to the fantasy that strategizing can be controlled by ever more complex rules without significantly increasing implementation costs for everyone. As I understand it from domestic applications on which I have worked, there are no legal impediments to maintaining bidder privacy during an auction, but probably all successful bidders must be revealed at some time after an auction is completed.

SEVEN

Psychology and Markets

We suffer more...when we fall from a better to a worse situation, than we ever enjoy when we rise from a worse to a better. Security, therefore, is the first and the principal object of prudence. It is averse to expose our health, our fortune, our rank, or reputation, to any sort of hazard. It is rather cautious than enterprising, and more anxious to preserve the advantages which we already possess, than forward to prompt us to the acquisition of still greater advantages.

Smith (1759; 1982, p. 213)

Psychology's Challenge to Constructivist Rationality

Researchers in psychology and behavioral economics who study decision behavior almost uniformly report results contrary to constructivist rational theory (Hogarth and Reder, 1987). This is reinforced by the manner in which papers are solicited by national and international meetings of societies devoted to this specialty. Thus, on October 25, 2005, the following email announcement was sent:

The IAREP-SABE joint meeting on "Behavioural Economics & Economic Psychology"...in Paris, July 5–8, 2006. This Conference aims at providing a platform to the fast growing number of economists, psychologists, neuroscientists and other social scientists who wish to discuss, rigorously but open-mindedly, their latest research in this field. *Relevant topics include all domains in which tenets of economic theory have been seriously and systematically challenged*, along with all concerns of economic psychology (emphasis added).

It was not always so, but the focus on what are called "anomalies," beginning in the 1970s, converted the emerging discovery enterprise into a deliberate search for contradictions between reports of behavior and standard decision theory. This has been documented by Lopes (1991):

Prior to 1970 or so, most researchers in judgment and decision-making believed that people are pretty good decision–makers.... Since then, however, opinion has taken

149

a decided turn for the worse, though the decline was not in any sense demanded by experimental results. Subjects did not suddenly become any less adept at experimental tasks nor did experimentalists begin to grade their performance against a tougher standard. Instead, researchers began selectively to emphasize some results at the expense of others.... The view that people are irrational is real in the sense that people hold it to be true. But the reality is mostly in the rhetoric (p. 66, 80).

This imbalance in the mainstream judgment and decision research program is echoed in social psychology by the review and critique in Krueger and Funder (2004).

Well before the 1970s, some economists had challenged work in decision theory under uncertainty. They disputed it as a theory that was unable to account for certain observations, such as individual purchases of lotteries and insurance, requiring explanation. Such economists sought to redefine the domain of the utility function to get a better representation (Friedman and Savage, 1948; Markowitz, 1952). Also, externalities in choice had long been recognized as potentially important qualifications of standard theory. It was the neoclassical hypothesis of a particular definition of self-interested agents, however, that led to the strongest theoretical results. Tractability is what motivated the hypothesis, not its plausibility, and therefore tractability was a prominent and easy target of criticism by the many who took the hypothesis at face value.

Daniel Kahneman has been widely associated with the idea that people are "irrational" in the narrow constructivist context used in modeling decision under uncertainty. In fact, he describes his empirical findings contradicting rationality, in this particular representation of the SSSM, as having been easy, thanks to the implausibility to any psychologist of this rational model.[1]

Unfortunately, the popular press – by focusing on the newsworthy rhetoric of behavioral economic studies – has often interpreted the contributions of Kahneman as proving that people are "irrational," in the popular sense of stupid, so elementary and transparent are their "errors," in the popular sense of mistakes. In the Nobel interview (see note 1), Kahneman seems clearly to be uncomfortable with this popular interpretation and is trying to correct it. Unless I have missed something important, neither psychologists nor behavioral economists have plainly sought to clarify that confusion. The pejorative meaning attached to words that define

---

[1] See the Nobel Foundation interview of the 2002 Nobel laureates in economics at http://www.nobel.se/economics/laureates/2002/kahneman-interview.html.

theoretical concepts and hypotheses has been all too prominent in professional interactions, and this allows easy public access to the muckraking fray in public discussions. Thus, "errors," in the sense of deviations from predictions, are referred to in the psychology and behavioral science literature as "cognitive errors," meaning mistakes as deviations from what "should be" observed based on theoretical constructions of optimal, or rational, behavior. Yet it may in fact be the theoretical constructions that are mistaken.

The "cognitive error" research program uncritically accepts the undoubted and undoubtable "truth value" of standard theory as a representation of optimality, and therefore deviant subjects are interpreted as making mistakes transparently contrary to their own rational self-interest. In science, however, observed deviations from theoretical predictions normally mean that either the theory is "wrong" – that is, it needs modification as a predictive theory, which may mean altering the original conception of optimality – or, as in tests of decision theory, if the tests are robust, human decision making is flawed. Observed deviations from predictions should not be identified with any particular unassailable interpretation of the meaning of the deviation if such confusion is to be avoided. As Kahneman (2002) has noted in the Nobel interview, if human decision making is as flawed as indicated by these representations, it is hard to understand our species' survival and occupation of the planet over the last 2 million years.

Kahneman here raised a natural question, one which the Kahneman and Tversky program did not investigate: To what kind of "optimal" decision-making process, if any, have human beings adapted? This question is at the heart of the two kinds of rationality underlying this book. After the 1970s, as noted by Lopes, this question was not addressed by mainstream cognitive psychologists, whose main business has been attacking the SSSM rather than developing a positive theory of why the alleged deficiencies in human decision have not blocked human success. Gigerenzer and his coworkers have prominently contributed much to correcting that imbalance by probing the sensitivity of the tests to context and procedures. See his critique of the Kahneman and Tversky program of narrow norms and vague heuristics in their debate: Gigerenzer (1996); Kahneman and Tversky (1996); see also Gigerenzer (1991); Gigerenzer and Selten (2002); Krueger and Funder (2004).

But looking beneath the rhetoric, psychologists, to their credit, generally have maintained an intensive program examining the contradictions between observed behavior and the predictions of classical models of choice, bargaining, and competition, and have moved the argument from an account of theoretical possibilities and anomalies to deeper empirical investigations. For example, Siegel (1959) and Fouraker and Siegel (1963) reported both confirmations and contradictions, and used the pattern to propose improved models. These were among the important contributions defining work prior to the 1970s but which were ignored in the rush to produce anomalies and to revise earlier work in the discredited "behaviorist tradition" (B. F. Skinner's aborted attempt to build a mentalist-free objective science of psychology).

Similarly, in prospect theory Kahneman and Tversky (1979) have proposed modifications in both the utility and probability weighting functions of standard expected utility theory, and thus revised the specification of optimality in expected utility theory. Research strategies that focus on the study of errors – as "mistakes" rather than deviations from predictions – can distort professional beliefs, to say nothing of popular representations, if the rhetorical emphasis is on the failures to the exclusion of the predictive successes of the theory. New theory needs to be able to embrace the old theory where it is accurate, and improve its performance where it is inaccurate.

Kahneman and Tversky's (1979) most widely recognized contributions in prospect theory were in empirical tests demonstrating the relevance of two propositions going back to Adam Smith, as expressed plainly and eloquently in the text quotation for this chapter: the idea that the theory applies to changes in wealth (that is, income) relative to the individual's current asset state, and that people choose as if they are risk-preferring in losses and risk-averse in gains. This much is not inconsistent with the axioms of standard expected utility theory, which requires only that the prizes of choice can be ordered, and therefore applies either to wealth or changes in wealth, since either can be ordered. Also, the theory does not preclude risk-preferring preferences so long as utility, kinked or smooth, is bounded.

The risk measurement experimental results (originally reported in 1994) by Berg et al. (2005) and subsequently corroborated by Isaac and James (2000a) have introduced an entirely different set of complications, and fundamental challenges to choice under uncertainty: The measurement of risk aversion for individuals varies with the type of market institution or procedures used in extracting risk measures from decisions. In fact, the measure of risk changes sign from demonstrating risk aversion in first-price sealed-bid auctions to showing risk-preferring behavior in English auctions. Thus, context shows up as a significant determinant of human perception and decision across alternatives that are formally identical in terms of probabilities and payoff outcomes. It has long been known that differing procedures for measuring what has been called "risk aversion" as traditionally defined exhibit very low correlations. In a recent such demonstration, Eckel and Wilson (2004) ask, "Is trust a risky decision?" But the inconsistencies they report between alternative measures suggest that the question should be, "Is any decision, X, a risky decision?" Holt and Laury (2002) provide impressively concise systematic comparisons of risk aversion under various reward conditions within a fixed set of measurement procedures. This does not address the problem of instability across procedures. Similarly, bidding in first-price auctions is impressively consistent with risk aversion and internally consistent across many variations (Cox et al., 1988), but says nothing about the stability of risk aversion across alternative means of measuring it.

All these particulars – impressive in the small picture – lose meaning in the whole.

Thaler (1980) proposed to extend the gain/loss asymmetry in prospect theory to explain the gap between willingness to pay (WTP) and willingness to accept (WTA), termed an "endowment effect" (or, more accurately, an ownership effect) that has surfaced in direct opinion measures of buying and selling prices. Support for the existence of such a gap has been mixed, particularly in the context of markets (Franciosi et al., 1995), to be examined later in this book, and recent contributions by List (2004) provide market field data showing that the choices of inexperienced consumers are consistent with prospect theory, but experienced consumers learn to overcome the postulated "endowment effect."

Plott and Zeiler (2005) study the WTP-WTA gap for both commodities and lotteries. They argue that if decisions are explained by preferences – in particular, a WTA-WTP gap attributable to a postulated "endowment effect" – then the observed effect of the decisions (the gap) should persist when the procedures (context) of decision are altered, an argument that is a critical part of this book and also Smith (2003). The phenomena should not disappear and reemerge as one varies the procedural context. Plott and Zeiler (2005) find, however, that this variability prominently characterizes the WTA-WTP gap. But their statement in this particular application completely parallels that of risk aversion in decision theory: Think of risk aversion as a gap between actual behavior and risk-neutral behavior. In Berg et al. (2005) and Isaac and James (2000a), the measurement of risk aversion (the gap) depends on the procedures, or the institutional context in which risk aversion is measured to the point that the sign of the gap shifts from risk-averse to risk-loving behavior. If risk attitude is a basic property of preferences over uncertain payoffs, why is it so sensitive to the procedural context? This class of issues is of priority importance; in Chapter 10, I return to the effect of procedural context as applied particularly to the choice between cooperation and equilibrium in one-shot two-person extensive form games.

Returning to the Kaheneman and Tversky (1979) results and whether the prizes are wealth or income, I intend no defense of the standard theory of decision making under risk; quite the opposite. The weakness of the theory is that it places *so little restriction* on choice. Hence, the prizes to which the theory is best applied have long seemed to be inherently a subject for empirical determination, with the objective of tightening up the wide breadth of the theory. If applied to wealth, the theory starts to infringe on preference theory across time, recognized from the start as especially difficult modeling terrain where tractability severely strains plausibility. The application to wealth was rejected empirically by Binswanger (1980) in his large-stakes tests using farmers in India.

The claims that people are not good Bayesian intuitive statisticians has been considered one more source of human cognitive errors, or mistakes in acting in their own best interests. But such predictive errors can just as strongly speak to weaknesses in the theory as to flaws in human judgment under uncertainty. An alternative to the Bayesian hypothesis is that human decision makers are adapted and attuned to decision making in uncertain environments that are not characterized by the balls-in-urns sampling paradigm – a constructivist simplification obviously driven by the demands of tractability. Is the world of decision like an urn, provisionally postulated by an individual to contain only black and red balls, but in sampling the individual proceeds to draw a yellow ball? If so, it is appropriate to "overweight" the sample evidence (as illustrated by the "representativeness heuristic") relative to prior evidence by real people, whose brains are conditioned to remember and process contextual information that carries information on the set of states and not only sampling data on a fixed set of states. In this example, a Bayesian has no updating rule conditional on observing a yellow ball because it is an event with zero prior probability. The SSSM postulates that the decision maker, with *probability one*, has an exhaustive description of the states of nature, and that all samples will be realizations only on those states. Given a sample yellow ball, the Bayesian must abruptly start over with a reformulated event space, a re-specification of "priors," and start sampling and updating again – until he draws a green ball! Yet the behavioral economics focus has been on the overweighting of samples relative to priors as one of the many "mistakes" in human judgment and decision rather than a possible clue that *the human brain might be trying to solve a different and more relevant problem* than the one posed by the Bayesian model. Does subject performance reflect a human failure in not satisfying the predictions of the theory, or is it evidence of human adaptation to a more relevant environment of decision than the one defined by the bounded rationality of the theorist? I think a good working hypothesis is the following: We owe our existence to ancestors who developed an unusual capacity to adapt to surprises, rather than to process sample information optimally from a stable historical environment in which all states were specifiable and presumed known.

Perhaps the Bayesian apparatus can be reformulated, by reserving some of the prior probability mass for a portion of the event space from which sample evidence may identify unanticipated outcome states, then update the apparatus when a new partitioning of the event space occurs. If life involves learning the composition of nature's urns, then successful samplers will want to update their decision-making formula when they encounter surprises. There are many good studies defending Bayesian decision as traditionally drawn: Gigerenzer (1991); Koehler (1996); Cosmides and Tooby (1996); Hoffrage et al. (2002); also see El-Gamal and Grether (1995). And such studies are relevant if and where event spaces are historically stable.

In principle, as I see it, experimental market economics and behavioral economics are complementary. Experimental economists study market performance (market rationality), incentives in public good provision and small group interactions, and other environments with dispersed individual valuations, whereas cognitive psychologists study the performance consistency (choice rationality) of individual decision making. If the objects traded are prospects, the appropriate valuations are their "cash values," whether based on expected utility, prospect theory (Kahneman and Tversky, 1979), or some other representation. Thus Plott and Agha (1983) study experimental markets in which speculators have the capacity to buy in a market with certainty of demand and supply and resell in a second market with demand uncertainty – in effect, the resale values are simple two-outcome gambles in the second market. They report convergence in the second market to a CE defined by demand and supply based on the expected values of the gambles – there is no risk-averse gap here. (Also see Plott and Turocy, 1996.) But the connective interface between rationality at the individual level and the market level and how institutions modulate the interface is yet to be fully explored and understood. It appears that markets may yield equilibrium outcomes given induced cash values, whatever the cash values – rational, irrational, or nonrational – that are provided by individuals. If those cash values are not individually stable or consistent, this may just mean that market-equilibrium-predicted outcomes are subject to uncertainty, not that they fail to support specialization.

Individual choice either in the sense of commodities and services based on preference ordering or under uncertainty is not where the action is in understanding economic performance and human achievement. The fundamental theorem in economics has not changed: Wealth is created by task specialization across individuals, groups, populations, regions, and climates; and specialization is determined by the depth and breadth of the market – personal or impersonal exchange systems. What are important about individual choices are the decisions that cause people across time and generations to change tasks, locations, and directions in an effort to better themselves in response to market prices. This does not require individuals to use their gains from specialization and trade for self-interested economic ends in the sense of always choosing dominant outcomes. To focus on the "rationality" of choice in the narrow frames defined to give it precision does not provide guidance toward a better understanding of this specialization process. People can make a lot of cognitive "errors" on the way to creating a new market – such as in off-the-shelf software programs allowing new firms to specialize and create much new wealth, and in the high failure rate of new firms trying to manage new technologies, out of which only a few discover how to do so and succeed in creating great long-term wealth for the social economy. The main work of socioeconomic systems is in specialization and the exchange systems that make possible the wealth they create, not the minutiae of choice and preference representation. The functioning of these systems is far beyond the field of vision of the individual, but it should not be beyond the vision of economic science.

Psychology, Economics, and the Two Forms of Rationality

Curiously, the image of economists and psychologists as protagonists obscures their underlying agreement on foundations. Both rely upon strict constructivism: To the extent that markets are rational[2] or irrational,[3] this

---

[2] An example is the double auction markets discussed previously.
[3] Experimental asset markets bubble and crash on the long path of experience to equilibrium. (Smith et al., 1988; Porter and Smith, 1994). It is only through experience (a third experimental trading session), not a reasoning process, that subjects in these experiments come to have common fundamental expectations. Nor can it be said that they are confused except from the perspective of a backward-inducting theorist, as I will show in the appendix

determination derives directly and only from the rationality or irrationality of agents. Thus, even a " ... monopolist ... has to have a full general equilibrium model of the economy" (Arrow, 1987, p. 207). Also see the previous discussion on racetrack market efficiency and the inference that the bettors must, ipso facto, have considerable expertise. Expertise in the sense of economic or probability sophistication on the part of all individuals is clearly not a prominent characteristic of subjects, whether they are agents or racetrack bettors. Yet the default explanation for market rationality is assumed to derive entirely from individual rationality: Markets cannot be rational if agents are not fully rational in the sense in which we have modeled it as theorists.

Individual rationality is a self-aware, calculating process of maximization. Here is a particularly clear statement of decision as rational constructivist action as seen by psychologists: "Incentives do not operate by magic: they work by focusing attention and by prolonged deliberation" (Tversky and Kahneman, 1987, p. 90). But this is an inference based on how the theorist, not the subject, models decision problems, and it ignores the demonstrated capacity of subjects to find equilibrium outcomes by repeat interaction in market experiments with no cognitive awareness of this capacity. Moreover, subjects are capable of achieving better outcomes in two-person anonymous interactions than if they applied traditional game-theoretic principles.

Predominantly, both economists and psychologists are reluctant to allow that naïve and unsophisticated agents can achieve socially optimal ends without a comprehensive understanding of the whole, as well as their individual parts, implemented by deliberate action. There is no "magic"; no room for the Gode and Sunder zero intelligence traders (see Chapter 4) who in simple environments reveal that at least some of the rationality in markets is encapsulated in an institution; no room for the results of hundreds of single- and multiple-market supply-and-demand experiments in which subjects with private information on values and costs achieve unintended equilibrium predicted outcomes. The phenomenon, not being constructively explicable with the tools of economic theory, is in effect denied: Even a monopolist must have a full general equilibrium model of the economy to behave rationally.

Consequently, psychologists test the rationality of individual decisions largely by asking for subject responses to choice problems to discover how the subjects "reason." Economists, subject to the identical vision (how do agents

to Chapter 10. For a new study of subject experience and asset bubbles, see Dufwenberg et al. (2002), and for a study showing for the first time that bubbles can be reignited with experienced subjects who had converged to fundamental value, see Hussam et al. (2006).

consciously think about economic choice?), are critical of the question-response survey methods used in cognitive psychology: The stakes are zero or too low, and it is claimed, without citing the experimental evidence on stakes, that the subjects are too unsophisticated, inexperienced, or untrained to allow a serious researcher to find out how "real agents really think." Many psychologists appear to find irrationality everywhere, and many economists appear to see the findings as everywhere irrelevant. To both economists and psychologists, how agents think indeed exhausts the core of investigation.

The use of cash or other reward medium in decision behavior experiments is listed by Hertwig and Ortmann (2001) as one of the key differences between psychology and economics experiments. The controversy over paying subjects, however, is being eroded as neurobiologists – including those who are informed on animal behavior models – enter these research programs and, following the traditions in experimental economics and animal research, pay subjects salient rewards (Thut et al., 1997; Breiter et al., 2001; McCabe et al., 2001a). This research is leading to a better understanding of how the brain encodes rewards.

Finally, psychologists and economic theorists have identified rationality almost entirely as expected utility (including expected profit, payoff, wealth, income, gain) maximization. But is this the concept of rationality that best explains either behavior or what agents seek and want? One alternative that is particularly appealing and leads to some very penetrating results is the theory of economic survival developed by Roy Radner and his colleagues, which I shall discuss in the next chapter.

In point of fact, opinion surveys can provide important insights. Sometimes survey findings can be tested more rigorously with reward-motivated choices in the laboratory or the field and may be found to have predictive content (for example, the choice asymmetry between losses and gains in wealth). But sometimes what people actually do completely contradicts what they say, and sometimes you cannot find out by asking because the agents themselves do not know what they will do or are doing. For example:

- Comparisons of risky choice under low- and high-monetary stakes have shown that actual reward levels sometimes do and sometimes do not have a statistically significant effect on decision; also, the qualitative conclusions from hypothetical choice response surveys are not always refuted by studies using very high stakes. Kachelmeier and Shehata (1992) report a statistically significant increase in risk aversion (and decrease in risk preferring) with reward levels, but no difference between low-reward and hypothetical-reward choices based on buying and selling prices for gambles. Binswanger (1980, 1981) studied lottery choices by farmers in India with payoffs ranging from hypothetical up to two to three times the subjects' monthly incomes; the hypothetical results were not consistent with the motivated choices in which risk aversion increased with the payoff stakes (but risk aversion was not related significantly to wealth levels). The recent comparisons by Holt and Laury (2002) agree with Binswanger's findings of risk aversion in income. In ultimatum games, a tenfold increase in payoffs does not significantly affect decisions (Hoffman et al., 1996a). Smith and Walker (1993b) found that reward levels do not change the tendency of bidders to "overbid" (as if risk-averse) in first-price auctions as the stakes are varied by factors of zero (hypothetical), five, ten, and twenty times normal reward levels, but the variance of the bids declines strongly with reward level. Payoff levels can make a big difference in behavior, but so also do context and experience. The general conclusion is that you are well advised to use incentive rewards, because it can matter substantially. (Also see Chapter 8.)

- Consider the double auction in classroom demonstration experiments. In debriefings afterward, students deny that there is any kind of quantitative model that could predict their market price and exchange volume, or that they were able to maximize their profits. However, a participant with a sealed envelope containing the predictions, provided in advance, opens it to reveal that this consensus is false. The dispersed private value/cost information is aggregated into prices that are at the equilibrium and each agent is maximizing his or her profit given the behavior of all others. Here there is indeed a kind of "magic," but only, I think, in not being well understood or modeled at the game-theoretic level of individual choice. In this sense, our bounded rationality as economic theorists is far more constraining on economic science than the bounded rationality of privately informed agents is constraining on their ability to maximize the gains from exchange in markets.

At the macromarket level, the classical Walrasian adjustment model does well in predicting convergence and cases of stable and unstable equilibrium, but the model is less reliable in predicting paths taken, including jumps across alternative unstable equilibria. See the outstanding summary by Plott (2001). The disconnection between the model and choice behavior is evident in the following: Walrasian dynamics makes ad hoc assumptions about price adjustments in response to excess demand, saying nothing about the corresponding payoff motivation of the agents who drive the price adjustment process. Walrasian dynamics is a story about the tatonnement mechanism in which there are no disequilibrium trades, whereas Plott's (2001) summary is about continuous double auction trading with a great many disequilibrium trades.

- In asset trading, participant survey responses reflect the disparity between the participants' information on fundamental value and their puzzling experience of a price bubble and crash generated on the long path to the rational expectations equilibrium in a stochastically stationary environment (Schwartz and Ang, 1989).
- Opinion polls administered to the Iowa Electronic Market (IEM) traders show the same judgment biases that psychologists and political scientists find in public opinion polls, but these biases need not interfere with the market's ability to predict the popular vote outcomes (Forsythe et al., 1992).
- In preference reversal survey experiments, subjects report many inconsistent choices: Gamble A is preferred to B, but a subject will sell A for less than B. Arbitraging the subjects' cash-motivated choices quickly reduces these inconsistencies (Chu and Chu, 1990, p. 906), and it has been shown that the inconsistencies are unbiased random errors under some, but not all, conditions (Cox and Grether, 1996); also see Sopher and Gigliotti, 1993, where choice intransitivity is studied directly and the errors are found to be random. But reducing inconsistency through experimental treatments, while of interest, does not mean that the subjects will generalize that "learning" to other contexts.

In the preceding bullet points, it is in private information environments – where the market is aggregating information far beyond the reach of what each individual knows, understands, or is able to comprehend – that the solicited opinions are so far off the mark. The surveys yield no useful understanding because the subjects have none to relate. In the complete information asset market, subjects are aware of its fundamental value structure, and come to have common expectations through an experiential process of repetition; that is, initial common information is not sufficient to induce common expectations.[4] They trade myopically and their expressed bafflement ("prices rise without cause") reflects this myopia. These

---

[4] This interpretation is consistent with asset-trading experiments using undergraduates, small business persons, corporation managers, and over-the-counter traders (Smith et al., 1988; Porter and Smith, 1994). Exceptions using inexperienced subjects, to my knowledge, have only been observed with advanced graduate students (McCabe and Smith, 2000).

comments suggest that much insight might be obtained from the systematic comparisons of the conditions under which survey results are robustly informative and the conditions under which they are not.

What Is Fairness?

The descriptor "fairness" has so many meanings in different contexts that I believe it is best to avoid the term entirely in experimental science except where it is explicitly modeled and the model tested in environments where subjects make decisions on the basis of the defining parameters of the proffered "fairness" model; then the descriptor "fair" and its ambiguity can be avoided altogether.[5] This is the way it was used in the utilitarian definitions by Franciosi et al. (1995) and subsequently by Fehr and Schmidt (1999) and Bolton and Ockenfels (2000). Of course, it is appropriate to use the descriptor if the purpose is to see how its instructional use might have an emotive affect on behavior, but this is not what is done in Kahneman et al. (1986; hereafter KKT; also see Kahneman et al., 1991). The emotive content of "fairness" is clear in the important work of Zajac (1995), who has also examined the rhetoric of fairness arguments as self-interest serving in the 2000 Florida election controversy, where each side charged the other with unfairness (Zajac, 2002). Also see Binmore (1994, 1997) on the social contract and the use of game theory to examine ethical issues.

KKT state "... the phrase 'it is fair,' is simply an abbreviation for a substantial majority of the population studied thinks it is fair" (1986, p. 201). Thus, KKT decentralize the definition to whoever is responding to a question that uses the phrase "it is fair." Response volatility will therefore compound variation in self-definition with variation in individual responses. This is not defensible instrument design. Here are some alternative concepts of fairness, each anchored in SSSM theories or cultural norms, which therefore may provide a more precise connection with principles – either constructivist or emergent:

- A utility for the reward of other(s) as well as self. This is the concept underlying the market tests of the effect of fairness in experiments reported in the next section.

---

[5] This is indicated by the observation that "fairness" is not a human universal across languages that is translatable. The linguist Wierzbicka (2006) examines the cultural baggage carried in many English words, showing that many of them are unique to English. A prominent example is the concept of "fair," which has no equivalent in any other European or non-Western language.

- Equality of outcomes. This is the imputation rule found by social psychologists to be preferred by people in situations where they have no knowledge or means of identifying differences in individual merit or in their contributions to the total to be apportioned to individuals.
- Equality of opportunity to achieve outcomes: to each in proportion to their merit or contribution to some resource to be allocated. This is the equity principle, the preferred imputation reported by social psychologists when individual contributions can be identified. It is also the "first harpoon" principle in Inuit, Arctic, and other cultures with likely much more ancient origins (see Freuchen, 1960, p. 53, quoted in Chapter 8 of this book; also Homans, 1967).
- The equilibrium market allocation principle: to each in proportion to his or her contribution to the net surplus of the group. This is closely related to the equality of opportunity to achieve outcomes. The achievement of such allocations and the fact that they are unintended are both confirmed by four decades of experimental market tests. Indeed, the fact that they are unintended can explain why the perceived fairness of the imputations is denied by people from the same population that confirms "fairness" in the sense of the equity principle.
- Property rights norms: Thou shalt not steal, and thou shalt not covet the possessions of thy neighbor. These historical – probably also as ancient as the "first harpoon" principle – Judaic laws are the property rights norms that support markets, and surely underlie the equity principle that emerges in social psychological findings.
- Reciprocity: It is "fair" behavior to return favors and "fair" to expect others to return yours. This principle also has historical Judaic-Christian roots in the Golden Rule: Do unto others as you would have others do unto you. These deep roots help to explain the principle's emergence in two-person extensive form games in the laboratory, even when subjects are anonymously paired.

As I interpret them, all these "fairness" norms with family and tribal origins enabled *H. sapiens* to evolve premarket exchange systems to support specialization in small-scale social economies long before cooperation through the extended order of markets was imaginable. It explains the human achievement of global migration from approximately fifty thousand to one thousand years ago, based on stone followed by metal technology. Although forms of limited personal exchange can be demonstrated in other primates – notably chimpanzees and capuchin monkeys – and further evolved in premodern humans, the takeoff was of more recent origins, and appears to be unique to *H. sapiens* (and perhaps Neanderthals). In game-theoretic terms, long histories of interaction and coevolutionary selection would have nourished reciprocity norms in which distinctions based on a dichotomy between the one-shot and repeated games would have been dysfunctional to the creation of wealth. If these things are true, there is little wonder that the norms survive in the two-person anonymous pairings that will be examined in Chapter 10.

Examples of Fairness

KKT provide many examples in which respondents are asked to rate the fairness, on a four-point scale, of elementary business actions in competitive environments. In one case, a hardware store raises the price of snow shovels from $15 to $20 after a snowstorm; 82 percent of the respondents consider this action either unfair or very unfair. Franciosi et al. (1995) substitute the words "acceptable" for "fair" and "unacceptable" for "unfair," and add one additional sentence to this KKT example: "The store does this to prevent a stock out for its regular customers since another store has raised its price to $20." Now only 32 percent rate the action unfavorably. This exercise suggests the possible sensitivity of survey results to emotive words and/or perceived "justification" in terms of impersonal market forces. It suggests the need not to extend again and again the range of examples using such frames, but to seek new insights by varying the instructional and procedural treatments and asking whether any new findings replicate or are subject to random variation. KKT's main interest, however, is in whether firm behavior is affected by community norms. Whether or not an action is "acceptable" would seem to be just as important in determining firm behavior as whether or not it is "fair." If the two terms map into different reported attitudes, then there is inherent ambiguity in specifying the effect on firm behavior, and the ambiguity of their program of exploring "fairness" is as open-ended as their definition.

Hurricane Andrew, with landfall in south Florida on August 24, 1992, was a newsworthy example for the *New York Times* of the divergence between popular conceptions of "fairness" and the perceived "unfairness" of unfet-tered free-market forces. According to this interpretation, the latter were allowed to reign during an unfortunate natural disaster, and to yield "unfair" price gouging in the market for plywood, the demand for which soared in the aftermath of the storm (Lohr, 1992). What companies took "unfair" advantage of Florida citizens? Home Depot? Louisiana Pacific? Georgia Pacific? None of the above:

What happened in the plywood market here after the storm is a classic example of fairness constraints at work. . . . The big companies (Home Depot, Georgia Pacific, and Louisiana Pacific) performed far differently (increasing the price only about half as much as the "market") than the price-gougers selling ice, water and lumber from the back of pickup trucks at wildly inflated prices in the first week after the hurricane hit. Classic economic theory, of course, defends these . . . (ibid., p. C2).

In fact, classical theory seeks to explain, not defend, the competitive market that operated out of the back of pickup trucks whose owners were responding to a temporary opportunity to gain by moving plywood supplies closer to where they were urgently needed. Buyers avoided the greater cost of losses by availing themselves of the protections of plywood. Theory states that it's all about the differential opportunity costs faced by pickup owners and homeowners and how both are made better off by their ability to respond to those differing opportunity costs in a temporary spontaneous market. Those who argue that people do not take account of opportunity costs cannot explain the roadside market. Modern economics, in the form of reputation theory, also explains the actions of the large firms that are specialists in making plywood and other building materials available through thick and thin. Their national networks of supply enable them, through transfers, to replenish stocks more quickly, price less aggressively, and build a long-term reputation and customer satisfaction for not "price gouging" (with free advertising provided by the *New York Times* article), but simultaneously reap supranormal profits. (Note that they raised prices, if only half as much as the competitive fringe of roadside entrants.) But one does not need a utilitarian fairness ethic to explain the repeated game (versus one-shot) nature of the long-run outlook of large suppliers. Of course, that theory may not be complete, and surely stands to benefit from a better articulation of consumer perceptions that might underlie reputation formation. Optimization theory predicts that if buyers believe that they can conserve their resources by complaining about price gouging being "unfair," then they will do it. The long-term result (hidden from the average consumer) may solidify customer loyalty and build the market power enjoyed by the large firms (is this why they are large?), reduce competition, and decrease realized, if not perceived, economic welfare – all under the cloak of "fairness."

Most customers of the local electric utility monopoly pay a fixed price per kilowatt consumed regardless of time of day, week, or season. They have always done so, and my conjecture would be that a suitably worded survey instrument would affirm that most people think this is only "fair." A similar instrument would surely show that people think it is "unfair" that, having paid $900 for their airline ticket purchased three days earlier, a seat near to them is occupied by the holder of a $400 ticket purchased two months earlier. Unfair to both of them is the fact that an on-line–purchase ticket holder, who had to take any route path and schedule available, paid only $150. Also "unfair" would be a similar story about differences in hotel room charges due to seasonal or time-of-week variations in occupancy.

Suppose I have all this opinion data. How do I use it to improve the predictive power of economic theory so that the theory still performs well where it has been successful, and to enable it to do better where it has not? That economic agents do not think like professional economists is well established (Smith, 1990). But how should this observation affect what we do?

In the next section, I report a static model, yielding either the competitive outcome or an alternative, motivated by this sort of survey data, and experimental tests of their alternative predictions. The model provides one illustration of how these questions might be addressed. The strong form of the KKT hypothesis is rejected, but a weak version ("fairness" may cause markets to be sluggish in their dynamic response to a change), not predicted or modeled by KKT, is supported by the different path of convergence to a new competitive equilibrium when buyers have information indicating that the new (equilibrium) state increases seller profits relative to the previous state. The results reinforce once again the predictive power of static equilibrium theory, but also the sensitivity of market dynamics to a particular context and information on how gains are being redistributed.

From the experiments, it is easier to see how people might hold a belief revealed in a survey, but that belief need not persist or be strong enough to change their myopically self-interested response in impersonal exchange. It also tells you, by implication, perhaps why someone might vote for a policy intentionally designed to change outcomes, but his or her market behavior creates outcomes contrary to those intentions. Here are possible examples: A minimum wage fails to raise all incomes by causing some to be unemployed and earn nothing; or a price ceiling fails to lower all prices by causing supplies to be diverted to a higher-priced underground market. We encounter Hayek's (1988, p. 18) distinction between "different orders" and the contrast in behavior between personal and impersonal exchange that has been an essential principle that we have learned from experiments.

Fairness: An Experimental Market Test

In their study of fairness, KKT articulate what is called a "descriptive theory." This methodology is driven by the untenable, and long-questioned, belief that general theories can be derived directly from observations if you just have enough data (see Chapter 13). "Perhaps the most important lesson learned from these studies is that the rules of fairness cannot be inferred either from conventional economic principles or from intuition and introspection. In the words of Sherlock Holmes in 'The Adventure of the Copper Beaches': 'I cannot make bricks without clay.'" KKT (1991, p. 234). Neither can a falsifiable or predictive theory of "fairness" be inferred from any amount of the KKT data. If N "fairness" rules are discovered by trial-and-error modifications in the survey questionnaires, you cannot reject the hypothesis that there is an N + 1st variation that will identify a new one; nor could such an observation be predicted. More data will not help, for the "fairness" concept as used here is a word that provides no effective means of modifying standard theory to correct for its predictive flaws.

In developing a descriptive theory of the "reference transaction," KKT state that what is considered "fair" may change: "Terms of exchange that are initially seen as unfair may in time acquire the status of a reference transaction" (KKT, 1991, p. 203). This opens the way without commitment for the adaptation of "fairness beliefs" to changes in the competitive equilibrium. Although the competitive model is the one that has static predictive content, its prediction is silent as to how long it will take to respond to a change in parameters. KKT's arguments are not predictive, but they tell a story about beliefs that explain why markets might be sluggish in responding to change. How good is their story?

Franciosi et al. (1995) state a preference model of optimal choice that allows for a utilitarian trade-off between private consumption and KKT's description of "fairness." For example, the utility of two commodities (x,z) is given by $u(x, z) = z + ax - (b/2) x^2 - \alpha x [(\pi/\pi_0) - 1]$, in which the seller's profit, π, relative to a reference profit, π_0, appears as an "externality" in the buyer's utility function. The usual maximization subject to an income constraint yields the inverse demand equation: $p = a - bx - \alpha [(\pi/\pi_0) - 1]$. Thus, for $\alpha > 0$, any change in the environment that increases a firm's profit relative to the reference profit has an external effect that lowers the buyer's inverse demand for units, x. This is the strong form of the KKT argument. If $\alpha = 0$, then we have the standard self-interested maximization of utility. Consequently, Franciosi et al. (1995) can test the hypothesis, never using the loaded and suggestive word "fairness," that if subject buyers have a utilitarian concern for profits not being increased relative to a baseline, then after a change from the baseline this should alter the observed equilibrium relative to the standard predicted equilibrium with no external effect ($\alpha = 0$).

In a posted offer market giving KKT their best shot – sellers cannot see each other's posted prices, and therefore cannot knowingly emulate or undercut each other's prices – Franciosi et al. (1995) find that under information conditions such that $\alpha = 0$ (implemented by either no disclosure of seller information or by [marginal] cost-justifying disclosure to buyers), the market converges quickly to the new competitive equilibrium. When $\alpha > 0$ (implemented by profit π and π_0 disclosure to buyers enabling "fairness" to affect behavior), prices converge more slowly, but decisively, to the new equilibrium. Hence, under conditions most favorable to a permanent external "fairness" effect, the response dynamics are different, but not the equilibrium as predicted by the standard competitive model. There is no utilitarian demand effect of "fairness"; the discipline of the market swamps all but a transient "fairness" effect. If sellers can see each other's prices, I would predict that a much smaller "fairness" effect, if any, would be observed.

The concept of "fairness," as here implemented, does not lead to any important modification of the traditional static equilibrium model, but it does establish that the concern for fairness expressed in opinion survey results may account for temporary resistance to adjustment to a new equilibrium. Also, we note that people's unmotivated responses to survey questions do not anticipate their own motivated equilibrating adjustments. But this observation may say much more about the disconnection between the two distinct decision processes – conscious verbal response to questions versus operating in a real-time market context without awareness of the social gains it creates – than it does about the fact that the survey responses are unmotivated. The study of these disparate processes is much neglected in economics but also prominently in cognitive psychology.

EIGHT

What Is Rationality?

...personal knowledge in science is not made but discovered.
Polanyi (1962, p. 64)

In this chapter, I reconsider the standard models of "rationality" in economics, asking whether subjects' decisions, often judged in various contexts to be irrational from the standard perspective, might not be judged
differently when viewed from a different perspective, including that of the
subjects and the environment to which they have become adapted. A divergence between the predictions of a theory and the results of tests of the
theory implies that either the theory or the experimental test is inadequate.
Just as the experimentalist can get it wrong in designing a test, and returns
to the drawing board when testing problems or suspected design flaws show
up in the first experiments, so the theorist can get it wrong in deciding what
assumptions, criterion of action, and constraints are most relevant for the
decision problem faced by the individual. Science is about reducing error –
deviations between observation and prediction – and the two instruments
of science are experiment and theory (a third instrument is the technology,
or machines, of experiment, which I discuss in Chapter 13); reducing error
necessarily involves being prepared to adjust either or both of these instruments. If the data are persistent and robust to procedural and contextual
issues and imply that the behavior stems from ecologically rational actions,
we are well advised to reevaluate our constructivist model of rational action.
On these questions, also see the important contributions by Cosmides and
Tooby (1996) and Gigerenzer et al. (1999).

Three areas in which theory has been weak in prediction and in accounting for variability concern individual decision making under uncertainty – where contradictions abound, as we have seen in the previous

chapter – and in accounting for subject heterogeneity of decision across a wide range of different experiments. The first has been masterfully addressed in the economics of survival, and deserves careful attention; the second provides a human capital-effort approach to understanding better how subjective rewards and skill/experience might interact in a model of individual decision effort to account for subject decision variability. The third concerns the rationality of collections of individuals in market and nonmarket institutions and environments.

A distinct set of issues concern rationality in groups ranging from aggregating subjective individual estimates to markets; these are discussed only very briefly in the last section of the chapter.

Economic Survival versus Maximizing Utility

In a series of papers, Roy Radner and his coworkers have developed models of economic survival, demonstrated contradictions between the implications of models of optimal survival and those of utility or profit maximization, and derived wealth valuation functions that are convex below a critical wealth level, and concave above, a property of the individual discussed by Adam Smith.[1] This important exercise explores alternatives to the standard utilitarian definition of rationality, and – central to the themes in this book – introduces a rationale for the difference between the predictions of the traditional constructivist models of choice and observations from the ecology of choice. Are we biologically and/or culturally adapted for choosing sequential investments to maximize the probability of survival or of expected utility (profit)? Kahneman and Tversky (1979, 1996) and others have provided numerous instances in which people violate the conditions of expected utility (including profit) maximization. But if it is simply not optimal, in the sense of criteria for survival, to maximize expected profit, then it is not prima facie a cognitive "error" to violate this constructivist model of behavior.

Maximizing the Probability of Survival

In the simplest model of survival, let an investor engage in sequential investment decision making in a capital market, or let a manager choose among

---

[1] Radner's (1997) Schwartz lecture provides a nontechnical exposition of the research program on which this section is based. Also, see Majumdar and Radner (1991, 1992); Dutta and Radner (1999).

real investment projects for her firm. Mathematically, the arithmetic is more compact if the choices among opportunities are made in continuous time.

In any short interval of time, h, let the investor's earnings be normally distributed with mean hm and variance hv, where m and v are fixed. It is acceptable to think of m as the yield ("drift" in continuous time) and v as the risk (or "volatility") of the earnings process from investment decision, where m can be positive, negative, or zero, but v by definition can never be negative, and is interesting only if strictly positive. Think of (m, v) being chosen continuously in time by the investor from a given (dense) set of investment opportunities, $S(m, v)$. In the constant returns model, the investor cannot increase the scale of investment, and is therefore special, but the model leads to interesting insights. The investor also faces a fixed rate of payout, which is best to think of as her maintenance or consumption, hc. Let $w > 0$ be the investor's current wealth; if she fails (goes bankrupt, never to rise again) at any time $t = T$, $w = 0$, then the probability that she survives forever is given by

$$P(w; m, c, v) = 1 - \exp[-2(m - c)\,w/v], \text{ if } m = 0, \text{ otherwise } P(0) = 0.$$

When $c = 0$, and m and v are fixed, this equation describes the classical Gambler's Ruin problem. The investor or firm is assumed to choose (m, v) from the set S to maximize the probability that she will survive forever.

Notice from the expression for P that the earnings process is under optimal control if the investor maximizes the ratio $(m - c)/v$ in the investment opportunity set, $S(m, v)$, represented by the closed convex curve shown as an ellipse in Figure 8.1. Hence, for any v, the yield m must be on the upper boundary of S, say $m = B(v)$, between $v = v'$ and $v = v''$ as shown. Also note that the ratio $(m - c)/v$ is maximum at M/V, which occurs at the point (M, V), where the slope of the ray from the origin to the upper boundary is at its greatest. Now let the largest value of m in the feasible set $S(m, v)$ be m^* and the corresponding value of v be v^*, or $m^* = B(v^*)$, shown in Figure 8.1. Since m must be at its maximum for each v, the optimal control policy becomes one of choosing the optimal risk, v, as a function of w. It can be shown that the optimal control has three important properties:

P1: *The optimal risk, v, decreases with the current wealth, w, so that the larger is wealth the smaller will be the optimal risk undertaken by the investor.*

This is indicated by the arrows moving from right to left on $B(v)$ in Figure 8.1; that is, movements to the left along the boundary are associated with increasing wealth and the choice of lower levels of volatility.

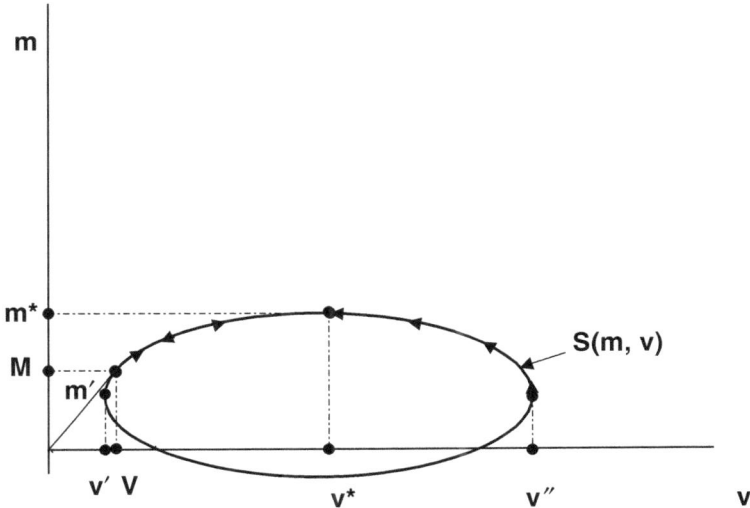

Figure 8.1. Economic Survival versus Profit Maximization.

P2: *If w is sufficiently small such that the optimal v is greater than v*, the investor appears to behave in the short run as if she is a risk-preferring investor (under utility maximization). The critical wealth w* below which this occurs corresponds to the point (m*, v*) on the upper boundary of S.*

Since v^* is the risk at which the yield is at its maximum in the feasible set, it follows that when w is sufficiently small, the optimal risk must be larger than v^*, and the optimal control will be on the boundary to the right of the point (m^*, v^*). Hence, along this boundary for $w < w^*$ the investor chooses the *maximum* feasible risk corresponding to the optimal yield. This behavior is completely contrary to the standard concept of an $(m\,v)$ efficient portfolio, which requires that the *minimum* v be chosen for any m. Note that as the investor's wealth approaches zero, the optimal v approaches the maximum risk v'' in S.

Only for $w > w^*$, corresponding to the segment of $B(v)$ that is to the left of v^*, does she choose the minimum risk corresponding to the yield. Radner (1997, pp. 198–9) lets $P(z)$ be the maximum probability of survival, where $z = \log w$. $P(z)$ increases with z and approaches unity asymptotically as z increases. $P(z)$ is S-shaped, convex below z^* and concave above, where the critical value $z^* = \log w^*$ corresponds to the critical wealth below which the investor appears to be risk seeking if one were applying the utility maximizing framework. The investor's behavior, however, is *not* due to risk aversion, a concept not meaningful to a survivalist. Any time wealth is below the critical value, then it is optimal for the investor to maximize risk in an effort to "get back into the game," and possibly escape the trap in which he will confront ruin. Choices that are not wealth maximizing can be rational for survivalists.

P3: *The optimal v increases with the payout rate, c, while the optimal m increases from m′ to m* and decreases thereafter.*

We note that P2, a result based on wealth, does not address the empirical finding that decision is more related to changes in wealth, or income, than to wealth. The empirical finding, however, concerns how people respond when asked about particular isolated prospects, chosen to put people in decision situations designed to reveal whether Adam Smith's proposition has validity: "We suffer more . . . when we fall from a better to a worse situation, than we ever enjoy when we rise from a worse to a better." But real people engaged in search and decision processes will want to avoid finding themselves in that position. In the framework of Radner and his colleagues, the investor is choosing from a large set of prospects, and the model characterizes the properties of choices on the boundary of that set depending on the investor's level of wealth relative to the critical w^*.

Maximizing Expected "Profit," or Discounted Withdrawals

Suppose the investor or firm does not wish to maximize the probability of surviving forever, but wants to maximize "profit," defined as the expected discounted value of all withdrawals – the consumption rate that is not available for reinvestment. Are the implications of the two behaviors different? In this model, the decision maker must choose a control policy for the yield and risk from the feasible set $S(m, v)$, as in the preceding formula, but also a withdrawal policy. A key feature of this "constant size" model is the cash or liquidity reserve that the investor maintains as protection against a run of poor yields that results in the firm's version of the Gambler's Ruin. This liquidity reserve is represented in equation (1) by w. It is not directly productive, but indirectly allows continuance. In terms of choosing (m, v), the applicable constraint set corresponds to the standard efficient portfolio subset (minimum v for each $m > m′$) of S shown in Figure 8.1. The key result is

P4: *The firm that maximizes expected profit (discounted total withdrawals) fails in finite time.*

Here is a succinct explanation as to why the proof works:

The larger the cash reserve, the greater is the protection that it provides, and hence the greater is the expected value of *future* withdrawals. However, this indirect (insurance) productivity ... is subject to decreasing returns. The larger is the cash reserve, the smaller is the *marginal* benefit ... from a further increase in the reserve, compared to the benefit of an immediate withdrawal. On the other hand ... in order to have a positive probability of surviving forever, the firm must accumulate a larger and larger cash reserve, without bound; but beyond a certain point such accumulation is no longer profitable" (Radner, 1997, pp. 203–4).

More recently, this model was extended to maximize distributions explicitly by the firm to investors and to allow the firm to fail, either if it is unable to obtain financing in the capital market or the firm obtains financing but goes bankrupt (Dutta and Radner, 1999). This model deals directly with the Alchian (1950) market selection problem, that firm success requires it both to attract investment financing and to earn positive profits. Two primary results are obtained. The first is proposition P4. The authors also establish the following:

There are, however, a variety of non-profit-maximizing behaviours that have a positive probability of never failing. Our second result shows that if there is *sufficient diversity*, in the set of potential firms, then some of the firms with a positive probability of survival will find financing in the capital market; consequently, *after a long time, most of the surviving firms will not, in fact, be profit maximizers*. Our conclusion, then, is the exact opposite of the market selection argument: firms that maximize profits are least likely to be the market survivors. These results hold also in the case that new cohorts of firms arrive periodically in the market (Dutta and Radner, 1999, pp. 771–2, emphasis in the original text).

As in the previous model, an important part of an entrepreneur's survival behavior is to violate the usual efficient mean-variance behavior by choosing the maximum mean return even if it has a high variance. Potentially, these results from the economic analysis of survival may have far-reaching consequences for how we interpret and study human risk-taking behavior. The SSSM – optimization of expected reward outcome under investment uncertainty – may be an inappropriate way to characterize the adapted human being.

Is It Rational to Be "Rational"?

In this section, I want to articulate a cognitive effort model of strategic decision cost to be balanced against the benefit from making better decisions, that is, reducing the error deviation from game-theoretic predicted equilibrium outcomes.[2] Hence, that which constitutes the best response by interacting players cannot be specified independently of individual costs and constraints on cognitive effort. Consequently, insofar as individual brains experience differences in such effort costs, optimal individual decision making is inherently *subjective* and there cannot exist a coherent concept of objective rationality. Think of this property of individual brains as part of the internal order of the mind – a mental adaptation that need not be any part of the individual's consciously self-aware attention.

Subject to some of the usual regularity conditions, we show that increasing the rate at which outcomes are converted into monetary reward for the players (1) reduces the deviation of realized outcomes from the game-theoretic equilibrium prediction, and (2) reduces the variance of outcomes due to sampling variation in subject decisions. The model therefore makes testable comparative static predictions of the qualitative effect of varying the monetary rewards in an experiment. This adds to a considerable literature that was precipitated by influential comments from Herbert Simon some fifty years ago. (See my discussion of some history of game theory and experimental economics in Smith, 1992.)

Traditional economic models of decision making in experiments have tended to assume that only the monetary rewards associated with the experimental decision task matter, while psychologists (Tversky and Kahneman, 1987; Dawes, 1988) have tended to assume that such rewards do not matter or matter little. Important exceptions to this characterization of psychologists include Siegel (1959), von Winterfeldt and Edwards (1986), Kroll et al. (1988), and others. In fact, Siegel seems to provide the earliest of many experimental studies surveyed in Smith and Walker (1993a), in which payoffs were systematically varied as an experimental design treatment. A common reaction of economists, when the predictions of theory are not supported by experiment, is to hypothesize that the payoffs were not large enough for the predicted outcomes to matter. Specifically, this may be a consequence of the problem of flat maxima, or low opportunity cost of deviating from the optimum, an issue revisited more recently by von Winterfeldt and Edwards (1986) and by Harrison (1989). To my knowledge, these problems were first postulated and explored by Siegel (1959, 1961) in the context of a game against nature, by Siegel and Fouraker (1960) in the context of bilateral monopoly bargaining and by Fouraker and Siegel (1963) in bargaining and in duopoly and triopoly markets. I am sure, however, that a more careful search of the literature would show that there were precursors.

---

[2] This is an edited and shortened version of Smith and Szidarovsky (2004).

Standard theory, however, predicts that decision makers will provide optimal decisions regardless of how gently rounded are the payoffs as a function of decision, provided only that there is a unique optimum. This means that the theory contains serious specification error and needs modification to take account of the common intuition that the size of payoffs, as well as payoff opportunity cost, ought to affect decisions. This intuition implies that there must be something else in agent utility functions besides monetary payoffs. (Subjective costs were discussed as a qualification of the theory of induced valuation in Smith, 1976, 1982a.) Smith and Walker (1993a, 1993b) take a decision cost approach in reformulating the problem; that is, a problem of balancing the benefit against the effort (mental labor) of reducing "error," the latter defined, *not* as a mistake, but as the deviation between the agent's actual decision and the optimal decision in the absence of decision cost. Thus, "when the theory is properly specified there should be nothing left to say about opportunity cost or flat maxima; i.e. when the benefit margin is weighed against decision cost, then there should be nothing left to forego" (Smith and Walker, 1993a, p. 245; also see Hertwig and Ortmann 2001, 2003, for additional evidence and elaboration in favor of these conclusions).

Of course, "proper specification" may involve far more than decision cost, but this seems like a reasonable hypothesis with which to start the modeling effort. That is, introducing decision cost is just one of many strategies to account better for the failures of the SSSM, and in the present case provides interesting and satisfying results. Smith and Walker's proposed model is consistent with the experimental data they surveyed showing (1) several studies in which the predicted optimum is at the boundary of a constraint set and in which increasing rewards shift the central tendency of the data toward the optimum; (2) in many studies, an increase in rewards reduces the variance of the data relative to the predictions of the theory; and (3) increases in subject experience often improve performance at rates that are equivalent to large increases in monetary rewards. Camerer and Hogarth (1999) have extended the Smith and Walker (1993a) survey and also argue that both ability and effort (or as they state it, a "capital-labor") theory of cognition are important, as is suggested by the pronounced effect of experience on performance in the Smith and Walker study. Ortmann and Rydval (2004) offer important empirical evidence on this capital labor framework, suggesting that cognitive ability differences appear to account for a larger part of performance than incentive rewards. No amount of reward and effort will make a superior pianist out of someone with inadequate music ability. Also, where ability has been leveraged by accumulated learning and experience, less effort is needed to achieve a given result.

There is an important technical gap in the Smith and Walker model: It is formulated in terms of a game against nature. Although many of the studies that they survey fall into this category, others are interactive market, bargaining, or oligopoly experiments conducted under private information on payoffs; that is, each agent knows only his or her own payoff and nothing about the payoffs of others. But the theoretical analysis proceeds as if there is complete information. As I have emphasized, it is well known that providing subjects private (as opposed to complete) payoff information yields the condition most favorable to realizing a noncooperative equilibrium (Smith, 1982a; McCabe et al., 1996, 1998).[3] Hence, in interactive decision, the equivalent of complete information *analysis* is conveyed to others by the choices made by each. But modeling that process, as was attempted in the Hurwicz program, has eluded formal efforts.

In this section, we modify the Smith and Walker model to allow for interactive decision in the n-person case and state conditions that permits one of the main theoretical results to be extended to the interactive case as follows: Where standard theory predicts an interior optimum, then increased payoff reduces the variance of decision error.

Literature Background

The challenge of modeling the response to monetary reward in experimental tasks that are realistically expected to exhibit subjective costs of executing the task was first examined formally in the context of predicting Bernoulli trials by Sidney Siegel (1959, 1961). Siegel attributed his motivation to an idea developed by Herb Simon:

... the necessity for careful distinctions between *subjective* rationality (i.e., behavior that is rational, given the perceptual and evaluation premises of the subject), and *objective* rationality (behavior that is rational as viewed by the experimenter) ... in the explanation of observed behavior.... If we accept the proposition that organismic behavior may be subjectively rational but is unlikely to be objectively rational in a complex world then the postulate of rationality loses much of its power for predicting behavior. To predict how economic man will behave we need to know not only that he is rational, but also how he perceives the world – what alternatives he sees, and what consequences he attaches to them.... We should not jump to the conclusion, however, that we can therefore get along without the concept of rationality (Simon, 1956, pp. 271–2).

---

[3] Motivated by the general experimental findings, Kalai and Lehrer (1993) model private information as a Bayesian game in which players have a diffuse prior on each other's payoffs, showing convergence to Nash. This is an important development, but is much beyond the scope of what is summarized here.

The model below gives force to Simon's observation that predictability under objective rationality "loses much of its power" under subjective rationality. Earlier he had noted "that there is a complete lack of evidence that, in actual human choice situations of any complexity, these (rational) computations can be, or are in fact, performed . . . but we cannot, of course, rule out the possibility that the unconscious is a better decision maker than the conscious" (Simon, 1955, p. 104).

The model we develop is also very much in the spirit of Siegel's original work, inspired by the first Simon quotation, and leads to the consequence that subjective decision cost implies that objective rationality is not definable. But the last clause in the second quotation is entirely relevant as there should be no presumption that the subjects are aware of any implicit weighing of cost versus benefit as their brains automatically expend increased effort when the reward centers signal greater value in doing so. If subjects implicitly take account of the effort cost of decision, then of course the subject's unconscious decisions are indeed better than the conscious, rational, constructivist decision analysis predictions of the theorist/experimentalist, and thus can be said to be *super rational*, and not the result of a "mistake."

Modeling Net Subjective Value

Consider a two-person experimental game with outcome $\pi^i (i = 1, 2)$, which is converted into a monetary reward using the conversion constant λ ($\lambda > 0$). Let $C_i(z_i)$ be the subjective or mental cost of cognitive effort z_i. The subjectively experienced utility of player i can therefore be written as

$$U^i = (1 + \lambda)\pi^i(y_1, y_2) - C_i(z_i) \tag{1}$$

where y_1 and y_2 are the decision variables of an interactive Nash game. When no cash reward is offered, then $\lambda = 0$, and the formulation (1) allows that there is some self-satisfaction in maximizing the paper profit $\pi^i - C_i$. This is supported by the common observation that subjects make reasonably good decisions – they do not choose arbitrarily or randomly – when there is no monetary reward; that is, they feel impelled to try their "best" although they do still better with a monetary reward (Smith, 1976).

Let (x_1^*, x_2^*) denote the Nash equilibrium computed by the theorist when net payoffs are π^i, and the subjective decision cost term in equation (1) is zero.

Now write each subject's decision in the error deviation form:

$$y_i = x_i^* - s_i \xi_i(z_i), \tag{2}$$

where s_i is a random variable and ξ_i is some function of i's effort, z_i. Thus, $\xi_i(z_i)$ is i's production function for reducing error by application of cognitive or physiological resource effort. Note that this production function could be written $\xi_i(z_i;$ cognitive ability) in recognition of the important Ortmann and Rydval (2004) finding that cognitive ability also enhances the brain's internal effort calculus.

Neither z_i nor $\xi_i(z_i)$ is observable, but we do observe their effect on y_i and the deviation error, $x_i^* - y_i$ and proceed to test hypotheses based on these observations. Under this assumption, the utility of player i can be rewritten as follows:[4]

$$U^i = (1+\lambda)\pi^i(x_1^* - s_1\xi_1(z_1), x_2^* - s_2\xi_2(z_2)) - C_i(z_i). \tag{3}$$

In (3), assume that

$$C_i'(z_i) > 0 \quad (i = 1, 2), \xi_i(z_i) > 0, \xi_i'(z_i) < 0.$$

The discrepancy between the actual equilibrium (y_1, y_2) based on the subject's experience, and the equilibrium (x_1^*, x_2^*) constructed by the theorist, is characterized by the componentwise discrepancies

$$\varepsilon_i = x_i^* - y_i = s_i\xi_i(z_i). \tag{4}$$

Simple calculation shows that

$$E(\varepsilon_i) = E(s_i)\xi_i(z_i) \quad \text{and} \quad Var(\varepsilon_i) = Var(s_i)\xi_i^2(z_i), \tag{5}$$

and therefore it can be shown that (see Smith and Szidarovszky, 2004)

$$\frac{dE(\varepsilon_i)}{d\lambda} = E(s_i)\xi_i'(z_i) \cdot \frac{dz_i}{d\lambda}, \tag{6}$$

is positive if $E(s_i) < 0$, negative if $E(s_i) > 0$, and vanishes if $E(s_i) = 0$. In the latter case, the subject's error in cognition is unbiased. Notice that $E(s_i) < 0$ indicates that y_i is above x_i^* in the average and since the derivative on the left in (6) becomes positive, in the expectation; the discrepancy ε_i becomes smaller as λ increases. Similarly, if $E(s_i) > 0$, then y_i is below x_i^* in the average, and since the derivative on the left becomes negative, the discrepancy

---

[4] The functions $\xi_i(z_i)$ and $C_i(z_i)$ have been assumed here to represent cognitive processes, but they can just as well be used to model the explicit cost and productivity of the time of paid consultants or others assigned a strategic decision task. In this way, the stakes in a field application of game theory can be weighed against the cost and productivity of observable effort.

decreases again, in the average, if λ increases. These conditions open a rich theory of biased cognition error, its effect on expected observational error, and its interaction with payoff levels (and as indicated previously, cognitive ability), a subject that will not be pursued here. Similarly (see Smith and Szidarovszky, 2004),

$$\frac{d\,Var(\varepsilon_i)}{d\lambda} = Var(s_i)2\xi_i(z_i)\xi_i'(z_i) \cdot \frac{dz_i}{d\lambda} < 0 \qquad (7)$$

showing that the variance of the discrepancy in the average must decrease if λ increases. This result corresponds to a principal characteristic of the data reported in (Smith and Walker, 1993a, 1993b). Note that to get this empirical prediction, the multiplicative form of the error equation (2) was used.

Smith and Szidarovszky (2004) also offer conditions such that all these results extend to the case in which there are n players.

Examples from Experiments

In Cournot duopoly, y_i, is output by firm i. With linear demand and zero production costs, profit is $\pi^i = [a - b(y_1 + y_2)]\,y_i$, $i = 1, 2$, where $a > 0$ and $b > 0$ are the linear demand parameters. Using data from Fouraker and Siegel (1963) for linear Cournot duopoly experiments, Smith and Walker (1993a) report mean square deviations from the optimum of 7.2 under low payoffs and 5.5 under high payoffs. Using data from various versions of bilateral bargaining (simultaneous move versus sequential move and private versus complete information on cash payoffs), the mean square deviations vary from two to twenty-seven times greater under high compared with low payoffs (ibid., p. 257). Double auction supply and demand experiments with perfectly elastic supply and perfectly inelastic demand (yielding a competitive equilibrium at the supply price boundary) exhibit large differences, period by period in both the mean deviation and the mean square deviation from equilibrium, when comparing weak with strong payoff motivation (ibid., p. 259). Smith and Walker (1993b) also published new first-price auction experiments that varied rewards, by letting λ = zero, one, five, ten, and twenty times the normal amounts paid to individual subjects, and reported a significant reduction in the mean square error for linear bid function regressions as λ is increased. They also report a large reduction in the mean square error linear bid function when subject experience is increased while controlling for payoff level. Experience is interpreted in the preceding model as lowering decision cost, that is, as familiarity with a task increases, decisions become routine and require less attention from higher cognitive resources. The results in Smith and Walker (1993b) parallel and reinforce earlier findings by Jamal and Sunder (1991), showing that reward saliency (payments based on trading profit instead of a fixed payment for participation) in double auction experiments speeds convergence to a CE more prominently with inexperienced subjects than with experienced subjects. Hence, rewards are most efficacious when a new task requires mindful attention, but once the task process has been "learned," that is, internalized to autonomic brain function circuitry, subjective decision cost is lowered. These findings directly support the view that experience is identified with a process in which there is a shift from mindful "focal" to "subsidiary" awareness in the brain; see the important discussion by Polanyi (1962, pp. 55–65, passim). In the internal order of the mind, this is the brain's way of conserving scarce reasoning and attention resources.

Monetary Incentives: Further Discussion

For further important evidence and discussion of the preceding conclusions, see the survey by Camerer and Hogarth (1999), Hertwig and Ortmann (2001, 2003), and Guala (2005, pp. 231–49). Reward levels can and do (and sometimes do not, or even can perversely) make a difference, and the experimental economics community has been cautious, and I think justified, in following the principles of induced valuation in laboratory research programs, but also making comparisons with econometric and other field empirical results wherever and whenever the occasion arises, as in, for example, Roth and Sotomayer (1990) and Deck and Wilson (2004). This is because experimentalists have long focused on motivation, subject sophistication, and the relevancy of laboratory to field observations. (See Fouraker and Siegel, 1963, and the treatment of' "parallelism" in Smith, 1976, 1980.) Moreover, in a variety of different contexts, experimentalists have done comparative total and incremental reward–level studies to examine the sensitivity of results to stakes and relative stakes in particular research contexts.

It is no longer acceptable for anyone to argue either the extreme position that rewards matter little or do not matter, or the opposite extreme that rewards cannot ever be high enough in laboratory experiments, without first studying what experimentalists have done and said about it going back nearly fifty years to the work of Siegel (1959, 1961), Siegel and Fouraker (1960), Fouraker and Siegel (1963), and many others. Of note in this respect we have the following statement by R. Aumann:

... the Gale-Shapley algorithm had in fact been in practical use already since 1951 for the assignment of interns to hospitals ... in the real world – when the chips are down, the payoffs are not five dollars but a successful career, and people have time to understand the situation – the predictions of game theory fare quite well (Roth and Sotomayer, 1990, p. ix).

This book is of course filled with examples of ecologically rational outcomes in both the laboratory and the field, with and without "successful career" motivation wherein there is no way in which it is possible that you can argue that they "understand the situation" in the sense of game theory. Aumann here is fostering the error that all rationality must be deliberate and based on the type of "understanding" that characterizes constructivist models of behavior.

Again, to quote Aumann, "It is sometimes asserted that game theory is not 'descriptive' of the 'real world'. . . . To back up such descriptions, some workers have conducted experiments using poorly motivated subjects, subjects who did not understand what they are about and are paid off by pittances; as if such experiments represented the real world" (ibid., p. xi). On this point, Camerer (2003) says, "Aumann is alluding to an earlier generation of experiments in the 1960s and 1970s which were not sensitive to subject comprehension and incentives" (p. 23).

These assertions are much too casual and uninformed to be credible: The experimental literature is replete with examples in which game and economic theory (especially in markets and certain auctions) do remarkably well at the levels of compensation and comprehension common in the experimental laboratory, and anyone who knows the Siegel work previously cited is aware that motivational and subject sophistication issues were examined extensively by "an earlier generation" in the 1950s and 1960s.[5] Camerer, however, correctly concludes:

If you are smitten by the elegance of analytical game theory you might take the data as simply showing whether the subjects understood the game and were motivated. If the data confirm game theory, you might say, the subjects have understood; if the data disconfirm, the subjects must not have understood. Resist this conclusion . . . by inferring subjects understanding from data, there is no way to falsify the theory (ibid., p. 22).

See Chapter 13 for more on this topic.

---

[5] Siegel's research included replicating undergraduate bargaining behavior experiments with General Electric executives (Siegel and Harnett, 1964).

Rationality in Collectives and the Sense of Number

Suroweicki (2004) calls it "the wisdom of crowds," and you can think of much of this book in those terms. Since crowds are not always wise, the topic is really the wisdom of crowds when they are indeed wise, and also about the conditions and circumstances that account for the failure of collectives to be wise. We have already seen a host of examples in which collectives in markets get things "right" in the sense of efficiency without the individual participants being aware that they are achieving a rule-governed emergent order: equilibrium in multiple market supply and demand, the hub and spoke airline service network, equilibrium adjustment of price and quality in some asymmetric information markets, market-driven retail gasoline prices that induce similar patterns in wholesale prices, efficiency in electric network wholesale markets, dynamic inefficiency in electric retail distribution in the absence of free exit and entry in the marketing of energy, prices that anticipate election outcomes in the Iowa Electronic Market, racetrack betting, and many other examples. In asset markets, however, traders are observed to bubble and crash on the long path to a rational expectations outcome.

Surowiecki (2004) summarizes many examples in which collections of individuals make numerical, comparative, or ranking judgments that are good aggregate estimates of the true values:

- *Weight guessing.* During the course of an English Country Fair, eight hundred people examined a fattened ox on display and bought tickets on which to enter their estimates of its weight after it had been slaughtered and dressed. Prizes were awarded to the best estimates in this wager market. Francis Galton was able to access the ticket entries. (See ibid., pp. xii–xiii, 275, and passim for references and a summary.) Using the sample mean as the aggregation principle for estimating the expected weight of the ox, Galton found that the collective estimate was 1,197 pounds for a dressed recorded weight of 1,198 pounds.
- *The "jellybean" experiment.* In this famous exercise, you pass a jar of jellybeans among a large group of people, asking each privately to write down their estimate of the number it contains. Again, the estimates seem "surprisingly good"; for example, with no motivation and a much smaller sample ($N = 56$) than with Galton's ox, the mean estimate is 871 for a jar containing 850 beans (Treynor, 1987, pp. 50–3). Estimating numbers of items behaves rather like estimating the weight of one item.

- *Estimating the value of a jar of coins.* Similarly, I have found that the mean of individual estimates of a jar containing coins in all denominations of 25 cents or less appear commonly to be a good estimate of the true total value of the coins in samples of forty or more estimates. Here the estimates are weighted averages of the number of items.
- *Ranking.* Also, groups of people are good at estimation in ranking objects by weight and ranking piles of buckshot by their numerical composition (Surowiecki, 2004, pp. 5, 276).

In these examples, the individual estimates are poor; for example, it is rare for any individual's estimate of a number to lie between the mean and the true number. How can collectives of unpracticed people make reasonably accurate estimates? Counting or computing is impossible in these examples. But other experiments enable us to understand why groups are able to make good numerical estimates.

Brain imaging studies have established that the brain modules used in numerical calculations are different from those employed in making magnitude comparisons. The former have to be taught; the latter are learned by animals and children without instruction (Dehaene, 1997, pp. 218–20, 223–6).

I suggest that what people rely upon in these group examples are their innate natural skills in "numerosity" – defined as the *perception of number* as a property of a set. Numerosity also characterizes an adaptive development that has been found in animals. Hence, the "accuracy" discovered in these examples follows in a straightforward way from what has been learned from studies of numerosity in individuals:

- *Numerosity in animals.* Let an untrained chimpanzee choose between two alternatives: a pile of seven pieces of chocolate and a pile of six. Most of the choices are for the larger pile. The error, however, varies with the numbers used and reflects effects due both to difference and magnitude. Holding constant the difference between the two piles, the estimation error increases as the compared numbers increase. The error increases moving from two versus one, to three versus two, to four versus three, and so on. Chimps can also discriminate on average between fifty and forty-five but not between fifty and forty-nine. Similar numerical discrimination has been observed in birds, rats, and other animals.
- *Numerosity in chimps.* Chimps, unlike other animals, can also choose as if they "know how to add" – that is, they can discriminate between a tray with two piles of four and three items each (seven) and a tray with five and one (ibid., pp. 17–31; for references, see pp. 255–6).

- *Numerosity in humans.* Humans can discern and identify one, two, or three dots in nearly the same flat response time, 600 to 700 milliseconds, with negligible error. But with four dots, the response time jumps to about 900 milliseconds and shows 8 percent error; five dots take 1,100 milliseconds, with 22 percent error; six dots take 1,300 milliseconds, with 38 percent error; and so on. Hence, beyond three, the response time and error rates increase rapidly. There is a discontinuity moving from three to four items. As with animals, humans show difference and magnitude effects when discerning which of two sets of dots is numerically larger. Thus, following "Weber's (scalar) law," if a person exhibits a 90 percent error when discerning sets of thirteen and ten, their error rate will be 90 percent comparing sets of twenty-six and twenty, thirty-nine and thirty, and so on (ibid., pp. 67, 70–2).
- *Piaget's errors* (see ibid., pp. 41–50): Piaget believed that children have no inborn preconceptions of number and must construct it from years of close attention. Thus, he found that three-to-four-year-olds judge two equally spaced rows of marbles to be "equal," but when four marbles are displayed in a row that is longer than a row of eight closely spaced marbles, their response is to indicate the longer row when asked, "Is it the same thing or does one row have more marbles?" Subsequently, however, more carefully controlled experiments showed that such findings were affected by both context and motivation. In particular – bypassing interaction through language – if you offer even a two-year-old the choice between four M&Ms in a long row and eight in a short row, most take the row of eight M&Ms (Mehler and Bever, 1967).

The experimental study of numerosity in individuals enables us to see why "crowds" can do so well when estimating large numerical quantities and ranking objects: If

A. the individual errors (as deviations from true value) increase with the size and complexity of the estimate;
B. and the errors are distributed approximately independently;
C. then we have the prospect that the sample mean will approximate the true value as the sample size increases; of course, the size of the sample needed for good performance may be much larger for more difficult (larger error) tasks.

The results listed in the preceding examples are "surprising" only because we are unaware, and do not expect, that individual estimates have the properties that have been established for them in experiments and that the law of large numbers is operating.

What are the characteristics of groups that enable good information aggregation outcomes? Suroweicki (2004, p. 10) offers four:

- *Diversity.* This characteristic enables a collective to embrace the range of skills and other attributes – information, opinion, viewpoint, market transactions – required by the various cognitive tasks in estimation or in market transactions.
- *Independence.* Outcomes can go astray if there is too much dependence on or correlation between the individuals' opinions and information states. For example, laboratory stock market bubbles appear to be strongly fueled by self-fulfilling but unsustainable common expectations that are extinguished by subject experience (Smith et al., 1988); similar accounts have been applied to stock markets where experience changes with the demographic turnover of investors (Renshaw, 1988). A new study, Hussam et al. (2006), shows that bubbles can be "rekindled" with the same experienced subjects after they have converged to fundamental value pricing by altering the dividend yield and increasing their cash endowments. Hence, "learning" is not robust to these changes in the environment.
- *Decentralization.* Judgments, values, impressions, and other information to be processed are dispersed in the collectives desiring to utilize that information. Each person has his or her own experiential history that constitutes each person's local information.
- *Aggregation principle.* A procedure, rule, mechanism, or institution exists for processing private knowledge and yielding group outcomes.

Giving people common but irrelevant information in an estimation task can be shown to bias individual estimates by introducing too much correlation in otherwise independent judgments. Thus, in a variation on the jellybean task, Treynor (1987) cautioned people to be aware of the air gap in the top of the jar, and that the jar was plastic, not glass, which made it thinner. This instructional treatment increased the error in the mean estimate by 15 percent. Similarly, if estimations are made sequentially rather than simultaneously, this can introduce dependence and "cascade" effects that reduce accuracy, much as in the asset-trading bubble experiments.[6]

---

[6] As I write (August 22, 2005), although share markets are well behaved, we are experiencing a housing market that has been claimed to have very high price-to-rental-property-earnings ratios.

Surowiecki (2004) provides many other examples of rationality failure in small crowds such as committees and advisory councils, ranging from the Bay of Pigs disaster (masterminded by smart people who thought too much alike), to the Challenger disaster, and to the general problem of evaluating intelligence. The bottom line is that when things go wrong, there are often too many "experts," insufficient diversity, and far too much correlation in the information aggregation process. This is why a "crowd" can outperform a few respected "experts." Also, experts often get their reputations in narrow specialties and are no better in solving problems outside their area of expertise than a random citizen off the street. But we expect them to have an absolute advantage over less accomplished people on all topics. These unrealistic expectations were part of the reason why Hayek (1974), in his banquet speech celebrating the 1974 Nobel awards, said that "if I had been consulted whether to establish a Nobel Prize in economics, I should have decidedly advised against it" (p. 1).

Market Rationality: Capital versus Commodity and Service Flow Markets

I want to close this chapter by talking about capital markets, and some important conjectures and observations about their contribution to human welfare and how and why they are so different from markets for goods and services. In both the lab and the field, there is compelling evidence of the capacity of flow supply-and-demand markets for goods and services to serve their function as institutions that support specialization and to capture gains from trade by discovering efficient and stable prices that are also responsive to fundamental changes in supply cost and demand values over time. In contrast, stock market prices are routinely subject to endogenous day-to-day volatility, unrelated to new information, and over longer time spans are prone to produce bubbles followed by crashes in prices. Quite apart from their noisy short-run behavior, stock markets are far more volatile and unpredictable than are existing goods and service markets because their underlying economic function is to anticipate the products and services of the future. I say "goods," not "commodities," because the latter, associated with raw material assets, are, like shares, sometimes volatile in anticipation of future shortages or surpluses from changing supply or demand. But these episodes are short-lived for basic raw materials compared with their long history of development and declining long-run inflation-adjusted prices.

Claims of stock market "bubbles and crashes" are not new; such expansions tend to be fueled by emergent new technologies:

- In the nineteenth century, the steam engine allowed the steamship to replace the square-rigger, and the railroad to replace the mule team and the stagecoach. Railroad expansion outran the shipping needs of interregional trade, as profitability turned to losses, bankruptcies, and consolidations. But out of this nineteenth century expansion, long-term value was created and retained for the entire economy.
- At the turn of the twentieth century, many new technologies emerged. Telephone, electricity, petroleum, and automobiles sustained a wave of investment and development. There was overexpansion in response to high profitability followed by declining margins, losses, bankruptcy, and consolidation, but long-run value was created and not lost to the economy. Bankruptcy typically allows the assets of failed managers – human and physical – to be reallocated to managers who launch a new attempt at making the business a success.
- A century ago, there were legions of small companies started by entrepreneurs hoping to achieve success manufacturing automobiles. A third of them were electric, and most of the others were experimenting with steam engines. Ultimately both failed: Electric-powered cars had very little range, and steamers took too long to heat up. The only good roads were in cities, built after the success of Henry Ford's high-clearance Model T. A long-shot won the race: the gasoline engine, which Ford combined with the assembly line to produce a low-cost car. Even so, electric car production did not peak until 1912. No one in advance could have picked the winner.
- The ball point pen is today an almost invisible but classic example of innovation. This midcentury invention sold initially for over $10 – in October 1945, New York's Gimbles sold its entire stock of ten thousand pens for $12.50; it was a very profitable new product and inspired a rush of entry, falling prices, losses, and consolidations. But the pen remains, yielding continuing long-term value of which we are not even aware, except that we are all a little richer as a result. When today we buy a 60 cent BIC pen, we are unaware that it is far superior to its expensive upstarts over a half-century ago.
- The 1990s brought an unprecedented volume of initial public offerings (IPOs) of shares by new companies. This "bubble and crash" was fueled by new electronic, communication, computer, biological,

and pharmacological technologies. The residual long-term value creation was suggested by the post-crash national income data: Output increased with little increase in employment as the economy produced more for less. This productivity increase has continued into the new millennium.

- America Online, Amazon, Google, Overstock, and so on were successful among a deluge of business failures. Internet technology has enabled new specializations by an enormous increase in the "extent of the market." Through Worldstock, Overstock has expanded into remote rural villages of poverty in Afghanistan, Nepal, Lebanon, and Ghana to expose the products of skilled artisans to the mass internet market (see the Overstock web site).

In these and thousands of other examples, people are conducting trial-and-error experiments in attempting to discover how new technologies can be managed to satisfy user demand and be profitable. How can the individual pain be eliminated, and the long-term value achieved, with a policy fix that avoids the risk of doing great harm to the selection process itself? We don't know. Here is the problem: If you limit people's decisions to make risky investments in which they have an equity stake, in an attempt to keep them from harming themselves, how much will that reduce our capacity to achieve major technological advancements? The hope of great gain by individuals fuels hundreds of experiments in an environment of great uncertainty as to which experiment, which combinations of management and technology will be successful. Business history in periods of rapid technical progress suggests that the failure of the many is a crucial part of the cost of sorting out the few that will succeed. After a wave of innovation and a bubble burst, managers know a lot about what did not work, and even a little about what did work. Moreover, rationality in this process is not about predictable processes of drawing balls from urns with known or even stable event structures – managing entirely new technologies to create profitable products is always a trial-and-error process of exploring new and untried event spaces.

PART THREE

PERSONAL EXCHANGE: THE EXTERNAL
ORDER OF SOCIAL EXCHANGE

...we must constantly adjust our lives, our thoughts and our emotions, in order to live simultaneously within different kinds of orders according to different rules. If we were to apply the unmodified, uncurbed rules [of caring intervention to do visible "good"] of the ... small band or troop, or ... our families ... to the [extended order of cooperation through markets], as our instincts and sentimental yearnings often make us wish to do, *we would destroy it.* Yet if we were to always apply the [competitive] rules of the extended order to our more intimate groupings, *we would crush them.*

> Hayek (1988, p. 18, emphasis in the original)

I charge you not to be one self but rather many selves ... the giver who gives in gratitude, and the receiver who receives in pride and recognition.

> Gibran (1928, pp. 105–6)

One of the most intriguing discoveries of experimental economics is that (1) people commonly behave noncooperatively in small and large group "impersonal" market exchange institutions; (2) many (typically about half in single-play games, depending on payoffs and the game structure) cooperate in "personal" exchange (two-person extensive form games); (3) yet in both economic environments, all interactions are between anonymous persons. In Part III of this book, I summarize some of the most compelling evidence of cooperation in personal exchange – in the field as well as the laboratory – and review some of the test results designed to discriminate among the more prominent predictive hypotheses for modeling this cooperative behavior. The unifying principle that relates the two-person studies with the study of markets in Part II is the interpretation of the former as a voluntary exchange in which there are gains from the exchange, but no external enforcement of property rights in those gains. That interpretation, sometimes called reciprocity, will be challenged by various tests and compared with alternative social preference interpretations of why people in such circumstances often cooperate contrary to their immediate incentive to defect.

Whatever might be the most useful way to model and explain cooperation, unaided by third-party enforcement of property rights, my working hypothesis throughout is that it is a product of an unknown mix of cultural and biological evolution, with the biology providing abstract function-defining potential, and culture shaping the great diversity of emergent but functionally similar forms that we observe. But to motivate the whole exercise in thought, I will begin in Chapter 9 with a discussion of some persistent cross-cultural social practices from business, law, anthropology, and American economic history.

How might a social rule (practice, norm) emerge, become a cultural fixture, and be widely emulated? I will use a parable to illustrate how a rule for "bargaining in good faith" might become established.

In bargaining over the exchange price between a buyer and seller, suppose the seller begins by announcing a selling price, the buyer responds with a lower buying price, the seller reduces his asking price, and so on. In this concessionary process, it is considered bad form for the buyer (or seller), once having made a concession, to return to a lower (or higher) price. To do so violates a principle of "bargain in good faith" (see Siegel and Fouraker, 1960). How might this cultural norm come about? One can suppose that those who fail to bargain in good faith would be less likely to be sought out by others for repeat transactions. Such behavior raises transactions cost by increasing the time it takes to complete an exchange. Trading pairs would be expected to self-select, tending to isolate the more time-consuming bargainers, and it would take them longer to find those willing to tolerate the time cost of bargaining. Such practices – inherently economizing in this parable – might then become part of a cultural norm, powerful enough to be codified ultimately in contract law and in the evolution of stock exchange rules requiring the bid-ask spread to improve between successive contracts in the double auction. This thread of development from bargaining in good faith to contemporary double auction rules is but one of the countless examples of possible institutional change lost to business history. (For comparisons of the effect of these improvement rules on volatility and equilibrium convergence in double auction experiments, see Smith and Williams, 1982). Proposition: In this manner, collectives discover law in the form of those rules that persist long enough to become entrenched practices. In this example, the emergent rule reduces transactions cost, leaving open the classical question of how equilibrium can be characterized in bilateral bargaining. (See also Tullock, 1985; Frank, 1988.)

NINE

Emergent Order without the Law

... all early "law-giving" consisted in efforts to record and make known a law that was conceived as unalterably given. A "legislator" might endeavor to purge the law of supposed corruptions, or to restore it to its pristine purity, but it was not thought that he could make new law.... But if nobody had the power or intention to change the law ... this does not mean that law did not continue to develop.

Hayek (1973, p. 81)

In one work (*Theory of Moral Sentiments*) we have a theory of the way in which "sympathetic transfers of feeling" set limits to the assertion of individual interests and promote social harmony: partly by creating moral sentiments in the minds of individuals which directly modify their conduct, and partly by causing society to evolve a legal system that expresses the moral sentiments common to the mass of mankind, and imposes restraints which not every individual would always impose on himself. In the other work (*Wealth of Nations*) we have a theory of the way in which individual interests, thus limited, themselves promote economic adjustment and harmony. The two treatises therefore give us complementary halves of [Adam] Smith's social philosophy.

Overton Taylor (1930, p. 233, quoted in Caldwell, 2003, p. 356)

Rules and Order

The early "law-givers" did not make the law that they presumed to "give"; rather, they studied social traditions, norms, and informal rules and gave voice to them, as God's, or natural, law.[1] The common lawyer, Sir Edward

---

[1] This is illustrated in Sumar with the beginning of writing and in the earlier Judeo-Christian oral tradition of the great "shalt not" commandments. Law as an expression of especially deep cultural insight is also plain in the teachings of Jesus. Thus:

Do not think that I have come to abolish the law of the prophets; I have come not to abolish but to fulfill. For truly I tell you, until heaven and earth pass away, not one letter, not one stroke of a letter, will pass from the law until all is accomplished. Therefore, whoever breaks

Coke, championed seventeenth century social norms as law commanding higher authority than the king. Remarkably, these forces prevailed, paving the way for the rule of law in England. What allowed the rule of "natural" or found law to prevail in England, " ... was the deeply entrenched tradition of a common law that was not conceived as the product of anyone's will but rather as a barrier to all power, including that of the king – a tradition which Edward Coke was to defend against King James I and Francis Bacon, and which Matthew Hale at the end of the seventeenth century masterly restated in opposition to Thomas Hobbes" (Hayek, 1973, p. 85; also see pp. 167, 173–4). For a thoughtful examination and critique of some of Hayek's arguments, see the recent contribution of Hamowy (2003).

Polanyi (1962) emphasizes that the rules of the practice of art, including political practice, are hidden, inexplicitly known, and passed on through tradition:

A society that wants to preserve a fund of personal knowledge must submit to tradition.... [T]he doctrines of political freedom spread from England in the eighteenth century to France and thence throughout the world, while the unspecified art of exercising public liberty, being communicable only by tradition, was not transmitted with it. When the French Revolutionaries acted on this doctrine, which was meaningless without a knowledge of its application in practice, Burke opposed them by a traditionalist conception of a free society (pp. 53–4).

The cattlemen's associations, land clubs, and mining districts in the American West all fashioned their own rules for establishing property rights, enforcing and practicing them. The brand on the hindquarters of his calf was the cattleman's indelible ownership signature on the equivalent of his property "deed," which was enforced by professional gunmen hired through his cattle club, just as today's Mafioso do (except that the Old West gunmen were making law, not breaking it[2]); squatter's rights were defended ably

one of the least of these commandments, and teaches others to do the same, will be called least ... but whoever does them and teaches them will be called great in the kingdom of heaven. For I tell you, unless your righteousness exceeds that of the scribes you will never enter the kingdom ... (Matthew, 5.17–22; Metzger and Murphy, 1994).

"The relation of Jesus' message to the Jewish Law was a great concern to the followers with a Jewish background.... Many Jews esteemed the prophets less than the law" (Matthew, note 7; Metzger and Murphy, 1994). Hence, what endured was the cultural message revealed in emergent law and expressed, not made, by the ancient prophets or by Jesus.

[2] These voluntary private associations for sharing the cost of a common good – monitoring and policing property rights – were subsequently undermined by statehood and the publicly financed local sheriff as the state-recognized monopoly law enforcement officer. This observation contradicts the myth that a central function of government is to "solve" the free-rider problem in the private provision of public goods. Here we have the

(possession is nine points of the law?) by the land clubs composed of those brave enough to settle wilderness lands in advance of late-coming veterans of a series of wars (1775–1848) who were seeking to exercise their military land script claims[3]; mining claims were defined, established, and defended by the guns of the mining clubbers, whose rules were later to become part of public mining law (Anderson and Hill, 1975; Umbeck, 1977). For over a century, the Maine lobstermen had established rights, using threats, then force, to defend and monitor exclusive individual lobster fishing territories in the ocean (Acheson, 1975). Eskimo teams hunting polar bear awarded the upper half of the bear's skin (prized for its long mane hairs used to line women's boots) to that hunter who first fixed his spear in the prey (Freuchen, 1960). Extant hunter-gatherers have evolved sharing customs for the products of communal hunting and gathering. For example, the Ache of Eastern Paraguay share the volatile products of the hunt widely within the tribe, while the low variance products of gathering are shared only within the nuclear family (Kaplan and Hill, 1985; Hawkes, 1990).

The following quotation from Freuchen (1960) concerns the "first harpoon" principle. Freuchen managed the first Danish trading post at Thule in Arctic Greenland a century ago. He came to know the Inuit people well, married an Inuit, and had two children by her:

"A (polar) bear is so constructed that it does not like to have spears in it," say the (Inuit) Eskimos. As if to prove what they say, the bear – as they run right up to the beast with their incredible courage and hurl their puny weapons at it – takes the spears that have lodged deeply in its flesh and breaks them as if they were matchsticks. . . . According to custom, all the hunters present are to get parts in the quarry, in this case both the meat and skin. There are three pairs of trousers in

reverse: The incentive of the cattlemen's clubs, which emerged prior to statehood by private association to solve a public good problem, was to free-ride on the general taxpayer, assigning the county sheriffs their erstwhile task of enforcing property rights in cattle. The same free riding occurs with school busing programs in which private costs are paid by public budgets; once a justifiable public budget is provided for education, there is a rent-seeking temptation to incorporate various private goods into that budget – transportation and various social service programs – although the Constitution has required religious activity to remain separate.

[3] The U.S. government had no important sources of revenue (other than excises on alcohol and tobacco, import duties, and land sales) until the twentieth century, when the income tax was introduced. Being rich only in land, soldiers were paid by the government in scrip – or "land office money," as it became known – allowing them to claim land after serving in the military. This policy began with the Revolutionary War, followed by the War of 1812 with the U.S. invasion of Canada, a series of Indian Wars (in part to make land safer for settlement and the discharge of previously issued scrip claims), down to the Mexican War.

a bearskin. If there are more than three hunters present, the ones who threw their spears last will usually be generous enough to leave their parts of the skin to the others. The hunter who fixed his spear first in the bear gets the upper part. That is the finest part, for it includes the forelegs with the long mane hairs that are so much desired to border women's kamiks (boots) with.

... So the hunter measures with his whip handle from the neck down, and marks the length of his own thighs on the skin and cuts off at that mark. The next hunter does likewise with the next piece, and the third one gets the rest (pp. 53–4).

Note that the Inuit "first harpoon" principle is an incentive rule that rewards with greater compensation the increased risk and cost of being the first to harpoon the dangerous prey. It is an equal opportunity rule, not an equal outcome rule, that evolved in this ancient culture. Any member of the hunting team is free to go first, pay the risk cost, and collect the higher revenue. All others, whose contributions cannot be differentiated, share equally in the revenue that is left. That deep human ethic surfaces in laboratory experiments showing that when there is no way to differentiate individual contributions to the joint effort, people tend to support the equal outcome rule. But when individual contributions to the group can be differentiated, then they prefer a rule that rewards in proportion to individual contributions – more to those who sacrifice more for the group.

But Freuchen now continues, and gives us some insight into cultural change, adaptation and institution (rule and practice) modification in the light of a great increase in potential revenue over that which prevailed in the environment under which the "first harpoon" rule traditionally emerged:

Since the white men like to have a whole bearskin with head and claws on it, a rule was made in the Thule district that if a hunter who had a "first harpoon" needed the skin to sell in the shop, he had only to say: "I skin with claws." Then he had to give his mates parts in the meat only, not in the skin. Often it is the intention of a young man to get himself income easily and quickly, just by saying these four little words. But they are difficult to pronounce. For they show that he isn't able to catch foxes enough to trade with. In addition, they make his friends trouserless. Often a remark is heard about "the good sense in keeping the skin whole!" and "This is a nice skin. Here is a man who uses his good luck when it comes along!" Such sarcasms do not fail to hit home. But then the young man breaks out in a laughter: "Naw, now I must laugh. At last, there is something amusing to tell others! Game mates think that the bear was to be skinned with snout and claws so as to rob them of the much desired trouser skins. Of course, it was a joke – the skin belongs to everybody!" With that, he feels relieved and can – since he ... [had "first harpoon"] – go up to the next two in line and measure the length of their thighs with his whip handles to transfer it to the bear skin. There is much satisfaction in this gesture (ibid., pp. 53–4).

We see in Freuchen's description a modification of the "first harpoon" norm in which the person now has the right to *propose* that "I skin with claws," but the proposal's acceptance requires consent from game mates, perhaps unanimity, but we are not told. Consent allows flexibility; for example, a particularly reputable "first harpooner" might have his proposal accepted immediately, and an upstart's proposal might be rejected until his reputation is established.

Also note that the statement, "At last, there is something amusing to tell others!" reveals a recurrent theme in Freuchen that the Inuit were indefatigable gossipers, an important mechanism of restraint.

Ellickson Out-Coases Coase

Using the rancher/farmer parable, Coase (1960) argued that if there were no costs of transacting, then in theory efficiency could not depend on who was liable for damages to crops caused by stray animals. Legal liability gives the rancher an incentive to employ cost-efficient measures to control straying cattle. But if the rancher were not liable, then in a world of zero transactions cost, victims would be led in their own interest to negotiate a settlement paying the rancher to undertake the same efficient control measures induced by legal liability. In so doing, trespass victims save the cost of crop damage, assumed to be more than the cost of cattle control – otherwise, it is inefficient to control them. The externality is internalized by market negotiation incentives.

Curiously, the Coase Theorem – that with complete information and the absence of transactions cost, efficiency does not depend upon the locus of liability – was controversial. It was clearly intended as a kindly spoof of oversimplified theories that, in particular, ignored transactions (including information) costs.[4] The real problem, addressed brilliantly by Coase, was to deal with the question of efficient liability rules in a world of significant transactions cost. He then proceeded to use the transactions cost framework to examine the problem of social cost in a variety of legal precedents and cases.

---

[4] Later game-theoretic formulations have allowed that with two or more alternatives there may exist "standoff equilibria" that stall agreement in Coase bargaining (see Myerson, 1991, p. 506). These untested esoteric cases may limit extensions of the Coase Theorem, but do not detract from its essential message that the locus of liability was irrelevant under the conditions then being posited in the literature.

Coase (1974) also noticed that the lighthouse was frequently cited by theorists as an example of a "pure" public good. As was his style (to challenge the casual parables of theory that finessed certain costs by fiat), his response, in effect, was, "Well, let's see what people have done who actually build and operate lighthouses, and use the services of lighthouses." It turned out that early lighthouses were private enterprises, not government services, that provided an emergent solution to a public good problem (read "capital investment" problem), and the alleged inevitability of free riding was solved by the lighthouse owner, who contracted with port authorities to collect lighthouse fees when ships arrived portside to load or unload cargo. Note that before the lighthouse is built, all are potentially excludable from its use. The builder, by offering to build it if port authorities contract to collect fees sufficient to fund the investment, provides a means of assuring payment before it is built, a return on the investment, and in effect enabling a private benefit/cost calculation. The government should go through the same exercise to measure investment worthiness. Increased property security has a market value in reduced marine insurance premiums, no matter who builds the lighthouse. The argument that the lighthouse, once built, can serve new users at zero marginal cost does not serve good design principles. Think of the relevant marginal cost as the discrete capital investment cost of the lighthouse as a supplement to port services; if the resulting benefits exceed the marginal capital cost, there are gains from the trade. This error – that the lighthouse, once built, should be made available at marginal cost – has application to fixed costs in the theory of the multidivision firm. If in a firm with capital inputs shared across different divisions and products, one is concerned about incentives that will ensure that lumpy inputs of capital will compete with other resources to enhance profitability, then you must avoid thinking that all sunk costs are irrelevant. This thinking is entirely parallel to the lighthouse problem; that is, in an environment in which knowledge is dispersed, how do we ensure that we have a mechanism for funding shared capital goods that is superior to alternative uses of the resources that are required?

Originally, Shasta County in California was governed by "open range" law, meaning that in principle ranchers were not legally liable for damages resulting from their cattle *accidentally* trespassing on unfenced land. Then, in 1945, a California law authorized the Shasta County Board of Supervisors to substitute a "closed range" ordinance in subregions of the county. Dozens of conversions have occurred since this enabling law. Under a closed range law, the rancher is strictly liable – even if not negligent – for damage caused by his livestock. Elickson (1991) out-Coased Coase by, in effect, asking, "given that this county applies the polar legal rules used in Coase's illustration, how do neighbors in Shasta County actually handle the problem of stray cattle?" The answer:

> Neighbors in fact are strongly inclined to cooperate, but they achieve cooperative outcomes not by bargaining from legally established entitlements,[5] as the parable supposes, but rather developing and enforcing adaptive norms of neighborliness that trump formal legal arrangements. Although the route chosen is not the one that the parable anticipates, the end reached is exactly the one that Coase predicted: coordination to mutual advantage without supervision by the state[6] (p. 4).

Thus, Shasta County citizens, including judges, attorneys, and insurance adjusters, do not have full working knowledge of formal local trespass law.[7] Citizens notify owners and help catch the trespassing animal; use mental accounting (reciprocity) to settle debts (for example, if a rancher's cattle have strayed, he might tell the victim to take some of his hay, or if your goat eats my tomato plants, you offer to help me replant them); use negative gossip, complain to officials, or submit informal claims for money (but not through the services of a lawyer) to punish deviant neighbors; rarely use lawyers to seek monetary compensation; share the building of fences, most often by a rule of proportionality (you pay more if you have more animals than your neighbor); ignore fence law as irrelevant; and do not change fence obligations with the planting of crops. Hence, as will be examined experimentally in Chapters 11 and 12, people use long-ingrained principles of positive and negative reciprocity to achieve workable systems of social exchange. Finally, contrary to the Coasian parable, the main cost of trespass is not from crop damage, but from highway collisions that kill animals and damage property.

---

[5] These are the outside options or threat points in game theory.

[6] The same results emerge in laboratory experiments reported by Hoffman and Spitzer (1985) demonstrating a case of parallelism between laboratory and field in which the stakes can be presumed to be much higher in the field; for a critique and extensions, see Harrison and McKee (1985a).

[7] Contrary to the beliefs of most of these people, under open range the animal owner is fully liable for intentional trespass, trespass of a lawful fence, and trespass by goats whatever the circumstances. Thus most people including legal professionals are wrong in believing that no liability was carried by the animal owners under open range. Note also that it appears that the unruly behavior of domesticated goats had long been recognized in pastoral norms and was now captured in codified law.

TEN

The Effects of Context on Behavior

Science believes itself to be objective, but is in essence subjective because the witness is compelled to answer questions *which the scientist himself has formulated.* Scientists never notice the circularity in this because they believe they hear the voice of "nature" speaking, not realizing that it is the transposed echo of their own voice. So, ... science ... is left with a fragmented world of things which it must then try to put together.

Bortoft (1996, p. 17)

The intelligent way to be selfish is to work for the welfare of others.

Dali Lama (Sobel, 2005, p. 428; attributed to Smuts, 1999)

Game theory is for proving theorems, not for playing games.

Reinhard Selten (quoted in Goeree and Holt, 2001, p. 1404)

Introduction and Elementary Theoretical Background

As in market exchange studied in Part II, cooperation has also emerged, if unexpectedly, in anonymous two-person extensive form games in laboratory experiments. This behavior is still actively being investigated.

   Although the behavior was found to be contrary to the traditional view (underlying the predictions) of "rational" prescriptions, further examination suggests that the results are not inconsistent with our examples of spontaneous order without externally imposed law in the evolution of behavior in human cultures in which rights to take action ("property rights") emerge endogenously by mutual implicit consent. Hence, there is no necessary contradiction between what we reported in the previous chapter (and in Part II) and what we will be reporting in this chapter and in the next two chapters following. There is, therefore, a direct parallel between what Ellickson found in testing the Coase predictions in Shasta County and what was discovered independently in testing the predictions of theory in two-person games.

Stated informally, here are the elementary assumptions and propositions in game theory that underlie the test results we report in Chapters 10 through 12. Examples include the ultimatum and dictator games in this and the next chapter, and the trust games of Chapter 12. The reference to Myerson (1991) provides corresponding formal treatments:

1) Given a choice between sums of money A and B, where A is greater than B, a rational person will always choose A, *whatever the circumstances* (including whoever supplies the money) (domination) (ibid., 1991, pp. 26ff).

2) If two people are to interact in a sequence of moves, each will use "backward induction," first looking ahead to analyze the sequence of future choices before deciding what to do in the current situation. At each decision point in the sequence, each person will choose rationally as in (1) and each will assume that the other will choose rationally as in (1) (sequential rationality; subgame perfection; backward induction) (ibid., pp. 154ff).

3) In repeated (in theory, infinitely) versus single-round ("one-shot") games, single-round games predict the dominant choice (1). Repeated games are said to favor "cooperation." This is the Folk Theorem in game theory because it has been thought to be intuitively plausible, but rigorous theorems have not been forthcoming (ibid., pp. 154ff).

The experimental literature strongly supports the Folk Theorem: Repeat interaction favors departure from the choice of strictly dominant strategies and outcomes and people achieve joint cooperative outcomes superior to the dominant choice equilibrium, although both parties have incentives in any round of play to defect from choosing to cooperate (See Chapter 12 and Camerer, 2003).

What is unexpected based on this constructivist framework of thinking is that we would observe "rampant cooperation" in single-round, "one-shot" interactions. But the concept of a single-round versus repeated interaction is an abstract theoretical construct; it is not somehow given in nature or a common property of our daily social interactions.

Perspectives on Interpreting Results

As we shall see, even when "strangers" are anonymously paired in games played only once, half or more of the subjects fail to choose dominant outcomes, electing rather to eschew defection strategies to ensure that their paired counterparts do not suffer a reduced payoff.

I think that the most comprehensive understanding of cooperation follows from postulating that subjects bring their social exchange experience with them into the laboratory, and do not automatically follow the game-theoretic abstract principle that they are in a one-shot game in which it never pays to cooperate. There should be no suggestion that reciprocity in this sense is a deliberate calculation, but simply something that comes naturally (or fails to under the conditions that apply) as part of an acquired style over a long history.

In the references that follow, many have interpreted the single-play experimental data to mean that the game-theoretic dominant choice version of the self-interest hypothesis must be rejected, *and* that we must perforce assume that people have other-regarding preferences. But in repeat interactions, people do not have to have other-regarding preferences to engage in other-regarding behavior. As Adam Smith put it, "Kindness is the parent of kindness," and as the Dali Lama formulates it, "The intelligent way to be selfish is to work for the welfare of others." Both "kindness" toward others and working for the welfare of others are sophisticated forms of taking actions that individuals may find consistent with their long-term self-satisfaction – your reputation is valuable. Therefore, when we observe a violation of the game-theoretic prediction that people will choose equilibrium-dominant, narrowly self-interested outcomes, either we reject this research hypothesis in favor of utilitarian preferences for other as well as own payoff, or we reject the implicit auxiliary hypothesis that the framework of subjects' decisions corresponds to the game-theoretic conception of a single-play game shorn of a past and a future between the players.

Building on the interpretation that the driving force behind cooperation is primarily (I would *not* claim exclusively) reciprocity, this also enables one to obtain a more comprehensive understanding of the principle that specialization does not require markets as such; specialization requires only that there be exchange-based norms of sharing, reciprocity, and other mutual self-help mechanisms – it requires only the "I owe you one" norm (see Crockett et al., 2006, for evidence in a new context). Hence, specialization and trade, which have roots in ancient human social groupings, can be appreciated as being with us still as an integral part of the two distinct worlds of personal and impersonal exchange in which we constantly live. In cross-cultural anthropological studies of behavior in some two-person games, the results are consistent with the hypothesis that greater cooperation in these games is associated with more participation in markets (Henrich, 2000; Henrich et al., 2005).

A window on this sociality, and on any tendency of subjects to bring the baggage of their repeated game life experience into the laboratory, is to study variations on context in single-play games. In these protocols, since context encodes past experience, we can examine the sensitivity of subject beliefs about whether the game they are playing is a single-play exception to the "repeated game of life." Hence, the issue is not cognitively impenetrable, as claimed in Henrich et al. (2005, p. 840), whose experiments do not systematically examine the issue.

This chapter discusses experiments that have examined the effect of context on behavior in ultimatum and dictator games. In particular, I want to address the use of the totality of instructions and procedures as treatments; if they are powerful treatments, this fact should not be viewed as a source of embarrassment, but as a truth and a means of examining the stability of behavior in one-shot games across different contexts. We have learned that people solve decision problems by drawing on their context-laden experience not by thinking about them as rational theoretical constructs with abstract fundamental properties that are independent of context. When we test those constructs, we are asking whether our reductions are valid predictors of behavior. If they are valid, as we have seen in tests of the supply-and-demand theory of markets, then our constructs reasonably capture the essential properties of that social system. If they are not, as in a large class of two-person interactions, we need to inquire as to how actual decisions are affected by various elements that are omitted from our reductions.

This perspective, that people solve decision problems by drawing on their contextual and social experience, has implications far exceeding what can be examined here. I do, however, want to apply this perspective to the question of subject understanding, or "confusion" – a construct entirely based on our models – as it has arisen in public good and asset-trading experiments where departures from theoretical predictions also have been particularly prominent. This is provided as an aside in the Appendix because the perspective there concerns larger group interactions than is the focus in the main chapter, although the issues are not dissimilar.

In the next chapter, I will build on the discussion of context to examine behavior in the investment trust game, behavior that is attributed to reciprocity or, as claimed in some of the literature, to some form of other-regarding utilitarian motives.

How Does Context Matter?

Two decision tasks, represented theoretically by the same abstract game tree, may lead to different responses when they occur in different contexts. Why is this? And why should we consider it important if our primary interest is in understanding the economic and strategic basis of human decision as we have formally modeled it in abstract games?

The answer to both questions is to be found in the process by which we perceive the external world and is therefore central to understanding human decision and how it changes over time – what we somewhat casually call "learning." As it turns out, experimental economic science is confronted with "entanglement" between our abstract constructivist models of human interaction and the context in which they are presented; this "entanglement" is a poorly understood property of how our brains work.

The more common view can be illustrated by some references to Camerer (2003), one of many who could be cited, who – too arbitrarily, I believe – define the issue much more narrowly than I will deal with it here and in the data that I will discuss.

- "Since Schelling's (1960) work on 'psychological prominent' focal points in coordination games ... it has been well understood that the way in which strategies are described could focus expectations on them and affect the way people play ... [and] ... the psychology of decision making shows that the way options are described or 'framed' can influence choices" (Camerer, 2003, p. 74). By "context," I will mean all aspects of the instructions and procedures, and the information carried thereby, and not only how the game is "described." As I see Schelling's work and critique, "context" is by no means a change of description only; it also says something much deeper about how our human mental equipment operates in decision making, and is more than a side issue or a nuisance to be recognized but not integrated into the science. Schelling spoke astutely as an economist, not as a psychologist, thinking outside of the traditional formalizations on how one might align or interpret propositions from game theory in terms of human information processing and decision.
- "Changing the way games are described can have modest effects.... There is little doubt that describing games differently can affect behavior; the key step is figuring out what *general* principles (or theory of framing) can be abstracted from labeling effects" (ibid., p. 75). The effects are not modest from the broader perspective to which I am referring. I will present ultimatum data in which instruction and procedures alone change the offered percentages of the total stakes from 44 percent to 28 percent, a decline of over a third, and dictator game data in which the change in offered percentages is from 23 to 0 percent of total stakes. (Some of this variation is evident in the summary tables

in ibid., pp. 50–8, 73, but it is not emphasized in reviewing the results and is in fact inconsistent with the interpretations that are offered as in the preceding quotation.) These procedural effects are easily as strong as those that are called "structural" (ibid., p. 75). It is correct, however, for Camerer to call for scholars to strive to explicate them in terms of general principles, which many of us are trying to do. Many if not most of these considerations are emerging as strongly linked to the deep sociality in humans as reflected in our cultural and biological heritage.

Anonymity as a Treatment Procedure

In defining context, one of the most common laboratory procedures is to ensure that subjects interact anonymously and that all know this fact. Why? The inspiration comes from the insight that the standard model of nonrepeated game theory must be about strangers without a history or a future who interact once, and anonymity has been the accepted means by which experimenters have sought to be true to this prescription. Rosenthal (1981) emphasizes the theory connection, but the experimental practice is more than two decades older than this and is importantly due to early work by social psychologists, particularly Sidney Siegel, whose pioneering contributions to experimental economics and methods influenced its subsequent development.

Moreover, anonymity has long been justified in small group experiments as a means of controlling variables that reflect the complexities of natural social intercourse. Thus, in their American Academy of Arts and Sciences prize–winning study of bargaining, Siegel and Fouraker (1960), state "... that such variables should either be systematically studied or controlled. ... We have chosen to control these variables at this stage of our research program, with the intention of manipulating them and studying them systematically in future studies" (p. 23).[1] That such interpersonal social considerations are important is well documented: Face-to-face interaction completely swamps subtler procedural effects in yielding efficient cooperative outcomes, which is an indication of why it may be the preferred popular way to bargain in the world (Hoffman and Spitzer, 1985; Radner and Schotter, 1989; Kagel and Roth, 1995). The original justifications given for studying two-person interactions under anonymity have evaporated with

---

[1] Siegel's untimely death a year later in the fall of 1961 arrested the further development of this program, which was unquestionably of great personal interest to him and which I believe he would have pursued with the energy and imagination that was his style.

the discoveries it engendered, and as variations in the degree of anonymity became a tool for systematically investigating the social influence that people bring with them into these environments and that affects decision even under anonymity.

Camerer (2003) embraces the original argument suggesting that anonymity "is used to establish a benchmark against which the effects of knowing who you are playing can be measured" (pp. 37–8), but the latter experiments hardly exist except in the early studies of face-to-face bargaining. This line of development has prominently *not* been pursued. Experimentalists continue studying the benchmark, ad infinitum, and it has become the standard protocol almost everywhere. I believe it is a defensible methodology, but not on the ground that it serves as a benchmark for comparisons that never are made. The problem is that with the use of natural language ("cheap talk"), people, even strangers, in face-to-face contact often successfully avoid all too easily the strategic principles that game theory was devised to analyze, a problem that simply cannot be adequately examined only within the confines of game-theoretic principles. Game theory may often be relevant in face-to-face interactions, but the theory does not predict the circumstances under which this will occur. These have to be preselected external to the theory.

But out of studies based on the early study of decision under anonymity there has emerged unintentionally a body of experimental knowledge that I believe provides much more important reasons for studying anonymous interactions than the "benchmark" justification: First, it is the anonymity condition that has provided the greatest scope for exploring the human instinct for social exchange and how it is affected by context, reward, and procedural conditions related to the social experience that we appear to carry with us everywhere; second, we have discovered that the deeply entrenched innate habits of cooperation are so strong that they persist even under the strong contrary condition of anonymity, a fact that game theory has been unable to account for except by the recent recourse to altering the utility function so that it provides an ex post fit with the discovered anomaly. This recourse is strange, for surely any such social preferences do not arise from anonymity. These considerations suggest the need for a broadening of the data set to examine the extent to which specific social considerations get imported through subjects' prior experience into the environment of "one-shot" games.

With anonymity, of course, a supposed cognitively aware strategist has an opportunity to eschew all social habits and norms and pursue her own immediate undercover interest. But for many, their sociality bleeds through the anonymity environment, although significantly less so in games using "double blind" conditions where your decision can be known by no other person (not only the experimenter, but all third parties) and thereby reveals the dimensions of the effect of our social experience on decisions in these one-shot encounters. Studying the effect of context, including degrees of anonymity, on decision gives us controlled access to a window on the hypothesized ubiquity of human sociality.

Perception, Context, and the Internal Order of the Mind

Hayek (1952)[2] was a pioneer in developing a theory of perception, which anticipated recent contributions to the neuroscience of perception and is particularly helpful in understanding why context is important. This theory merits systematic study. It is natural for our calculating, inferring minds to suppose that experience is formed from the receipt of sensory impulses that reflect unchanging attributes of external objects in the environment. Instead, Hayek proposed that our current perception results from a relationship between external impulses and our past experience of similar conditions. Categories formed in the mind are based on the relative frequency with which current and past perceptions coincide. Memory consists of external stimuli that have been modified by processing systems whose organization is conditioned by past experience (ibid., pp. 64, 165).[3] More recently it has been accurately noted that there is a "constant dynamic interaction between perception and memory, which explains the...identity of processing and representational networks of the cortex that modern evidence indicates.... Although devoid of mathematical elaboration, Hayek's model clearly contains most of the elements of those later network models of associative memory..." (Fuster, 1999, pp. 88–9).

Thus if a person's past experience is predominantly in repeat social interaction, then what is in memory may be called upon for application in a one-shot game in which the theory we have constructed is conditioned on the auxiliary hypothesis that the individual will easily assimilate what abstract game theory assumes: that there is a sharp and essential difference between single- and (infinitely, discounted) repeat-play protocols for dominance choice players. The important point here is *not* "that reciprocal

---

[2] *The Sensory Order* was not published until 1952, when Hayek revised a manuscript, originally written in the 1920s, titled in English translation *Contributions to a Theory of How Consciousness Develops* (noted in correspondence to me by Bruce Caldwell).
[3] The interdependence between perception and memory is revealed by the different descriptions of the same event by two eyewitnesses (Gazzaniga et al., 1998, pp. 484–6).

behavior in anonymous one-shot experiments is due to subjects' inability to adjust properly to one-shot interactions" (Fehr and Fischbacher, 2002, p. C6). We have known since the original experiments of Fouraker and Siegel (1963) that subjects do not behave the same under repeat and single play when they have complete information on payoffs. McCabe et al., 1996, and Chapter 12 of this book summarize the effects of different matching and repeat interaction protocols on behavior in trust games. Subjects adapt to their outcome experience in repeat play under differing such protocols. Rather, the question missed by Fehr and Fischbacher is whether subjects understand the distinction between one-shot and repeated interactions that is specified in the abstract theory, and whether that distinction has any relevance to human interactive behavior as we observe it in or outside the laboratory. I would claim that in human experience the prospect of repetition is commonly open to two people interacting for the first time. Whether there is repetition depends importantly on whether they experience mutual cooperation, and I would conjecture that it also depends on there being synergistic gains from the exchange as we have seen in the investment trust game as a two-stage dictator game. These circumstances have been studied by McCabe et al. (2002, 2007). Subjects should not be presumed to bring to a one-shot game the "understanding" assumed by theorists in the concept of single play, but they have no difficulty slowly adjusting their choices to the outcomes they experience when such a game is repeated, as we shall see in Chapter 12.

Hayek's model captures the idea that, in the internal order of the mind, perception is self-organized within a flexible developmental structure: Abstract function combines with experience to determine network connectivity and expansion. Loss can occur either from lack of function or lack of the stimulus of developmental experience. Block or distort sensory input, and function is impaired; impair function by brain lesions or inherited deficiency, and development is compromised.

Built into your brain's development structure is the maintained hypothesis that the world around you is stationary. Look at the wall and move your eyes back and forth, keeping your head still. The wall does not move. Now press your eyeball with your finger through the eyelid from the side. The wall moves as you jiggle your eyeball. Why the difference? When you flex the eye muscles and move your eyes back and forth, a copy of the signal goes to the occipital cortex to offset apparent movement of the wall so that your net perception is that of a stationary wall. This stabilizing self-ordered adaptation for seeing the world as stationary has in practice a minor cost: It makes you vulnerable to certain optical illusions of motion. Moving your eyes back and forth between the tunnel gate and your airplane as it docks, you ambiguously "sense" that either the gate or the plane, or both, is in motion. The ambiguity is resolved only when the gate, or plane, stops, or you focus on a third stationary object for reference.

Hayek's model is consistent with the hypothesis that the mind is organized by interactive modules (circuits) that are specialized for vision, language learning, socialization, and a host of other functions. In this view, mind is the unconscious product of coevolution between the biological and cultural development of our brains that distinguishes us from other primates. It is what made reason possible. Our folk predilection for believing in the "blank slate" concept of mind (Pinker, 2002) makes plain that this interpretation of mind derives from our subjective direct perception of control and learning, much as did once the idea that the earth is flat. A great puzzle concerns how it is that we can sometimes escape from our subjective folk perceptions so that the falsifying indirect evidence, based on reason, becomes part of our "felt" experience – that is, how do we come to abandon the subjective "flat earth" experience and quite unconsciously come to think of the world as round? In such cases, constructivist rationality somehow succeeds in re-forming our perceptions into an ecologically rational adaptation, but we do not know how this occurs. In Polanyi (1962, pp. 55–65), this means that the individual has moved his awareness of some phenomenon from being focal to being subsidiary, exactly as the pianist acquires the skill of playing without thinking about the notes.

Hayek's theory – that mental categories are based on the experiential relative frequency of coincidence between current and past perceptions – seems to imply that our minds should be good at probability reasoning. But Tversky and Kahneman (1983) have famously emphasized the opposite (their data are actually more mixed, but they were into highlighting "anomalies"). For example, people violate the most elementary rule of probability reasoning: the "conjunction fallacy." But what is "probability" for people who are untrained in probability theory and its abstract axioms or rules? Consider the following Linda (or the bank teller) problem: Linda is thirty-one-years-old, single, outspoken, and very bright. She majored in philosophy. As a student she was deeply concerned with issues of discrimination and social justice, and also participated in antinuclear demonstrations.

Which of the following two alternatives is more probable? Linda is

A. a bank teller.
B. a bank teller and is active in the feminist movement.

A is the correct choice since the state of being a bank teller must be more inclusive than the conjunction of bank teller and any other nonempty set of characteristics with positive probability mass, such as being a feminist activist. Hence: Probability (A) > Probability (B). But 85 percent choose B. The Kahneman and Tversky wording of the problem tricks the minds of most people into thinking that the descriptive content of *B* is better associated with the problem statement.

So what happened to the idea that we naturally think in terms of outcomes relative to a class? Or is it that our thinking is actually poor at defining the classes correctly, which does not sound like a good adaptation? Gigerenzer (1993) helps us to understand the resolution of this anomaly by restating the question slightly, leaving in all the diverting descriptive pitfalls in the original problem statement. The question is aligned with our hypothesized ecological experiential learning, rather than the abstract – I am going to say constructivist – notion of "probability." Following the preceding problem description, Gigerenzer states the following:

There are one hundred people who fit the description above. Which of the following two alternatives is more probable? Linda is

A. a bank teller.
B. a bank teller and active in the feminist movement.
Most people now choose *A*.

Left to its ecological devices, when problems are presented in terms of relative frequency of outcomes in classes, our brain seems to do tolerably well, but when the untrained mind is required to apply constructivist event probability principles of the kind that are well suited to proving theorems, the mind is likely to make errors. Those logical reasoning principles are not the autonomic brain's way of calculating, estimating, and managing to get through our day-to-day tasks. (Tversky and Kahneman, 1983, noted earlier this disappearance of conjunctive error, but decision failure has been the remembered theme, as in Lopes, 1991.)

The Significance of Experimental Procedures

The data reported here will demonstrate how social context, as reflected in various experimental protocols, can be important in the interactive decision behavior we observe. This possibility follows from the autobiographical character of memory and the manner in which past encoded experience interacts with current sensory input in creating memory. I will be reporting the results of decision making in single-play, two-person, sequential move game trees. Subject instructions avoid using technical and role-suggestive words such as "game," "play," "players," "opponent," and "partner" (except where variations on the instructions are used as systematic treatments to identify their effect; see Burnham et al., 2000, for a study comparing the effect of using "partner" and "opponent" in a trust game). Rather, reference is made to the "decision tree," "decision maker 1" (DM1), "decision maker 2" (DM2), "your counterpart," and so on. The purpose is to provide a baseline context – one that attempts to avoid potentially emotive words that might trigger unintended meanings. But we must always keep in mind that this purpose may be very difficult to achieve, and we should never assume without evidence that it has been achieved.

With respect to the point I am emphasizing here, it is not meaningful or helpful to talk in the pejorative about "experimenter effects." There certainly can be instructional and procedural effects, including the presence or absence of an experimenter, what the subjects know about what the experimenter knows or does not know (as in double blind behavioral experiments), and what he or she does or does not do. The point is that all of the elementary operations used to implement these two-person experiments are potential treatments that may have a significant effect on observed outcomes. The purpose of identifying these effects is to understand why they are important to the individual and how we are to interpret people's responses in abstract games, not to control or eliminate them as a nuisance. It is also important to realize that a great many experiments are very robust with respect to instructions and procedures. One example is the class of double auction market experiments; another is the asset "bubble experiments" widely replicated under varying instructions and protocols – including subjects providing all their own money to "buy into the game." A notable feature of the instructions in the "bubble" experiments is the heavy-handed emphasis on calculating and reminding the subjects at the beginning of each period what is the current fundamental dividend or "holding value" of the shares, information that they summarily ignore. This is an example showing clearly that "jawboning" by the experimenter simply does not influence the robust phenomena that are established by these experiments (Smith et al., 1988).

Thus, I do *not* mean that this baseline context is "neutral," a concept that is not clearly definable, once it is understood that context effects are defined by autobiographical experience. The effect of instructional variation on decision is an empirical matter, concerning which the abstract theory has nothing to say, and any particular set of instructions must always be considered a treatment unless the observations are found by experiment to be robust to changes in the instructions. Implicitly, this view is revealed in the strong tendency, and justifiable fussiness, of experimentalist economists to maintain uniform instructions across replications and treatment changes. In spite of this good intuition that testing and context may be entangled, few studies have systematically compared the effect of different instructions (and procedures) as socioeconomic "treatments" that might affect observations.

All observations, however, must be seen as a joint product of the treatments derived from the theory, implemented by particular parameters that define precise hypotheses that we test, and the protocols and procedures used to implement the test. This is not unique to laboratory observations, but a characteristic also of all field observations and the whole of science, as we shall see in Chapter 13. It is therefore important to understand how procedures as well as different parameterizations (games, payoffs) affect behavior. Procedures are to be understood as setting the perceptual stage for the subject that affects how he or she sees the immediate world of the experiment, as per the past experience on which the subject is drawing. Consequently, if important, procedures must be studied and ultimately related to our interpretation of strategic decision making. But to study them, it is necessary to look for inspiration outside of game/economic theory. Abstraction is essential to the mechanics of theory but cannot be assumed to be sufficient for understanding the range of behavior that is manifest in tests of theory.

Overview of Experimental Procedures

In the experiments I will be discussing, subjects are recruited in advance for an economics experiment. As they arrive near the appointed time, each registers and receives a show-up fee, and, in many but not all of the experiments we report, people are assigned to a private computer terminal in a large room with forty stations. Commonly there are twelve people, well spaced throughout the room, in the experiments reported. After everyone has arrived, each person logs in to the experiment, reads through the instructions for the experiment as the experimenter reads them aloud, responds to instructional questions, and learns that he or she is matched anonymously with another person in the room, whose identity will never be known, and vice versa. This does not mean that a subject knows nothing about his or her matched counterpart. For example, it may appear evident that he or she is another "like" person, such as an undergraduate, or part of some business group with whom one may feel more or less an in-group identity.

Obviously, each person imports unconsciously into the experiment a host of different past experiences and impressions that may be associated with the current experiment. If the subjects all gather in a room together before being assigned private space, if it is a very large or very small group, or if there is a disproportionate number either of men or women, such variations should not be assumed to make no difference, because they are all part of the unconscious initializing experience of the subject. The safest path is to either hold constant such conditions or systematically vary them until one has learned how robust subject behavior is in any particular two-person game.

The Ultimatum Game Example

Consider the ultimatum game, a two-stage, two-person game with the following abstract form: For each matched subject pair, the experimenter makes a fixed sum of money, m, available (for example, m will be ten $1 bills, or ten $10 bills); player 1 moves first offering x ($0 \leq x \leq m$) units of the money to player 2, player 1 retaining $m - x$; player 2 then responds either by accepting the offer, in which case player 1 is paid $m - x$, and player 2 is paid x, or by rejecting the offer, in which case each player receives nothing (see Güth et al., 1982, for the pioneering study of ultimatum game behavior).

In the following, I report ultimatum results from four different instructional and procedural treatments (contexts) that have the same underlying abstract game structure. In each case, imagine that you are player 1 (see Hoffman et al. [1994; hereafter HMSS] for instructional details):

- "Divide $10." The instructions state that you and your anonymous counterpart have been "provisionally allocated $10" and randomly assigned to positions. Your task as player 1 is to "divide" the $10 by filling out a form that will then go to your counterpart, who will accept or reject it.
- "Contest entitlement." The twelve people in the room each answer the same ten questions on a general knowledge quiz. Your score is the number of questions answered correctly; ties are broken in favor of the person who first finished the quiz. The scores are ranked from 1 (highest) through 12 (lowest). Those ranked 1 through 6 are informed that they have earned the right to be player 1 and that the other six will be player 2.
- "Exchange." Player 1 is a seller, and player 2 is a buyer. A table lists the buyer and seller profit for each price $0, $1, $2, and so on, up to $10,

charged by the seller, and the buyer chooses to buy or not buy. The profit of the seller is the price chosen; the profit of the buyer is ($10 – price). Each receives nothing if the buyer refuses to buy.

- "Contest/exchange." This treatment combines "contest entitlement" with "exchange"; that is, the sellers and buyers in "exchange" are selected by the contest scoring procedure. In one version, the total money available is ten $1 bills, and in the second it is ten $10 bills.

Whatever the context, there is a game-theoretic concept of equilibrium (subgame perfect, Selten, 1975) that yields the same prediction in all four treatments: Player 1 offers the minimum unit of account, $1 ($10) if $m = \$10$ ($100), and player 2 accepts the offer. This follows from the assumptions (1) that each player is self-interested in the narrow sense of domination; (2) that this condition is common knowledge for the two players; and (3) that player 1 applies backward induction to the decision problem faced by player 2, conditional on player 1's offer. Thus player 1 reasons that any positive payoff is better than zero for player 2, and therefore player 1 need only offer $x = \$1$ ($10).[4]

One difficulty with this analysis is that, depending on context, the interaction may be interpreted as a social exchange between the two anonymously matched players who in day-to-day experience read intentions into the actions of others (Baron-Cohen, 1995). Suppose the situation is perceived as a social contract as follows: If player 2 has an entitlement to more than the minimum unit of account, then an offer of less than the perceived entitlement (say, only $1, or even $2 or $3) may be rejected by some players 2. Such action may serve player 2's long-term interest across many interactions as a life style, and be imported into this environment. Player 1, introspectively anticipating these possible mental states of player 2, might then offer $4 or $5 to ensure acceptance of his offer. Alternatively, player 1 might enjoy (get unexplained utility from) giving money to his counterpart – that is, be altruistic. The point is simply that there are alternative models to that of subgame perfection that may qualify predicted choices in the ultimatum game, and these alternatives leave wide latitude for the possibility that context will affect the behavior of both players by invoking habits from their autobiographical experience in social interaction.

---

[4] This game also can be interpreted as having multiple Nash equilibria, but exploring this avenue will not help us with the problem I want to address here. Also, this interpretation provides no insight into the behavioral differences observed in ultimatum and dictator games.

Abstract game theory can embrace these alternatives through the artifice of "types" – utilities or perhaps social norms or belief states such as sharing, trust, trustworthiness, and reciprocity, but the original game-theoretic tradition focused on utility types. More promising, I believe, are approaches by Stahl (1993) and Holt (1999) in which "types" are based on thinking processes, breaking from the mold of utility typing, or the work of Houser in a series of contributions with several coauthors.[5]

The norms, rules, and morality that evolve in cultures have important features as cross-cultural human universals. That individuals enter an experiment with initialized mind characteristics that can be differentiated from others is demonstrated in repeat interaction games by the empirical discovery that a person's first decision provides a measure of "type" that has explanatory power in subsequent play.[6]

Observe that in "divide $10," the original $10 is allocated imprecisely to both players. Moreover, a common definition of the word "divide" (*Webster's*) includes the separation of some divisible quantity into equal parts. Again, random devices such as lotteries are recognized as a standard norm for "fair" (equal) treatment. Consequently, the instructions might be interpreted as suggesting, unconsciously, that the experimenter is engaged in the "fair" treatment of the subjects, cueing them to be "fair" to each other.

As an alternative, "contest" deliberately introduces a pregame procedure that requires player 1 to "earn" the right to be the first mover. This may cue some insipient pattern or norm of just deserts or entitlement based on performance in the pregame quiz; alternatively, for many this may be seen as an illegitimate reason for departing from a more pervasive and dominant norm of equal division.

In "exchange," the ultimatum game is imbedded in the gains from exchange from a transaction between a buyer and a seller. In an exchange, both the buyer and the seller are made better off, and buyers in our culture may accept implicitly the right of a seller to move first by quoting a price. Thus, in experiments reported by Fouraker and Siegel (1963) for single play in a bilateral monopoly exchange, the seller moved first by quoting a price, followed by the buyer choosing a purchase quantity. This game had the same two-stage structure of an ultimatum game except that there were variable gains from the exchange, and the surplus to be divided varied with price and quantity. As in the ultimatum game, the buyer was free to reject the seller's

---

[5] McCabe et al. (2001b); Kurzban and Houser (2005); Houser (2003); Houser, Keane, and
     McCabe (2004); Houser and Kurzban (2002); Houser et al. (2005); Houser et al. (2006).
[6] See the references in note 5 to the work of Houser.

Table 10.1. *Mean percentage offered by treatment in ultimatum games*

Context	Measure	$10 stakes			$100 stakes		
		Divide $10	Exchange	Exchange strategic prompt	Divide $100	Exchange	Exchange (grad students)[a]
Random Entitlement	Mean	43.7%	37.1%	41.7%	44.1%	NA	NA
	N	24	24	24	27	NA	NA
	(%Rejected)[b]	(8.3%)	(8.3%)	(12.5%)	(3.7%)	NA	NA
Earned Entitlement	Mean	36.2%	30.8%	39.6%	NA	27.8%	28.8%
	N	24	24	24	NA	23	33
	(%Rejected)[b]	(0)	(12.5%)	(2.9%)	NA	(21.7%)	(21.2%)

[a] The graduate students were visiting participants in an introductory workshop in experimental economics. These are new data not previously reported.

[b] Refers to percentage of the N pairs in which the second player rejects the offer of the first.

Source: Data from Hoffman et al. (1994) and Hoffman et al. (1996a, 2000). See the references for the statistical significance of pair-wise comparisons as discussed in the text.

offer by choosing a quantity of zero. The interesting result was a strong tendency to the equilibrium outcome, distinct from an equal split of the joint surplus created between buyer and seller in the exchange.

"Contest/exchange" combines the implicit property right norm of a seller with a pregame mechanism for earning the property right.

Table 10.1 summarizes the results from three different studies of ultimatum game bargaining with stakes of either ten $1 or ten $10 bills for each of N pairs of players, where N varies from twenty-three to thirty-three subject pairs.

Here are some findings from these results:

- *Finding 1.* Comparing "divide $10" with "divide $100" under the random entitlement, we observe a trivial difference in the amount offered between the low stakes (43.7 percent) and the tenfold increase in the stakes (44.1 percent). Also, there is no significant difference in the percentage rate at which offers are rejected, 8.3 percent and 3.7 percent respectively. (But such tests have very low power, so small is the typical number of ultimatum rejections.)
- *Finding 2.* Comparing the "divide $10/random" entitlement condition with the "exchange" entitlement, the offer percentage declines from 43.7 percent to 37.1 percent, and comparing the former to the "earned" entitlement, the decline is from 43.7 percent to 36.2 percent, both reductions being statistically significant. Even more significant is the reduction form 43.7 percent to 30.8 percent when the "earned" and "exchange" entitlements are combined. Moreover, in all four of

these comparisons, the rejection rate is null or modest (0 percent to 12.5 percent).

- *Finding 3.* When "exchange" is combined with an "earned" entitlement, the increase in stakes lowers the offer percentage from 30.8 percent for $10 stakes to 27.8 percent for $100 stakes, but this difference is within the normal range of sampling error using different groups of subjects and is not significant. Many will be surprised, however, to learn that this minuscule decline in the mean offer causes the rejection rate to go up from 12.5 percent to 21.7 percent. Three of the four subject players 1 that offer $10 are rejected, and one offer of $30 is rejected in the game with $100 stakes.

- *Finding 4.* The small proportion of the offers that were rejected (except when the stakes were $100 in the "earned/exchange" context and the mean offers declined to a low of 27.8%) indicates that players 1 read their counterparts well and, as the context is altered, normally offer a sufficient amount to avoid being rejected. The exception shows clearly that pushing the edge, even if it seems justified by the higher stakes, may invite a sudden escalation of rejections.

- *Finding 5.* Table 10.1 also reports an anticipated "tiny" variation on the "exchange" treatment in which only one slight addition is made to the instructions: The instructors "prompt" player 1 sellers to consider what they expect the buyer to do before making a decision. The motivation for this variation was to encourage sellers to think ahead, like a game theorist doing backward induction. Symmetrically, buyers were asked to think about what they expected the seller to do. Apparently – not having had the course – all this "prompt" did was to focus the sellers on the fact that they might have their offer rejected. Under the "random" entitlement, the mean offer actually *rises* from 37.1 percent to 41.7 percent; and under the "earned" entitlement, the mean offer jumps from 30.8 percent to 39.6 percent. What we seem to be observing here is the severe limits to which one can use language to induce in subjects an alleged greater sophistication in analyzing their situation. Far from helping, this example of a constructivist scheme for trying to make the instructions more "relevant" actually backfired. The rules of morality (culture) are indeed not the conclusions of reason, at least as we try to define it in this application of game theory. It's far better to let the subjects do their thing in alternative contexts, and try to learn from those contexts how the subjects see the situation and make the appropriate decision. If this is inimical to our preconceptions based on game theory, then it is important to reexamine the constructs of the theory.

These data indicate some of the ways in which context is important relative to the baseline "divide/random" treatment (which cannot be proclaimed "neutral") in the ultimatum game: The percentage offered varies by over a third as we move from the highest (44 percent) to the lowest (28 percent) measured effect. Like variation is reported in cross-cultural experiments: A comparison of two hunter-gatherer cultures and five modern cultures reveals variation from a high of 48 percent (Los Angeles subjects) to a low of 26 percent (Machiguenga subjects from Peru) (Henrich, 2000; Henrich et al., 2005). These studies have selectively reinforced the view (1) that there are cross-cultural, universal, "neutral" *procedures* yielding observations on ultimatum game behavior, and therefore (2) any variations are due to cultural differences.

The cross-cultural comparisons carefully attempted to control for instructional differences across languages, but this is inherently problematic, given the nature of perception. One cannot be sure that the instructions, translations, payoffs, or procedures for handling subjects (with little if any "education" as the developed world knows it – hence, the use of intensive experimenter-subject interactions) control adequately for context across cultures. In each culture, one needs to vary the instructions/procedures in a manner analogous to the preceding, but in a way that is considered relevant for the culture by those with the most local field knowledge; to observe the sampling distribution of outcomes; and then to compare the sample distributions across cultures. Instead of point comparisons, one asks whether the instructional sampling variations in each culture are comparable to those in other cultures.

These are extremely important subject populations to reach with their huge inventory of specialized knowledge and demonstrated capacity for adaptation to challenging environments – knowledge that is disappearing with economic development and exposure to market economies. They are a laboratory unto themselves deserving of careful attention to how protocols may interact with cultural variation. For further discussion, see Henrich (2004); Houser, McCabe, and Smith (2004); Ortmann (2005); also, the subsequent rather spirited ex post defense of procedures articulated in Henrich et al. (2005), and the comment thereon by Smith (2005b) and numerous others.

Exercises in deliberate variation in the defined context are especially important in cross-cultural studies where there must be no presumption, given the wide variation in the techniques needed to convey comprehension, that different experimenters are all implementing the same "abstract game." For example, it is not meaningful to continue to suggest that most "researchers stuck to entirely abstract explanations of the game, and experimental context ... " (Henrich et al., 2005, p. 805). What does that mean? How would you determine whether you had achieved an entirely abstract explanation without examining variations? How do you know in cross-cultural comparisons that the verbal descriptions of "division" and "sharing," and any illustrations used, have equivalent meaning across the cultures? There was much subject-experimenter interaction, including face-to-face interaction in at least some implementations, a slippery condition not commonly used or studied when these experiments are performed in Western cultures (see a photograph of such an interaction between an anthropologist and a subject in Papua New Guinea in Camerer, 2003, p. 72). Such variations need to be studied within each culture. By varying protocols, you discover empirically what is or is not a contextual "treatment" in any given culture.

These methodological concerns do not, however, detract from the Henrich et al. (2005) finding of many close cross-cultural parallel variations between subjects' behavior in the experiments and their corresponding routine behavior in cultural interactions across time in everyday life. But they ignore the implications of these parallel comparisons in interpreting their one-shot games as reflecting social preferences rather than simply an instance of repeat play of many such distinct events across time. For example, in some of "these societies, accepting gifts, even unsolicited ones, implies a strong obligation to reciprocate at some *future time*. Unrepaid debts accumulate, and place the receiver in a subordinate status. Further, the giver may demand payment at times or in forms ... not to the receiver's liking, but the receiver is still strongly obligated to respond. As a consequence, excessively large gifts, especially unsolicited ones, will frequently be refused" (ibid., p. 811; italics added).

This quotation powerfully supports the main points in this and later chapters: (1) human sociality derives from the emergent cultural "rules of morality" or norms that routinely invoke rewards and punishments to overcome and neutralize temptations to defect and choose short-run dominant outcomes that the individual otherwise might prefer; (2) it is a statement about culturally conditioning behavior by operating on the structure of payoffs and creating group cohesion and stabilizing community through altering expectations of individuals; (3) moreover, this sociality derives quintessentially from repeat cultural interaction across time and across many games in the blur of life, some one-shot, some repeated; (4) such characteristics of humans cannot be assumed to be abandoned at the door of the experimental laboratory, or the hut in the village, or the shelter in the forest, wherever human sociality is manifest; (5) the abstract theoretical idea of a one-shot game is in all societies an experiential fiction as people generally grow up accustomed to one interaction after another, some repeated and some not, but all experienced in an ongoing interactive stream, with the exception of certain tragic cases of social isolation, or deficiencies of inheritance, that lead to maladaptive sociopaths; (6) in one-shot game experiments under strict conditions of anonymity, subjects are afforded an unfamiliar environment that invites them to lean toward choosing dominant outcomes – some lean more or less than others, but the effect of their experienced sociality bleeds through to a hard core of cooperative behavior to varying degrees in different games; (7) consequently, as we have learned, cooperation in one-shot experimental games exceeds the predictions of twentieth century economic and game-theoretic reasoning.

As we shall see in Chapter 12, cooperation increases even more dramatically with repeat interaction between the same pairs, but although cooperation is high across randomly repaired (or distinct) anonymous pairs, their ability to apply the tools of their acquired sociality to sustain increasing gains from cooperation is less reliable.

Henrich et al. (2005, p. 797) is not about human sociality and how behavior is modified by that sociality, but rather the "empirical challenge to what we call the *selfishness axiom*," that is, maximizing one's own gain and expecting others to do the same. They reject the idea that our social norms are a means of overcoming the consequences of strict selfishness and changing the payoff structures through actions across time. (Yet their descriptions of "everyday life" in these societies implicitly recognize selfish actions that are overcome with norms of reciprocity and sharing that reward giving and punish obligation failures with low status in repeat interactions.) To achieve this narrow interpretation, the authors have to invoke, and not doubt, the extreme assumption that, while people in all cultures fail as dominant strategy game-theoretic players, people are resolute and completely dependable in accepting the game-theoretic concept of the one-shot game that fully controls on their past experience and all thought of repeat interaction by a subject once they enter an experiment and leave their everyday life behind.

But once a key assumption in game theory fails – that people always choose dominant outcomes whatever the circumstances – why would you rely so conclusively on its equally key concept of the one-shot game isolated from a history and a future based on constructivist assumptions? I suggest that history will always matter to the extent that what subjects bring to an experiment is encoded in past experience. What varies across different experiments is how quickly the effect of that initializing experience might decay relative to the immediate interactive experience in the experimental environment.

The instructional treatment effects reported in the preceding from the undergraduate population of subjects call into question the extent to which one can define what is meant by "unbiased" instructions simply by examination and discussion among designers. We need empirical guidelines until we have a better understanding of how to study the elements of language that define the structure of social context. If the treatment effects in HMSS had turned out to have been all statistically insignificant, then all the instructions studied would have been "unbiased," and one has reason to say that people, like game theorists, perceive that there is just one underlying "abstract" ultimatum game.

The main lesson is that, because of the nature of perception and memory, context can matter, and in the ultimatum game, the variation of observed results with systematic instructional changes designed to alter context shows clearly that context can and does matter.

Next, I will consider dictator game experiments, which are much more sensitive to variation in procedures than ultimatum games. In this game, there are treatments that yield the purely self-interested outcome within the same general subject population.

Dictator Games

The ultimatum game is converted into a dictator game by removing the right of the second mover to veto the offer of the first. Forsythe et al. (1994; hereafter FHSS) had the important insight that if the observed tendency toward equal split of the prize in ultimatum games is due primarily to "fairness" – a word that was used in the sense of a social norm of just division that is achieved when outcomes are equal – then it should be of little consequence whether this right is eliminated. But if it is the prospect of rejection, as an immediate or life-style strategic consideration that tempers the amount offered by player 1, then the outcomes should be materially affected by removing the right of rejection, which converts the ultimatum game into what is called the dictator game. By traditional game-theoretic criteria, the latter does not even qualify as a "game," since there is no player interaction in the sense of joint decision making. But as a comparison control in interpreting the ultimatum game results, a significant reduction in the mean percent offered in the dictator game would be consistent with the hypothesis that the large percentage offered in ultimatum games cannot be explained only by fairness. Alternatively, if outcomes are little changed, this would be a powerful indicator that "fairness" in the preceding sense is indeed an important driver of the standard ultimatum game results.

Comparing the results in Table 10.2's column 1 with those for "Divide $10, Random Entitlement" in Table 10.1, we see that the mean dictator offer is only 23.3 percent compared to the mean ultimatum offer of 43.7 percent. FHSS conclude that fairness alone cannot account for behavior in the ultimatum game. This is a correct conclusion, but the results raise a new question of equal or greater interest: Why are dictators still giving away nearly a quarter of their endowment?

This research puzzle was picked up by HMSS, who conjectured that such generosity might be, at least in part, a consequence of the incompleteness of anonymity. In all the games prior to the HMSS study, the members of each player pair were anonymous with respect to each other but not with respect to the experimenter, and perhaps many others in the subjects' perception, who knew or might find out each person's decision. Hence, they introduced a double blind treatment category (two versions, one expected to provide more secure privacy than the other) in which the protocol made it transparent through what the subjects saw and experienced that no one, including the experimenter, could possibly learn the decisions of any player. Data from the second version, Double Blind 2, are reported in Table 10.2. In this treatment, mean dictator offers decline to only 10.5 percent, less than half the 23.3 percent under single blind. Camerer (2003, pp. 57–8) reports these data and others showing that experimental procedures, not just "descriptions," have important effects. These results reinforce the importance of context, memory encoding, and decision. Moreover, the data shifts – by factors of two to three, implying that context is easily as strong as structure-theoretical variables, and, I suggest, reflects a deep structure of how the brain works to draw on its past experience.

Consequently, context – in this case, social connectedness or distance in the sense of whom or what others can know about any individual's decision – has an important effect on dictator transfers. These double blind procedures and treatment effects have been replicated by two other investigations (Eckel and Grossman, 1996; Burnham, 2003). Bolton et al. (1998) use a double blind procedure that differed from HMSS, and report failure to replicate

Table 10.2. *Dictator giving with and without gains from exchange and social history*

Treatments	Standard dictator game giving		Double blind[b] dictator giving under gains from exchange[c]					
History	Single blind	Double blind 2	Baseline; no history			Social history		
Player role	FHSS[a]	HMSS[b]	Sent Players 1	Returned Players 2	Yield Players 1	Sent Players 1	Returned Players 2	Yield Players 1
Percent Transferred	23.3	10.5 (Cox 36.3)[e]	51.6 (Cox 59.7)[e]	30.3[d]	−9.1	53.6	35.5[d]	+6.5
Percent Transferred By Top 50% Givers	38.3	21.0 (Cox 69.3)[e]	74.4 (Cox 92.5)[e]	49.4		49.4	55.8	

Notes:

[a] Data from Forsythe et al. (1994), replicated by Hoffman et al. (1996b).

[b] Data from Hoffman et al. (1994) for Double Blind 2.

[c] Data from Berg et al. (1995). Their procedures are different, but are nearest to those of Double Blind 2 in HMSS.

[d] Since the sender amount is tripled, if the receivers return an average of 33.3 percent, then the average amount returned will equal (pay back) the amount sent.

[e] Data from Cox (2004).

the HMSS results. Their failure to replicate the results suggests that in these tasks, procedures matter and interact with the double blind condition. They neglected to run a comparison control showing that they could replicate the HMSS results with their subjects and experimenters using the HMSS procedures, thereby establishing that their procedures constituted a treatment difference. Therefore, all that can be concluded is that their combination of experimenters, subjects, and procedures yielded results different from those of HMSS. Once you establish that procedures matter, as in HMSS, you cannot assume that there is a single abstract double blind effect independent of how it is implemented.

Thus, following up on HMSS, various double blind procedures merging into single blind are systematically explored in Hoffman et al. (1996b), who vary social distance by varying the instructional and protocol parameters that define various versions of single and double blind dictator games.[7]

---

[7] They also report their replication of FHSS and one variation on the FHSS instructions. Cox and Deck (2005, note 20) report that HMSS did not independently replicate the

Experimental studies have inquired as to whether emergent norms of cooperation and constructivist incentive schemes are substitutes, the latter crowding out the former. See Bohnet et al. (2001) and Fehr and Gachter (2000) for studies suggesting that they are substitutes (formal rules undermine informal cooperative norms), and Lazzarini et al. (2002) for results suggesting that they are complements – contracts facilitate the self-enforcement of relational elements beyond contractibility. See also Henrich et al. (2005), where it is found that in tribal societies, people who are more involved in market exchange opportunities exhibit less selfish behavior in ultimatum games, suggesting that there might be complementarities between market and personal exchange.

In evaluating institutions for solving common problems, Ostrom (1990, pp. 61, 90–102, 136, 178–81) has discussed the hazards of importing constructivist design criteria that fail to rely on the indigenous experience and wisdom of local customs and culture.

Finally, in this string of studies I want to summarize the intriguing recent findings of Cherry et al. (2002) in dictator games with low ($10) and high ($40) stakes under three different experimental treatment conditions:

1. Dictator endowments are provided by the experimenter, with single blind anonymity.
2. Dictators are recruited for a presession test where their endowments are earned based on their scores on seventeen questions taken from the Graduate Management Admissions Test (GMAT), with a score of ten or more correct yielding the endowment of $40, while a smaller score provided the $10 endowment, again with single blind anonymity.
3. The procedure is the same as (2), but subjects decide under double blind anonymity in which no one, including the experimenter, can know the earnings of individual subjects (the DB1 procedures in Hoffman et al., 1996b).

Table 10.3 compares the percentage of the dictators who give nothing to their recipient counterparts under the various treatments. First, it will be seen that the stakes, $10 versus $40, have only a modest effect on the percentage given. (As has been concluded in other contexts, if stakes are adequate, increasing them may have a minor effect on central tendencies in decision, although the variance of decisions will likely be reduced.) When

FHSS single blind experiments, but in fact the replication was conducted and reported in Hoffman et al. (1996b) but overlooked when cited earlier in Cox (2002).

Table 10.3. *Earned endowments and dictator game results: Percent of dictators who transfer zero dollars to their counterpart*

Treatment	$10 Stakes	$40 Stakes
Single Blind; Experimenter Gives Endowments	19%	15%
Single Blind; Subjects Earn Endowments	79%	70%
Double Blind; Subjects Earn Endowments	95%	97%

Source: Based on data reported in Cherry et al. (2002).

endowments of $40 are *given* by the experimenter, 15 percent give nothing; when dictators *earn* the endowments, the percentage of dictators who give nothing jumps to 70 percent. And the latter rises to 97 percent when combined with the double blind treatment. It appears that when a strong sense that the stakes belong to the dictator are combined with the condition that "no one can know" his or her decision, this all but eliminates dictator giving.[8] In effect, dictators are altruistic with respect to the experimenter's money, not their putatively "own" earned money.

These results have now been replicated and much extended, by Oxoby and Spraggon (2007), using twenty questions from both the GMAT and Graduate Record Examination (GRE) to study the effect of earned stakes of $10, $20, and $40 by dictators and, in comparison, by receivers. They report that 100 percent of the dictators allocated nothing to the receivers under all three stake conditions. More powerful in demonstrating human sociality in the property rights culture of their subjects are the results when receivers earn the stakes but dictators decide the final allocation between them. Dictators recognize the legitimate claim rights of receivers, by awarding them median levels of 20 percent, 50 percent, and 75 percent, respectively, of the stakes of $10, $20, and $40. This is in striking contrast with the award of 0 percent in all three cases by *dictators* who have earned the stakes.

This direction of research opens the gates of exploration of the question: How does the earned-endowment context vary with culture and the associated wealth-producing performance of those cultures? It is also particularly important to the understanding of personal exchange systems that the

---

[8] Such findings should not, however, be supposed to extrapolate to other games without further testing. For example, in public good games, it has been shown that these procedures for creating earned endowments do not have an effect on public good contributions (Cherry et al., 2005). These are empirical issues and you have to proceed on a case-by-case basis to find and establish regularities across environments.

earned stakes context be extended to the investment trust game discussed in the next chapter. (In the preceding results, I should note the voluntary trustworthiness of dictators when receivers earn the stakes, implicitly recognizing that dictators are "second movers" in the time sequence.) The essence of the investment trust game is that gains from specialization and trade are made possible by the interaction of senders and receivers – that is, "investors and entrepreneurs." There is not only the one-shot version of the trust game, heavily laden as it must be by the cultural norms that subjects import into the experiment, but how these norms jump-start them into further developing a relationship over time (see Chapter 12).

The dictator game data in Tables 10.2 and 10.3 cannot be understood in terms of a utility for money – own or other – independent of circumstances; if dictator giving is altruistic, why is giving so sensitive to context? And what is there left to say about dictator "altruism" when Cherry et al. (2002) reduce it to only 3 percent (0 percent in Oxoby and Spraggon, 2007) under double blind protocols, by requiring subjects to earn the endowments they are invited to share? It seems to be OPM (other people's money – in this case, the experimenter's money) and human sociality: To reduce the choice to near-dominant outcomes in the dictator game appears to require an environment in which people feel isolated entirely from any social context and use money that has been unambiguously earned by them.

These are powerful treatments; their use was motivated by considerations not part of game-theoretic analysis; they constitute significant road signs indicating that context, the ways in which stakes are acquired, and sociality have a structure easily as important as the payoff and move structures treated in classical game-theoretic formalism. In particular, some reappraisal and new inquiries are in order for the social utilitarian research program. The investment game (Chapter 11) run with an earned endowment protocol needs to establish the emergence of trust and trustworthiness in the investment and return decision. If dictator giving out of earnings is minor, then investment in the triple-return double-dictator game is largely in expectation of reciprocity. How much will investment be affected?

As I interpret it, the Dali Lama is referring to the means whereby an individual achieves true betterment for herself, but this form of other-regarding behavior is worlds apart from naïvely always choosing dominant payoffs as hypothesized in traditional game theory. Working for the welfare of others has meaning only in the framework of human sociality, which is inherently about human interaction over time. What makes experimental one-shot games significant is that even under such austere conditions, we get what game theorists might once have called "repeat-play-like" results. But it is questionable to suppose that what the subjects understand about one-shot games is what the theorist/experimentalist understands about them, namely that control for repeat interaction effects, *within an isolated experiment*, allows us to give a clear social preference interpretation to choice. If the degree of anonymity – for example, single versus double blind – makes a difference in giving, and if preferences are a credible property of individual outcome values, then why does the double blind treatment reduce the weight of "other" in preferences? The "other" has not changed and is still anonymous; the only change is whether third parties can know the decision. If double blind heightens the sense of privacy and isolation from human interaction (making it less true that "no man is an island"), then it seems natural to expect choices to be more narrowly self-interested.

Sobel (2005) provides an excellent formal discussion of the concept of social preferences, which can reflect context and explain cooperation and punishment behaviors in single-play games. He also offers a succinct reduced-form repeated-game model of long-term self-interested cooperation. Thus, individual i with current utility u_i chooses strategy s_i in a stage game to maximize $(1 - d) u_i(s) + d V_i(H(s))$, where $s = (s_1, \ldots s_i, \ldots s_n)$, d is a discount factor, H is the history of play, and $d V_i(H)$ is i's endogenous discounted value of continuation. Hence, the continuation value perceived by i may easily make it in his interest to forgo high current utility from domination because it spoils his future. Sobel rejects this model in favor of "interdependent preferences." Why? "Because laboratory experiments carefully [sic, try to] control for repeated-game effects, these results need a different explanation . . . for conventional repeated-game arguments to apply, the future must be important" (ibid., p. 411). Yes, but we have seen experiments in which potential futures are important – double blind, earned rights, and, in the next chapter, the mere suggestion of an undefined "Task 2." An alternative to social preferences is to enrich what is captured in $V_i(H)$. Why assume that history and continuation require repetition of the same stage game, rather than sequences of different games so long as agents perceive long-term benefits from continuation? Perhaps part of the answer will come from experiments studying single-play sequences of different stage games with same, random, and distinct pairs.

The key issue in applying any of these formalisms to the phenomena we study is to keep in mind that neither the subjects in our experiments nor their real-life counterparts think about these decision problems the way we do.

Even we may not consistently think that way because, as Hume (1739; 1985) states, "When we leave our closet, and engage in the common affairs of life, [reason's] conclusions seem to vanish, . . . " (p. 507).

APPENDIX: BEHAVIORAL DEVIATION FROM PREDICTIONS: ERROR, CONFUSION, OR EVIDENCE OF BRAIN FUNCTION?

In what sense should we expect subjects in experiments (or people in any decision situation) to understand the consequences of their decisions? For example, to borrow from a market example in Part II of this book, should we expect that a seller's privately induced marginal cost for a unit be understood by him as a minimum willingness-to-accept limit price, meaning that if he sells below that price it clearly proves that he has made an error? Not necessarily: This is not evidence of error, or "confusion," if the subject believes that he can gain in the end by driving his competitors to the sidelines. (I have heard this justification from people.) This may be infeasible, even foolish, in the environment of the experiment, but do you tell subjects this or let them discover it? Thus, in experiments designed to study the implications of predatory pricing, selling below marginal cost is interpreted by an influential body of theory as prima facie evidence of predation. But in that testing framework, such behavior, far from showing "confusion," is supposed to show "sophistication" and be theory confirming. Describing subject choices as "error" or "confused" can be read as a pejorative judgment of human decision competence based on our own theoretical, constructively "rational," and in this sense biased interpretations of subject behavior. It is easy to not guard against defaulting to our defensive analytical prejudice, and fail to stay open to Herb Simon's plea for "the necessity for careful distinctions between *subjective* rationality, and *objective* rationality. . . . To predict how economic man will behave we need to know not only that he is rational, but also how he perceives the world – what alternatives he sees, and what consequences he attaches to them . . . " (Simon, 1956, pp. 271–2). It is from that perspective that we can learn the most about human decision processes as distinct from decision constrained and interpreted *only by the models we test*. Ultimately it is those processes that we must model to gain insight into why game and economic theory has sporadic prediction success. After all, the institutions, businesses, and life practices whose abstract principles we presume to model by reducing them to game trees emerged in societies consisting of populations that included the same students and professionals we place in our experiments. If they produced the situations we model, it behooves us to better understand how.

Interpreting subject behavior as "error" or "confused" requires the experimentalist to interpret the state of mind of subjects from the perspective of reasoning based on formal analysis, a perspective not normally shared by the subject or even our own actual decision behavior.[9] Imagine Jane Goodall, after an intense period of observational study of a group of chimps, concluding that their behavior is in "error" or "confused.'" With justification, animal behaviorists would categorically reject this as unacceptable anthropomorphizing. Likewise, we as experimentalist-theorists must take care not to convey effectively an impression that we anthropotheorize people – that is, attribute to our subjects our specialized constructivist interpretation of decision-making behavior. The same holds for our subject economic agents in the field (the "real world"), because exactly the same considerations apply – context and its associated beliefs are at the heart of memory, perception, and subjective rationality. Of course, we seek not only to instruct, but first to learn what knowledge merits instruction.

I will not attempt in this space to survey the literature on what has been called subject error, misunderstanding, or confusion, but three studies are particularly valuable because they represent serious and informative explorations of this issue through careful designs and/or data analysis.

Lei et al. (2001; hereafter LNP) set out to better understand the circumstances as to why there is a robustly replicable tendency of inexperienced subjects to generate price bubbles and crashes in laboratory asset markets with well-defined fundamental dividend values. Although their study focuses on undergraduate subjects, it is important to note that all subject groups studied – small business persons, middle-level corporate managers, over-the-counter stock traders – tend strongly to produce the same pattern of results. Indeed, there is a huge literature on bubbles in stock markets around the world implying that it is a universal property of human behavior in asset

---

[9] Of interest in this respect is that McCabe and Smith (2000) report a departure from the theoretical predictions in two-person interactions even with graduate students trained in the theory. Theory knowledge may not be sufficient to ensure that intervention by the trained mind will trump social habits or norms in the autonomic brain. The theory mind is trained to do analysis for research, publication, and teaching, not necessarily for life and its real-time decision-making demands. Such training might be a way of thinking about life, but not necessarily living it since the rules of living are not the result only of constructivist rationality. But the same graduate students did not depart from the prediction of financial market theory in asset trading. Such a departure is very robust across a great variety of other subject groups, so here was a case in which their training mattered.

markets. In the laboratory, however, fundamental value is controlled and subjects participate in up to three distinct sessions; the results have established the finding that with repeat experience the bubbles recede to insignificance across three sessions. The original results, reported in Smith et al. (1988), were consistent with the proposition that a common dividend, and common knowledge (information) about the dividend, is not sufficient to induce common expectations (also see Porter and Smith, 1994). Experientially, however, over time the subjects arrive at prices that are consistent with common expectations and are equal to fundamental dividend value – what Nash called rational expectations equilibrium – however volatile is the pathway to that equilibrium. What was unexpected to all researchers was not that convergence was observed, but that the convergence process took so long in such a seemingly "transparent" environment. In earlier studies of much less transparent flow supply-and-demand environments, convergence was found to be unexpectedly fast. This seems to be because researchers implicitly tend to think that people solve problems by thinking carefully about them, then acting. Asset market bubbles were believed not to require much thought to be avoided, while achieving supply-and-demand equilibrium with dispersed private information was believed to be a very hard problem – unsolved formally in economics – and hence for naïve subjects to solve it seemed like magic.

From the perspective of finance theory, one readily can conjecture that if people buy at prices above the known fundamental value, it is because they expect (or hope) to resell to another speculator at a higher price. LNP seek to test this constructivist explanation of the thinking process underlying subject behavior. In the clever LNP experimental design, each participant is assigned one of two roles, a "buyer" or a "seller," fixed throughout a sequence of three sessions. A buyer is endowed with cash but no units of the asset, and can buy only against dividend realizations. A seller is endowed with units of the asset and the capacity to make sales, but has no initial cash, no ability to buy, and can only hold for dividends or sell for cash. It follows that if buyers naturally apply backward induction reasoning and seek to maximize profits, then they will never pay more than the maximum possible dividend value on which they have common information, since this yields a certain loss. Purchase for resale is not possible, so there can be no speculative demand for the asset.

The findings constitute a major blow to applying standard "rational" reasoning processes to explaining subject asset-trading behavior. A remarkably high fraction of total transactions (45.3 percent, 41.6 percent, and 27.5 percent in the three sessions) takes place at prices higher than the maximum possible future dividend realizations. LNP report a second experiment in which the instructions to subjects emphasize that participation is optional and that it may be in their best interest not to participate in an attempt to counter any natural "compulsion" toward active participation. (The subjects also can participate in an alternative stationary market in a commodity with only a one-period life.) This treatment yields a bubble only in the first session and substantially reduces the number of money-losing purchases. Hence, even with a heavy-handed instructional treatment, people still rely on their experience to adapt and converge to the predicted equilibrium. The effect of the instructional treatment is to speed up that experiential learning.

How do the authors describe their objectives and findings in these two experiments? They note that the bubble and crash phenomenon may be due not to speculation but to actual decision errors on the part of subjects; and they constructed an asset market, in which errors in decision making are the only way that bubbles can arise. By "error," they clearly mean that subjects deviate from what is predicted if they were to reason by applying the same principles used by the experimentalist-theorist – backward induction and profit maximization. In spite of this "error," across one series of three sessions the subjects adapt their behavior toward the constructivist rational outcome (with the instructional treatment, this adaptation is stronger), but this is also the qualitative pattern of adaptation in the standard asset market experiments replicated hundreds of times with a great variety of subjects – students, businesspersons, and professional stock traders. Even in the LNP "badly behaved" asset trading environment, subjects who do not consciously apply and act upon basic economic principles have the ability to adapt over time, materially improve their performance, and approach an ecologically rational state of equilibrium corresponding to our constructivist vision of rationality. Their beliefs – so poorly characterized by the LNP rational construct of reasoning – adapt by obscure and inscrutable processes that are not easily comprehended by scholars imbued with the usual economic training. If asset market traders deviate ("err") in how they think about their decisions relative to how we think they should think about them, this simply teaches us that they think differently about their decisions. In particular, they fail to backward-induct, maximize, or both, but this finding begs the question of how they are able to converge over time to the equilibrium we predict. Herb Simon would say that we are failing to ask, concerning our subjects, "how he perceives the world – what alternatives he sees, and what consequences he attaches to them ... " (Simon, 1956, p. 272). What LNP show us, however, is something important: how subjects do not perceive the world, its alternatives, and their consequences. Yet this helps not at all to understand how subjects *do* see the world.

Houser and Kirzban (2002) revisit a topic initially explored by Andreoni (1995): In the voluntary contribution mechanism (VCM) experiment, to what extent is the commonly observed decay over time in contributions to the public good due to individuals (1) discovering over time that their "altruism" in giving is not reciprocated, or (2) learning by experience that it is in their self-interest not to give to the public good? If it is the former, then the individual is learning to modify a strategy that failed to serve the individual's objectives; if it is the latter, then the individual is simply learning by experience the implications of what was stated in the instructions, that is, that individual payoffs from the public good depend only on total giving, not on whether the individual contributed to that total. In LNP's second experiment, subjects were also learning the implications of what was stated in the instructions.

Houser and Kirzban (2002; hereafter HK) compare contributions in two treatment conditions: one consisting of a collective of four persons, the other consisting of one person and three computer "players" or robots. Individual subjects in the latter condition were instructed that the contributions of the computer players were fixed independently of their individual contributions. Moreover, in this condition, on each round of play, the human subjects were reminded that, "This period, regardless of what you do, the computers will contribute_____ ... " (ibid., p. 1068).

HK find that (1) "confusion" was responsible for approximately half of all "cooperation" in VCM experiments, and (2) all of the decay over time in the level of contributions was attributed to reductions in "confusion." This interpretation assumes that "confusion" is essentially measured by the observation that a person gives to the public good when he or she cannot reap any possible future benefit and currently is giving up money by not allocating all his or her endowment to the private good. I suggest an alternative interpretation: Half the subjects do not consciously think about and analyze decision problems the way we economists do. They adapt their behavioral responses to their experience; over time, they "get the message" that we believe is transparent in the verbal information, but their brains require the real-time feedback of experiential reinforcement. Recall that the behaviorist tradition associated with B. F. Skinner was built on the idea that all "learning" could be so interpreted. The new cognitive psychology replaced that program with one based on the hypothesis that decision making was the result of cognitively-aware verbal-processing minds. It seems clear that both traditions have a demonstrated effect in this context.

This study and that of LNP make it quite plain that a substantial fraction of people – even college students with high minimum verbal proficiency – generally do not gain an operating understanding of the outcome implications of their decisions in a simple interactive decision situation based only on carefully worded instructions and examples. Over time, however, the data strongly indicate that individuals preponderantly come to have an operating grasp of their decision situation.

Both the HK and LNP studies tell us something important about cognitively heterogeneous minds and the importance of real-time adaptive experience for many people making decisions in unfamiliar environments. Much of the critique of experimental economics is based on the claim that the subjects are not "sophisticated," and the stakes "inadequate," and the implicit untested assumption that such claims are not true of people in the "real world." This is of course why experimentalists typically study repeat decision making and experienced subjects, test the effect of altering stakes, frequently recruit practitioners for subjects, and do field experiments and applications.

The HK paper nicely illustrates the importance of distinguishing between knowledge of "how" and language-based descriptions that aim at articulating the implications of a decision task so that people will be readied to learn "how." Even in college subpopulations, most people learn "how" by doing and practicing, not by simply thinking about it. People do not learn how to use computers by reading instructions and following them – they learn by doing, from others, by observation, and by trial and error. The HK paper reinforces the practice of experimentalists who in complex mechanism design and dynamic environments run long experiments and use highly experienced subjects from college, as well as professional and business populations. Language-based instructions are simply an entry-level device for starting the learning process. Decision making involves information acquisition, processing, and judging its outcome implications for action or choice according to preferences, and each part of that complex must be experienced-based to be meaningful.

Surely only academics believe that most of the world's operating knowledge is acquired entirely by reading. Real people seem to learn from experience – the school of hard knocks – not simply by reading instructions and then implementing what they read.

In a second paper, Kurzban and Houser (2005; hereafter KH) present evidence consistent with the hypothesis "that individual differences reflect genuine strategic differences as opposed to differences in, for example, amounts of confusion between participants" (p. 1807). They sought to avoid "confusion" by requiring all participants to complete a ten-question quiz

"... to insure that subjects understood the structure of the experimental environment" (ibid.).

The device of screening subjects based on verbal quiz responses is a common means of ensuring that they understand (that is, they are informed verbally and perhaps pass a verbal test) what the experimentalist intends to convey in instructions. I have done this myself, but I have never felt comfortable with the procedure. It leaves unanswered many gnawing questions: What are the characteristics of those omitted? Are they action learners, not verbal learners? Do some of them never learn? Suppose you gave the test, but let those who "fail" the test participate for a repeat trials experiment or return after experience. How would the poor test performers behave asymptotically? How obstinate are the outliers? We do not have good answers to these questions because it is difficult for us to think about decision making from the perspective of someone who does not have our professional perspective and training. Without a perspective on the thinking of untrained people, we are limited in our ability to test hypotheses that will enable us to learn how they solve the problems we give them. Rather, we strongly rely on that training in designing our experiments and then interpreting what people do, and in consequence we discover that they are "confused" and make "errors."

ELEVEN

Investment Trust Games

Effects of Gains from Exchange in Dictator Giving

Of all the persons . . . whom nature points out for our peculiar beneficence, there are none to whom it seems more properly directed than to those whose beneficence we have ourselves already experienced. Nature, . . . which formed men for their mutual kindness, so necessary for their happiness, renders every man the peculiar object of kindness, to the persons to whom he himself has been kind.[1] . . . No benevolent man ever lost altogether the fruits of his benevolence. If he does not always gather them from the persons from whom he ought to have gathered them, he seldom fails to gather them, and with a tenfold increase, from other people. Kindness is the parent of kindness.

> A. Smith (1759; 1982, p. 225)

But what will explain the explanation?

> Krutch (1954, p. 148)

A Celebrated Two-Stage Dictator Game

Berg et al. (1995, hereafter BDM) made an important and widely influ-ential modification of the dictator game that explicitly introduced a key characteristic of personal social interactions: gains from "exchange."[2] Their investment trust two-stage dictator game also uses a double blind (most comparable to the HMSS Double Blind 2) protocol: Dictators in room A send any portion of their $10 (0 to $10) to their random counterpart in room B. People in both rooms know that if $x is sent by anyone, it is tripled, so that the counterpart receives $3x. Thus, the most generous offer, $10,

---

[1] Notice that Smith is talking about reciprocity, but without using this word from our time. But Smith then goes on to talk about reputation formation, and negative and positive reci-procity in cultural responses, again without using those words, concluding that "Kindness is the parent of kindness."

[2] For an example of a recent extension and replication of the BDM findings, see Pillutla et al. (2003). Also see Ortmann et al. (2000).

yields a gain to $30. The counterpart can then respond by sending any part (0 to $3x) of the amount received back to his or her matched sender. Unlike the dictator game, the exchange itself increases the surplus available to both parties, and BDM ask whether this context is a significant treatment. Another feature of the BDM protocol takes on potentially new but yet to be fully explored dimensions in the light of the Cherry et al. (2002) results in Table 10.3[3]; BDM depart from the standard dictator protocol by asking those who are randomly assigned the role of players 1 to share their $10 show-up fee with players 2, who also received $10 show-up fees to which are added any amounts sent to them and tripled, but not returned.

The analysis of the game, however, is no different from that of the one-stage dictator game: By the principle of backward induction, player 1 can see that player 2's interest is to keep all the money received, and therefore nothing should be sent. The fact that the sender's transfer will be tripled is irrelevant. But it is not irrelevant if both players see the interaction as an exchange based on trust by player 1 and trustworthiness by player 2 (or if player 1 likes to give money to player 2 even under a double blind protocol).

Referring back to Table 10.2, sender players 1 now give 51.6 percent when the transfer is tripled, compared to 23.3 percent when it is not. This shows how the tripled pie shifts the distribution toward larger transfers by players 1. But on average, the senders do not quite break even: 30.3 percent of the amount received by players 2 is returned to players 1 (breaking even would be 33.3 percent, since x is tripled). In the "social history" treatment, the instructions and protocol are the same as described previously except that the second treatment group is shown the distribution of amounts transferred and returned for the first group. Comparing the social history with the baseline mean percent given and returned reveals the effect of being exposed to the decision data of the first group. Social history does not cause a reduction in transfers, which actually increases marginally from 51.6 percent to 53.6 percent. And the average percent returned actually increases from 30.3 percent to 35.5 percent, just above the breakeven level. If $S = \$$ sent, and $R = p(3S) = \$$ returned, where p is the percent returned, then the yield on the investment is $y = (R - S)/S = 3p - 100$. Thus, in Table 10.2, the baseline average return to players 1 is -9.1 percent. Moreover, with "replication" in the form of social history – that is, learning by observation from the experiments of others, like Alex the parrot – the average return is 6.5 percent. From social history, people do *not* learn that free riding occurs and that giving should be avoided. This is the most striking of the BDM

---

[3] I write this in October 2006, but this state of our knowledge is likely to change.

findings – a point largely ignored in the outpouring of replications and variations on BDM.

These results are not explicable by the canons of traditional game theory wherein agents were supposed to always choose dominant strategies on the assumption that their opponents would do the same. By introducing gains from the investment by player 1 – who can only receive a self-regarding payoff benefit if player 2 perceives the process as an exchange calling for payment for services rendered – dictator giving more than doubles. But the challenging results reported by Cherry et al. (2002) establish that the observed altruism in standard dictator giving is based on "other people's money" (OPM) in the particular sense that when dictators earn the stakes, their altruism declines to only 3 percent. Similarly, in the investment trust game, trustors may be investing OPM in this sense, and we do not yet have data on what they will do when they earn the stakes to be invested.

Keep in mind that everything I have to say about investment trust games is conditioned on the protocols that were used, which differed from the dictator game reported in Cherry et al. (2002). None of this matters in standard game theory because money is money, and it makes no predictive difference where the payoffs come from in isolated one-shot games. But if we use real people, they come from a world and return to a world that entangles them in a sociality laden with a past and a future, where it matters how you acquired your rights to act with money and who supplied it. If this contextual feature along with double blind and a host of others affects experimental play, then we cannot claim to have created the conditions that allow interpretation in terms of single-play game theory.

This precautionary warning about protocols may turn out to be unnecessary if the gains from exchange introduced by BDM dominate effects found to be important in the dictator game such as the OPM problem. In the original BDM protocol, the show-up fees constituted the stakes that funded investment by those randomly chosen to be the first movers. In this sense, the money was an earned payment for arriving, but this is unlikely to be equivalent to the Cherry et al. (2002) explicit protocol for earned stakes.

An alternative to the GMAT (or GRE) earned right to the endowment would be to recruit dictators to the investment trust game with the provision that they have to bring their own stakes. Or, following an unrelated experiment, invite subjects to stay for the investment game provided they risk $10 of the money just paid to them (tripled gains from specialization exchange provided by the experimenter). (Schwartz and Ang [1989] report asset market experiment with subjects who were required to each bring

$20 of their own money. They report price bubbles just like those observed when cash endowments are provided by the experimenter). Still a third alternative is to administer a GMAT/GRE-derived instrument to the entire group. The high-scoring subjects earn, for example, $40 and the right to be dictator; the low scorers earn $10 and the recipient position. It is now experimental open season on the effect of money source on dictators.

Reciprocity or Other-Regarding Preferences?

Double blind protocols in the dictator game have the intention of removing the individual from a social context, or at least of maximizing the social distance between the individual and all others. It provides tighter controls on any possibility of a future in single-play interactions. Why would your brain encode a concern for whether third parties know your decision if it never affected your future? As we have seen, double blind protocols are effective in the sense that they lower substantially the amounts given to others. This is consistent with the hypothesis that dictator altruism is about nurturing your sociality in interaction, and not only about a utilitarian unrequited kindness toward others. Moreover, even under double blind, when increased relative gains from exchange are introduced in the BDM two-stage version of the dictator game, we observe a corresponding large increase in the amounts sent and returned by dictators.

Cox (2004) has proposed a two-person social preference model as an alternative to the exchange model of this behavior, and an experimental design intended to decompose the BDM stage I transfers into motives of "trust" and motives of "altruism," and the stage II transfers into motives of "trustworthiness" and of "altruism." This approach seeks to maintain the traditional perspective of game theory – in particular, that a single-play game controls for cooperation as it emerges in repeat play – but allows that some dictators derive utility for both own and other payoffs. In effect, some subjects arrive with a propensity to derive subjective utility from giving money to anonymous others in single-play protocols as well as from the sums they keep. Others are motivated by considerations of trust and trustworthiness as in an exchange.

The "trust/altruism" decomposition was effected by comparing the distribution of the amounts sent by subjects in BDM stage I (trust), Cox's (2004) experiment A, with the distribution of the amounts sent by members of an independent experimental control group to their counterparts in a standard dictator game (altruism), Cox's experiment B, where any amounts sent are tripled, but their counterparts have no decision to make.

By direct analogy, Cox's "trustworthy/altruism" decomposition was effected by comparing the distribution of the amounts returned by subjects in BDM stage II (trustworthy), experiment A, with the distribution of the amounts sent in an independent dictator game control (altruism), experiment C, where the endowments in the dictator control were adjusted by amounts that correspond to the transfers received by subjects in the paired BDM stage II experiment.

This was a clever cross-subject comparison design, in which Cox sought to tighten comparisons and strengthen inferences about reciprocity in the BDM game. Recall that the two primary points of this as well as the previous chapter are that (1) instructions/protocols, which define context, can be important treatments, and that (2) we can learn about different protocol effects only by testing their robustness in any particular abstract game. I want to summarize the data reported by Cox from this perspective because they shed important new light on the strength of protocol effects.

In building on and extending important previous work of others, Cox had a natural check on the validity of his design and protocols: His BDM results should replicate those of others, and his dictator game results for BDM stage I comparisons should approximately replicate those of others after taking account of the effect of a tripling of any amounts sent by the dictators. Since Cox reports the first dictator game results with unequal endowments in his BDM stage II dictator control, no direct comparison check is possible, although it seems reasonable to expect the results to be generally consistent with ordinary dictator transfers with equal endowments – his unequal endowments were randomly allocated.

Cox (2000, 2004) reports two different series of results using the above A-B-C design comparison: A1-B1-C1 from Cox (2004) and A2-B2-C2 from his earlier working paper (Cox, 2000) (see Table 11.1). The differences are not attributable to sampling variation only, but (as it turns out) to a considerable change in instructions/protocols that are particularly interesting with respect to protocol/context effects. In Cox (2000), the instructions/procedures treatment included (at least) two behaviorally significant provisions that were removed in the Cox (2002, 2004) treatment: (1) In the A2-B2-C2 experiments, subjects' instructions informed them that the experiment in which they were about to participate was "task 1," and that there would follow a "task 2" experiment with instructions also to follow; and (2), subjects were also informed that a coin flip would determine which of the two experimental task earnings would actually be paid to them (Cox, 2000; 2002, pp. 333–9). Thus, "Each session contains an individual decision task ('Task 1') and a group task ('Task 2').... The individual task precedes

Table 11.1. *Dictator and investment game transfers*

Source; game	Number observations	Mean dollars sent	Mean dollars returned BDM, A; Sent: C
BDM; Investment DB[a]	32	5.16	4.66
Cox; A1, Investment DB[b]	32	5.97	4.94
Cox; A2, Investment DB[c]	30	6.00	7.17
Cox; B1, Dictator DB[b]	30	3.63	
Cox; B2, Dictator DB[c]	38	5.81	
Cox; C1, Dictator DB[b]	32		2.06
Cox; C2, Dictator DB[c]	30		5.10
HMSS; Dictator DB 1[d]	36	0.92	
HMSS; Dictator DB 2[d]	41	1.05	
HMSS; Dictator SB[d]	24	2.67	
FHSS; Dictator SB[e]	24	2.33	
FHHS-R; Dictator SB[f]	28	2.75	
FHHS-V; Dictator SB[f]	28	2.18	
Orma; Dictator SB[g]	NA	3.1*	
Hazda; Dictator SB[g]	NA	2.0*	
Tsimane; Dictator SB[g]	NA	3.2*	

Notes:
DB: double blind.
SB: single blind.
* Reported as percentages, here normalized on $10 for comparability with all other dictator results appearing in this column.
Sources:
[a] Berg et al. (1995).
[b] Cox (2004).
[c] Reported in Cox (2000) using different procedures than in Cox (2004); note that the mean dictator transfers are much larger in the 2000 study (B2 and C2) compared to the 2004 report (B1 and C1). As noted in the text, these differing results are explained by important context/protocol differences.
[d] Hoffman et al. (1994).
[e] Forsythe et al. (1994).
[f] Hoffman et al. (1996b).
[g] Henrich et al. (2005).

the group task in every session. At the end of a session, a coin is flipped in the presence of the subjects to determine which task has monetary payoff" (Cox, 2002, p. 333). This departure from previous replicable procedures in BDM and dictator games turned out to be very important as a new example of how different, seemingly "minor," protocol contexts affect behavior, while holding constant the game structure and payoffs.

As dictator game experimenters have learned, this is a very sensitive game to the protocol context, and any suggestion of a future in the subjects' immediate session horizon is potentially problematic. All "one-shot" game experiments are potentially sensitive to any implicit suggestion of a future, but this is particularly the case in dictator games. Here is an explanation of why: From the perspective of the subjects, the dictator game is an unusual laboratory experiment in that subjects are given the opportunity to transfer money to another person, who then has no decision to make! ("The Gods Must Be Crazy!") In fact, it is important for the dictators to know and be able to verify that human recipients actually exist.[4] It is this strangeness that led HMSS to run their experiments double blind to guard against any thought of a "future" in which the subjects would be subsequently judged by anyone, anywhere, for their decisions, or that their subsequent treatment would be affected by their decisions. The intention was to try to increase control over prior human socialization – whether someone knows or not a person's decision.

I should note, however, that the existence of a future "task 2" is perhaps less likely to affect the BDM investment game because decisions in "task 1" are to be made by both groups, and the stage 1 BDM dictators need to know the essence of BDM: that there is a stage 2 with the prospect that money will be returned. The stage 2 recipients, however, need only know there was a stage 1 to which they are to respond, not that there is anything to follow their response. But the dictator control experiments are another matter and it is important that they not expect a subsequent task.

How did all these considerations impact Cox's observations? To his credit, all the experiments summarized originally in Cox (2000, 2002) were rerun by him for Cox (2004) in response to receiving comments from those of us who noted the preceding serious problems with introducing an explicit "future" in the form of a "task 2." Consequently, we have the considerable benefit of learning from two different instruction/procedure treatments, both different from those in previously replicated double blind experiments. Instructive "treatments" can be unintended.

In Table 11.1, we observe that the original BDM results – $5.16 sent, and $4.66 returned – are lower but compare reasonably well with the second

---

[4] Many subjects know that psychologists routinely perform deception experiments, so you have to guard against them thinking that experimental economists also do deception experiments. See Hertwig and Ortmann (2003), who discuss deception as one of the hallmarks of psychology experiments and the problems this creates. This critique of psychology experiments need not be about morality in the sense of "moralizing," but an issue of experimenter credibility and of not contaminating the subject pool and the social community; it is about morality in the sense of property rights, of the performance of promises, and of guarding against the unintended creation of external costs.

set of experiments reported in Cox (2004) after he removed the preceding potentially contaminating features in the standard dictator and BDM game instructions/procedures – \$5.97 sent, and \$4.94 returned in experiment A1. The corresponding results before the procedural change were 6.00 sent and \$7.17 returned for experiment A2. Observe that the procedural change considerably affected the amount returned in A2 versus A1, but not the amount sent. Hence, in this controlled comparison, the amount returned was much affected by subjects who were aware that there was a "task 2" and therefore another step.

How were the dictator control experiments affected by the different protocols? After the procedural change, Cox (2004) records results that are in line with other double blind dictator experiments if we simply triple the transfers reported by others: His stage I control dictators transfer 36.3 percent of their endowments (B1 in Table 11.1), which is only a little more generous than triple those in the HMSS Double Blind 2 treatment (10.5 percent × 3 = 31.5 percent) (Table 10.2). Table 11.1 reports observations on mean transfers from several dictator and investment game studies. (Except for Henrich et al., 2005, the references report standard errors and all manner of statistical tests showing sampling variability, which I will not reproduce here.) Particularly interesting, however, is the fact that the observation, \$3.63 for B1, is substantially exceeded by a larger dictator average transfer from the first set of instructions/procedures, which yielded a 60 percent increase in the mean to \$5.81 for B2. Similarly and consistent with the B1-B2 comparison are the stage 2 dictator control experiments, C1 (2.06) and C2 (\$5.10), in which removing all suggestion that dictator transfers would be followed by another task overwhelmingly reduces the level of dictator giving.

This strongly points to the hazards implicit in the maintained hypothesis that single-play experimental games implement the concept of a single human interaction event shorn of a future and a past. The mere fact that there exists a subsequent undefined "task 2" materially increases dictator giving. The double blind condition, normally important in dictator games, helps to isolate a person's decision from a social context with a future and a past, but not when it is inadvertently coupled with the creation of an *expectation* that there is an impending "task 2." Moreover, these important findings of Cox are consistent with the many studies of the treatment effects of context reported in the previous chapter.

Social interaction across time is perhaps one of the most highly developed human characteristics whose emergence is deeply rooted in our earliest primate ancestors. Our brains "know" this even as our conscious mind uses reason to invent the wholly abstract concept of the "one-shot game," postulated to be devoid of contamination by our rich experience in the repeat interaction games of life, thereby enabling us to articulate a pure theory of single-play strategic interaction provided that this auxiliary maintained hypothesis is not subject to doubt.

The important lessons between the lines in the directive of "how to identify trust and reciprocity" are about experimental knowledge, procedures, and technique, not mathematical modeling and the mechanics of single-hypothesis testing shorn from the auxiliary hypotheses required to interpret a test's meaning. As for the methodology of identifying decisions that are motivated by other-regarding components of utility, what are we to make of the large variations in altruistic giving in experiments B1, B2, C1, and C2 "not conditioned on the behavior of others" (Cox, 2004, p. 262), but substantially conditioned on the context and therefore expectations of the behavior of others derived from social experience, as defined by the experimenter's instructions and procedures? (Context indirectly conditions decision on the behavior of others through our past socialization.) I think we have a right to expect utility functions of own versus other fixed monetary rewards to be tolerably stable across the situations that yield the outcomes, and not to be jumping all around in response to different contexts.

In contrast, positive (and negative) reciprocity in social exchange systems, which can only have meaning across time, are intrinsically behavioral hypotheses about context and relating past experiential episodes to the current task; it is about how the brain processes information, not only about the reward centers, although clearly the brain must be sensitive to both own rewards and own and other opportunity costs in order to interpret properly the intentions in moves and engage in meaningful interactions across time. One of the many issues to be examined in the next chapter's report on trust games will be experiments testing for the effect of what the first mover forgoes on second-mover behavior – a key consideration in reciprocity theory.

All these characteristics of human interactive decision are about the social brain and how it has been shaped by our ongoing repeat interaction with other humans from earliest childhood. It is only from this social context that Adam Smith can speak of nature, including culture, having formed men for their mutual kindness.

This and the previous chapter have focused on the large variety of contextual treatments that affect behavior in ultimatum, dictator, and investment trust games. These exercises enable us to examine the robustness of our ability to create in the laboratory an environment in which people interact without any influence from a history or a future beyond the particular interaction we study. As experimentalists, going back to Siegel and Fouraker (1960) and Fouraker and Siegel (1963), our technique for achieving this has been (1) to match subjects anonymously and (2) to limit the interaction to a single play only. All other elements that define context are among the initiating conditions that experimentalists have varied. I will summarize the import of these two chapters by discussing each of these elements:

- The most common baseline context supplements (1) and (2) by randomizing subjects to pairs and randomizing their assignment to the roles of player 1 and player 2. Clearly this is not required by the game-theoretic postulate that people are all identically motivated by self-interested dominance, as all such people are interchangeable. We routinely randomize to achieve independence in errors – in this case, subject deviation from the theory's specifications, such as the failure of strict dominance. Randomization helps not at all in escaping the effects of history. To the extent that subjects have a common experiential history of social interaction in repeated play with old and new associates, then that commonality is part of the background defining the initial mental state of subjects entering a laboratory experiment. The protocol and instructions then supplement this initial entering state of mind, resulting in outcomes that may or may not have a significant effect relative to some baseline.
- In the ultimatum game, we have seen that this initial baseline context – as further defined by contest entitlement, exchange, and contest/exchange – each yield significant deviations from the arbitrary random baseline. Subjects whose prior socialization disposes them to honor rights legitimately earned by equal opportunity procedures will offer less as players 1 without increased rejection by like players 2. The same enhancement of player 1's self-interest, and its acceptance by player 2, follows if the game is formulated as an exchange.
- If the context involves the experimenter increasing the ultimatum stakes, from $10 to $100, this fails to change ultimatum offer behavior significantly. However, under exchange, the rejection rate rises as buyers react negatively to the small increases in the seller's offer price.
- Prompting players in exchange to think ahead like game theorists about what behavior they expect from their counterpart significantly increases the amounts offered. In effect, prompted sellers expect buyers to behave less self-interestedly, not more so.
- Double blind controls for what third parties can know about the identity of players and their decisions. Sensitivity to what others can know

is about image and reputation, and necessarily derives from repeat interactive experience over time. Double blind protocols in dictator games significantly decrease transfers to anonymous counterparts relative to single blind protocols, reducing the amounts given by more than half.

- Cherry et al. (2002) show that HMSS did not go nearly far enough in documenting the effect of an earned rights context on decision. After the assignment of subjects to the roles of dictator and recipient, the Cherry et al. (2002) dictators are required to earn the stakes ($10 or $40, depending on their performance score) that they subsequently may voluntarily share with an anonymous other person who has nothing to do but receive whatever is offered by the randomly matched anonymous dictator. With an earned right to stakes, the percentage of dictators giving nothing increases from 19 percent for $10 stakes (15 percent for $40 stakes) to 79 percent (70 percent for $40); when in addition the transfers are double blind, 95 percent give nothing (97 percent for $40). Oxoby and Spaggon (2007) replicate and extend these findings.
- The dramatic effect of earned stakes on dictator giving does not, however, detract from the important contribution of Cox (2000, 2004) documenting the large effect on dictator altruism of instructions that inform dictators that after the first experiment, task 1, is completed, they will participate in an undefined task 2 to be explained later. Being informed that there is a future task significantly increases dictator giving. Cox's controlled comparisons demonstrate the great sensitivity of the alleged single-play dictator game to introducing an unknown future explicitly, a behavioral characteristic that subjects import from their cultural experience.

TWELVE

Reciprocity in Trust Games

... Humboldt quotes without a protest the sneer of the Spaniard, "How can those be trusted who know not how to blush?"
Darwin (1872; 1998, p. 317)

The questions you ask set limits on the answers you find. . . .
Grandin and Johnson (2005, p. 281)

The fundamental fact here is that we lay down rules, a technique, for a [language] game, and that when we then follow the rules, things do not turn out as we assumed. That we are therefore, as it were, entangled in our own rules. The entanglement in our rules is what we want to understand.
Wittgenstein (1963; quoted in Strathern, 1996, p. 65)

Introduction

Extensive experimental studies have established the truth that the amount of giving in ultimatum and dictator games is greater than predicted by standard economic theory based on an especially unsophisticated form of self-interest defined as always choosing dominant outcomes, whatever the circumstances. The replicable results from these games alone, however, are subject to overinterpretation in terms of social utility without testing this interpretation in less restrictive interactions.

Thus, "Since the equilibria are so simple to compute . . . the ultimatum game is a crisp way to measure social preferences rather than a deep test of strategic thinking" (Camerer, 2003, p. 43). You do not generally learn *more* about the wellsprings of behavior by constraining choice and suppressing opportunity cost as it arises in more relaxed environments. Far from allowing a "crisp" interpretation, the results in these decision environments confound the following motives: (1) self-interest based on dominant choice; (2) positive reciprocity; (3) negative reciprocity; (4) other-regarding utility or "altruism"; (5) the constant sum form precluding any effects based on gains from exchange that are created by the actions of the participants – the environment to which we may have become unconsciously adapted in ordinary day-to-day social exchange; and (6) gains to the individual whose accustomed horizon encompasses lifetime reputation-based benefits that are not captured in the exceptionally abstract idea of a single-play one-shot game, and who may not, dutifully and pliantly, operate from this perspective postulated by game theory. The single-play interaction between strangers with no history or future emerged from the need to make a conceptually clear distinction between a game and a super game, to enable constructivist thinking to be clear on the ease with which considerations of the past and of spillovers into the future can and should influence interactive choice. I defend this exercise without qualification in the world of theory, but it is quite another matter to take seriously the proposition that it is a world easily created with real people, and that it is "rational."

What Camerer means is that the interpretations are crisp given that all the game-theoretic auxiliary hypotheses are not subject to doubt. This and the previous two chapters report experiments that confront those auxiliary "maintained and inexplicit" hypotheses to much doubt. If one believes that one-shot ultimatum and dictator games control for most of the confounding elements in the preceding list, then of course it follows that other-regarding utility is the default motive that "explains" by elimination both first-mover largesse and second-mover rejections when the former is "unfairly" low, as revealed by the fact of rejection.

> The belief persists that the predictions of game theory are sharp and unambiguous compared with alternative less formal models that are believed to be inherently ambiguous compared with those of game theory. But this belief confuses exactness of conditional prediction with the imprecision of falsification tests in which a negative result can falsify any of the extra theoretical maintained hypotheses required to implement the predictive hypothesis, a topic on which game theory tells us nothing. This property does not of course somehow excuse vague theories whose predictions fail to command a consensus. In truth, the stronger the assumptions of a theory and the tighter its conditional predictions, the more it may be subject to "waffling" on whether the auxiliary hypotheses are satisfied.

As we have seen, context has a deep structure of "thinking" in the sense that it invokes access to forms of personal and tacit knowledge derived from experience, from social interactions, and from unconscious sources and processes – a nexus that is a driver in the equation of what I have called ecological rationality. This deep structure of experience rooted in human sociality is not the same thing as controlling for a "deep test of strategic thinking," to which game theory formally limits its analysis, its world view, and the hypotheses that it considers relevant. Indeed, "the questions you ask set limits on the answers you find . . . " (Grandin and Johnson, 2005, p. 281), and powerful evidence against the predictions of a theory require us to explore some of the theory's support structure that may blind our narrowly construed interpretations.

Wilson (2005) provides a fresh and illuminating perspective on how we might go about unpacking the complex concept of "reciprocity" in language meaning. Wittgenstein is his point of entry into the "language game" as it applies to this concept. Thus, "What determines . . . [human behavior] . . . , is not what one man is doing now, an individual action, but the whole hurly-burly of human actions, the background against which we see any action" (Wittgenstein, 1967, §567). Wittgenstein's "hurly-burly" is deeply embedded in our language and our sociality. Wilson uses semantic primitives to define "reciprocate," show that it is tautological, and that this circularity can be avoided by an appeal to the essential sociality of human interaction. Such a tautology is no stranger to science, or to scientists, however much the use of "reason" is touted. Sir Isaac Newton's first law of motion – derived allegedly by direct observation – that an object remains at rest or in uniform motion unless altered by a force impressed on it, is a tautology, since he earlier defined an impressed force as an action that if exerted on a body causes it to change from its state of rest or of uniform motion. And, of course, we have already seen that "fair," defined as a tendency to equal division, is quickly invoked by sober scientists as an "explanation" of equal division when it is observed. These ways of using words in ordinary language creep naturally into scientific discourse and our formal mathematical models; they become generative sources of "sophisticated triviality" (to borrow a term from Lakatos, 1978, vol. 2) from which escape can be difficult. There is a need always to ask "what will explain the explanation" (Krutch, 1954, p. 148).

The preceding confounding thesis has also been explored imaginatively by embedding the ultimatum game in an extended environment. Thus, in comprehensive work building on several previous papers, Schmitt (2003) studies ultimatum games with (1) asymmetric payoffs in which bargaining is over chips that have differential conversion rates into cash for the first movers (proposers) and second movers (responders), (2) outside options (disagreement payoffs) for either player, (3) both players having complete information on both players' chip values and outside options, and (4) players knowing their own chip cash values and outside options and having no information on the other's chip values or outside options. In all treatments, players are informed as to how much information on chip value and outside options has been given to the other player. Schmitt's results question the standard argument that ultimatum games support the hypothesis that people predominately seek "fair" outcomes in the sense of harboring other-regarding preferences (compare Camerer, 2003). Here is Schmitt's summary of relevant previous papers as well as her own paper on this particular finding:

Existing papers find that fair proposers seem to exist, but more in attempt(s) to be perceived as fair rather than true "other regarding" (preferences). Rapoport and Sundali (1996) find proposers make lower offers when the responder is uninformed, strategically lowering the proportion of the offer as the size of the pie increases. Guth et al. (1996) find that proposers pretend to be fair to uninformed responders by "hiding behind a small cake." When responders have no information regarding the distribution of the pie, Straub and Murnigan (1995) find that proposers (knowing that responders are uninformed) make *strategic* offers. The results in this paper are generally consistent with these findings. The results clearly suggest that the strategic component to offers may be a more relevant explanation than the altruistic or other regarding (preference) component. When players had incomplete information . . . offers that started at 44.5 chips (out of 100) in round 1 fell to 16.0 chips in round 10. When players had complete information offers only fell by as much as 3.3 chips. . . . In both information states, increases (decreases) in offers were due to high (low) rejection rates (Schmitt, 2003, pp. 70–1).

Schmitt's conclusions are reinforced by the findings of List and Cherry (2002) that innovate a different procedure (also used in the dictator games of Cherry et al., 2002, discussed in Chapter 11): Proposers are recruited separately from responders, and the former must earn their stakes, $400 ($20), depending on whether they correctly answer ten or more (fewer than ten) GMAT questions correctly. Each proposer participates in ten rounds of ultimatum game play, each round with a distinct responder (the repeat single-play condition); for each subject, one round is chosen at random to receive the actual payoff. Compared with other studies of the ultimatum game, where the stakes do not have to be earned, offers are lower in both the high- and low-stake conditions. Second, in both stake conditions, proportionately lower offers are rejected more often than higher offers. Also, unconditioned on the offer amount, subjects reject more offers in low-stakes than high-stakes games. Under high stakes, they report a reduction in the rejection rates over time, suggesting that people are gradually learning to play the equilibrium under the condition of repeat single play.

It is particularly difficult for me to see why proposers, who are referred to as "fair" in much of this literature, are such Machiavellian exploiters of their uninformed counterparts.[1] The experiments summarized in Schmitt (2003) expose the self-serving proclivities of subjects who know the size of the pie and know that the counterpart does not. This seems to be the dark side of utilitarian largess; the claim that social preferences are the driver of cooperation in single-play games is weakened at its foundation if responder ignorance is exploited by the first mover. These results, however, are consistent with the view that fairness arguments are frequently self-serving and self-justifying attempts to achieve an advantage, as in the Florida Supreme Court controversy (Zajac, 2002), in which each side argued that it represented the fair approach.

In this chapter, I report results that occur when subjects from the same populations that generate the outcomes reported in the preceding studies choose among low-cost altruistic, equilibrium, and cooperative outcomes with and without punishment options for defection in the same game tree. Also I will examine reciprocity from several perspectives, including direct comparisons with other-regarding utilitarian predictions. That reciprocity need not be "calculative" is indicated in experiments with repeat pairing over time between pairs rank ordered by their cooperative choices compared with pairings that are random; in neither case do subjects know the matching rules. The bottom line is consistent with the hypothesis that people are much better at producing outcomes in their more sophisticated longer-term ("intelligent") self-interest than is predicted by dominant strategy game-theoretic players. Utilitarian "fairness" models perform poorly by comparison, but we cannot yet rule out the possibility that there exists a

---

[1] Camerer (2003), in defending social preference explanations of cooperation, focuses not on what proposers are doing, but on the behavior of responders: "If Responders accept less when they do not know the Proposer's share, that is very strong evidence that rejections are an expression of preference when they *do* know the Proposer's payoff" (p. 78). But this interpretation requires subjects to behave in accordance with all of the game-theoretic assumptions except domination. Also, observe that not knowing payoffs disables preference, reciprocity, and reputation interpretations: Thus in repeat play the way to get strong support for equilibrium (domination) outcomes is to provide only private payoff information (McCabe et al., 1998).

small minority of other-regarding utility functions, as in the double blind and earned stakes dictator experiments.[2]

Trust Games without a Punishment Option

Figure 12.1 is typical of the "trust" game trees that we will study.[3] Play starts at the top of the tree, node x_1, with player 1, who can move right or down; a move right stops the game, yielding the upper payoff to player 1, $10, and the lower payoff to player 2, $10; if the move is down, player 2 chooses a move at node x_2. If player 2's move is right, player 1 receives $15 and player 2 $25. This is the cooperative (C) outcome. If, however, player 2 moves down, the payoffs to players 1 and 2 are respectively $0 and $40. This is the defection (D) outcome, in which player 2 defects on player 1's offer to cooperate. The subgame perfect equilibrium (SPE) is $10 for each. This follows because at node 1 player 1 can apply backward induction by observing that if play reaches node x_2, player 2's dominant choice is to defect. Seeing that this is the case, player 1's dominant choice is to move right at the top of the tree, yielding the SPE outcome.

Note in particular that if every player is exactly like every other player and is strictly self-interested, there is no room for "mind reading" or inferring intentions from actions, and no room for more sophisticated and subtle action in the "intelligent" self-interest. In effect, play is robotic. To illustrate, suppose that you have been through the standard economics course in game theory and that you are in the position of player 2 in Figure 12.1. Consequently, you expect player 1 to move right at the top of the tree. He doesn't. He moves down, and it is your turn. Surely he did not move down because he prefers $0 to $10, and expects you to defect. He must think that you think

---

[2] Camerer and Weigelt (1988) study sequential reputation formation in a "trust" or borrower-lender game based on externally induced borrower payoffs that may be either normal (default pays) or "nice" (repay the loan). Equivalent labels are "fighting" or "sharing" in entry deterrence. Camerer (2003, pp. 446–53). These studies of sequential equilibrium based on game-theoretic models, labeled "trust," should be clearly differentiated from the free play of choice in the experiments I will report here that are concerned with studying any home-grown tendency for cooperative outcomes to emerge in single- or repeat-play interactions in extensive form games.

[3] See McCabe et al. (1996) for a more complete report of trust games with and without punishment of defection and with a wide variety of matching protocols. In the versions of their games that we report here, I will use trimmed versions of their game trees in which certain nodes, chosen only rarely in most protocols, are omitted. This greatly simplifies and sharpens the exposition without sacrificing substance. But the skeptical reader is invited to refer to the original paper for the complexities omitted here.

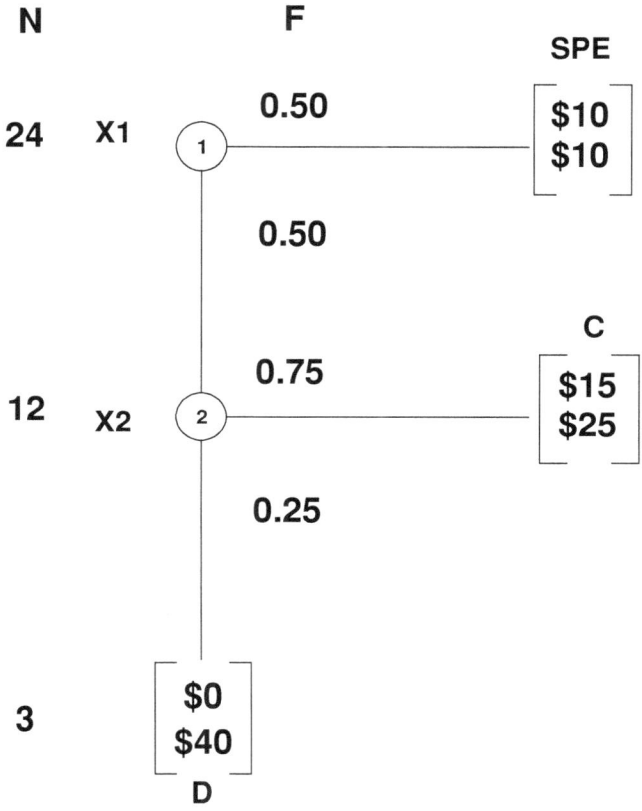

Figure 12.1. Invest $10 Trust Game: Frequency of Play
Notes:
N: Number of subject pairs by node
F: Frequency at which pairs move right or down
SPE: Subgame perfect equilibrium
C: Cooperate
D: Defect

that he wants you to choose C. Whatever else can he have in mind? Maybe he cannot do backward induction, or thinks you are not self-interested. So how are you going to respond? He is making it possible for you to increase your payoff by 150 percent, and his by 50 percent, compared to the SPE. He is not even asking for the larger share of the pie that his action has created. According to reciprocity theory, by choosing C, you will reciprocate his inferred intentions and complete an exchange – exactly in the same way that you trade favors, services, and goods across time with friends, associates, and perhaps even strangers in life, unless you are a sociopath.

According to Mealy (1995):

Sociopaths, who comprise only 3–4% of the male population and less than 1% of the female population, are thought to account for 20% of the United States prison population and between 33% and 80% of the population of chronic criminal offenders. Furthermore, whereas the "typical" U.S. burglar is estimated to have committed a median five crimes per year before being apprehended, chronic offenders – those most likely to be sociopaths – report committing upwards of 50 crimes per year and sometimes as many as two or three hundred. Collectively, these individuals are thought to account for over 50% of all crimes in the U.S. (see ibid., pp. 523, 587–99 for references and caveats; also Lyyken, 1995).

Without a conscious thought, you may often say, "I owe you one," in response to an acquaintance's favor. Player 1 has done a favor to you in giving up SPE. Do you owe him one? Apparently this English phrase translates easily into one with essentially the same meaning in many other languages, whether, Chinese, Spanish, Italian, Hebrew, German, or even French; as a cross-cultural human universal, this is unsurprising (Gouldner, 1960, reviews the literature on the norm of reciprocity, and Brown, 1991, treats it extensively as a human universal): If you are paired with another like national or some in-group member, then you don't have to have any altruism in your utility-for-own reward to know how to respond. Instinctively, you might choose C with hardly a thought, or, since your counterpart will never know your identity, upon closer reflection, you may think that it just makes no sense not to take the $40. Although you are not a clinical sociopath, here is an opportunity to cut a corner and no one can know. Alternatively, if you are player 1 in Figure 12.1, are you certain that you would want to play SPE?

These thought processes may explain why, in data reported by Corelli et al. (2000) comparing faculty with undergraduate subjects, the faculty take much longer, and earn less money, than the undergraduates in deciding whether to offer cooperation or defect. Yet given knowledge of game theory, and knowing that one's counterpart has the equivalent knowledge, what is there to think about?

In regard to this reciprocity analysis of the game, we should note that the game in Figure 12.1 was originally derived as a reduced-form version of the BDM game: Think of player 1 as sending $10, which becomes $30; player 2 can either split the $30 equally with player 1, giving the imputation C, or player 2 can keep it all, yielding the D outcome. But there is another difference, one of context. In the experiment using Figure 12.1, the subjects play an alleged abstract game, one that is not embedded in a BDM-type story about sending $10 upstairs, which becomes $30, and the receiver can either keep it all or split the gain made possible by the sender. But given the BDM outcomes reported previously we should not be too surprised that some subject pairs might end at outcome C.

The outcomes are shown in Figure 12.1 for twenty-four undergraduate subjects: Fifty percent move down (trust), and of these 75 percent "recipro-cate" (trustworthiness). Trustworthiness is here stronger than trust, which we will observe emerging later in repeat play versions of trust games with various matching protocols (Figure 12.2) and also reported by others (see Ashraf et al., 2006; also Schotter and Sopher, 2006, where trustworthiness increases trust, but not vice versa). I think this reflects the fact that in the extensive form, any cooperative tendencies by players 1 are tempered by uncertainty about the response of players 2, whereas player 2 does not have to make a decision until after player 1 has decided.

Why So Much Cooperation?
My coauthors and I have interpreted the outcome C as due to reciprocity, a hypothesis that of course may be wrong, and which we therefore have sought to test in a variety of experiments to challenge the validity of our own way of thinking.

Is It the Subjects? Undergraduates versus Graduates
In the preceding trust game, nearly half the players 1 forgo the "sure thing" derived from dominance and SPE, and three-quarters of the responses are cooperative. I have sometimes heard such results dismissed as a consequence of using naïve subjects. This dismissal has the logical implication that the original theoretical hypothesis is either not falsifiable – in a new test with less naïve subjects, a negative outcome leads to the conclusion that the subjects are still too naïve – or if the outcome is positive, testing stops, as an unpredicted consequence of the search for "sophistication." McCabe and Smith (2000) examined this explanation using advanced graduate students from the sample of subjects who participated in the $100 exchange/entitlement version of the ultimatum game reported in Table 10.1, which showed almost identical results for graduate and undergraduate students. The trust game comparisons are shown in Figure 12.2. In both groups, 50 percent of players 1 offer cooperation, while 75 percent of the undergraduate and 64 percent of the graduate student players 2 reciprocate. (To break even, a cooperative response must occur with probability 2/3. However naïve undergraduates

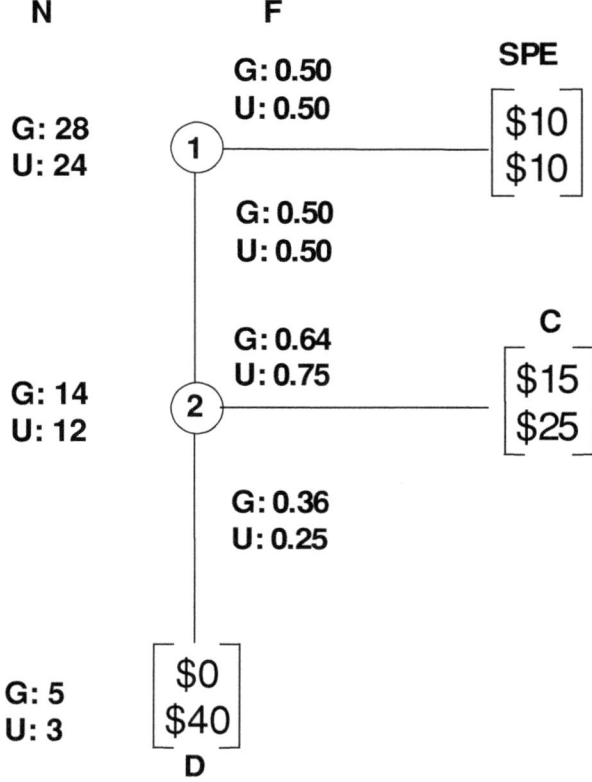

Figure 12.2. Invest $10 Trust Game Comparing Undergraduates (U) and Graduates (G)
Source: Based on data from McCabe and Smith (2000).

are alleged to be, these tests suggest that some graduate students with training in economic theory are capable of showing very similar behavior in this
extensive form game. In an unpublished sample of eighteen pairs of House
and Senate chiefs of staff, both Democrat and Republican, Mary Rigdon
has obtained unpublished results identical to those of the undergraduates
in Figure 12.1.

Machiavelli, Trust, and Cooperation: Mandeville's Knaves?
Mealy (1995) notes that sociopaths are inveterate nonreciprocators who
cheat in not returning favors from others; they radiate superficial charm
and are untruthful, exploitive of others, and lacking in remorse, shame, and
emotion. Normal individuals, however, may vary substantially in the extent

to which they follow social norms of reciprocity and exhibit some of these features of the sociopath as part of a continuum of personality traits; only the extreme tail of the distribution of those traits can be said to involve a personality disorder. In this spirit, Christie et al. (1970) developed an instrument – the Mach scale – to measure the existence of a subclinical trait they call "Machiavellianism." Since their work, hundreds of studies have used the Mach scale.

Gunnthorsdottir et al. (2002b) summarize some of these studies and note that the prediction seems clear: Players 2 in the trust game in Figure 12.1 with high Mach scores should be less likely to reciprocate, whereas the prediction is ambiguous in the player 1 position. Generally, the studies suggest that high Machs are cool, calculating, self-interested, "and thus appear to be true *hominess economici* who continuously test how far they can go" (ibid., pp. 54–5). They cautioned, however, that Mach score effects in the psychology literature all involve very large samples, yielding statistically significant, but very low (for example, 10 percent), R square estimates. Samples are costly (up to $40 per observation in Figures 12.1 and 12.2) and necessarily not large in studies of two-person games. To guard partially against the fact that any relationship between trust game choice and Mach score is likely to be weak, Gunnthorsdottir and his colleagues ran the Mach-IV scale on 1,593 freshman, thereby allowing a sufficient number of extreme scoring people to be selected for the trust game experiments. Even so, they approached the exercise with skepticism as to the predictive power of the measure.

Mach-IV consists of twenty responses; here are two examples:

- "Never tell anyone the real reason you did something unless it is useful to do so."
- "Most people are basically good and kind."

Each person responds on a seven-point scale from "Strongly Agree" to "Strongly Disagree." A high Mach would strongly agree with the first and strongly disagree with the second statement. The maximum raw score is 140 and the minimum is 20. Gunnthorsdotti et al. (2002b) recruited 266 subjects drawn from low scores, low to average scores, and very high scores (eighty-eighth percentile and above).

The subjects participated in the trust game of Figure 12.1, reproduced in Figure 12.3, showing the number of observations (pairs) in the low-to-average-score range and the number in the high-score range. Whenever an odd number of high Mach subjects signed in for a session (four out of twenty-eight total subjects), they were placed in the player 2 position,

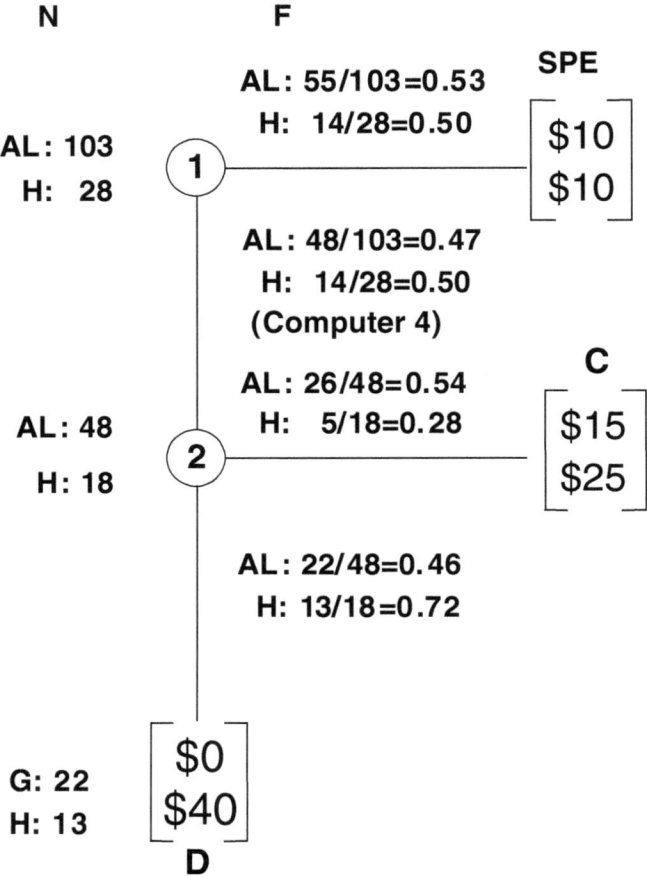

Figure 12.3. Invest $10 Trust Game Comparing High (H) to Average and Low (AL) Mach Scores. *Source:* Based on data reported in Gunnthorsdottir et al. (2002b)

and the computer played down for player 1. This allowed eighteen high Mach subjects to participate as players 2, with twenty-eight as players 1. The relative frequency of play for each score class is shown on the decision tree.

The cooperation rate for the low to average Mach scores (0.542) is nearly double the rate for the high scores (0.278). But there is little difference between the two classes in the player 1 position. Low, average, and high Machs all read their paired counterparts the same when they are in the player 1 position, and are about equally likely to "trust." But high Machs are much less likely to be "trustworthy" than those with low to average scores.

Is It Utility for Other Payoff?

Fehr and Schmidt (1999), Bolton and Ockenfels (2000), and others have proposed fitting preference functions to decisions that aim at accounting for behavior in a variety of experiments, such as ultimatum and dictator games. (See Sobel, 2005, for a summary treatment.) As we have seen in Table 10.3, these models perform poorly in dictator games when dictators are required to earn their endowments, and worse when, in addition, giving occurs under the double blind protocol (Cherry et al., 2002; Oxoby and Spraggon, 2006). It is the intrinsic properties of outcomes that are assumed to drive behavior. Intentions, as reflected in the move choices and conditioned by context based on experience, are assumed to be superfluous in the interactions between the parties. The preference approach identifies adjusted other-regarding ("fair men") utility types. The alternative identifies types who signal intentions, who are into reading move signals, and who risk misidentifying reciprocity versus defection types. The important testable distinctions are that the former are immune to instructional procedures and to path dependency, including the opportunity costs of foregone options.

In the data from these trust games, however, we have an identification problem because an altruistic utility interpretation of cooperation can be invoked in games like that in Figure 12.1. Thus, player 2 may move down because her utility for reward is increasing in both own and other payoff. Sobel (2005) erroneously states that "considerations based on reciprocity are largely retrospective" (p. 396). They need not be and indeed are not: Many direct prospective tests have been reported, and the results have shown consistent statistical support for reciprocity.[4] I will summarize some of these relevant findings in this and in subsequent sections of this chapter.

In Figure 12.4 is a trust game using a within-subjects design that enables us to distinguish subjects who deviate from the subgame perfect equilibrium from motivations of altruism, and those whose deviation from equilibrium derives from reciprocity interpreted as an exchange. The game starts at the top, node x_1, with player 1, who can move right, which stops the game, yielding the upper payoff to player 1, \$7, and the lower payoff to player 2, \$14, or move down, in which case player 2 chooses a move at node x_2. If the move is right, each player gets \$8. If player 2 moves down, player 1 can then move right at node x_3, yielding \$10 for each, or down, yielding \$12

---

[4] That Sobel's (2005) important survey would miss some of these results is not surprising in view of the large experimental literature on trust games and his focus on utility explanations.

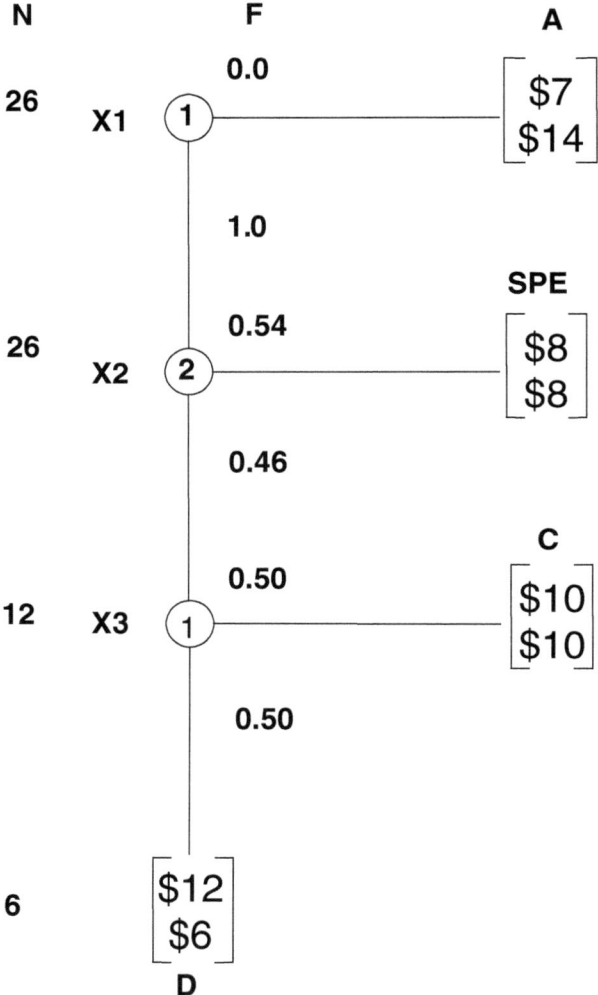

Figure 12.4. Trust Game with Altruism: Frequency of Play. *Source:* Trimmed version of tree and data reported in McCabe et al. (1996), with payoffs in dollars.

for player 1 and $6 for player 2. The subgame perfect equilibrium (SPE) is $8 for each player. This follows because at node $x1$, player 1 can apply backward induction by observing that if play reaches node x_3, player 1 will want to move down. But player 2, also using backward induction, will see that at node x_2 he should move right. Since right at node x_2 yields a higher payoff to player 1, at node x_1 player 1 will conclude that she should move down. Hence, the SPE outcome would prevail by the logic of self-interested players who always choose dominant strategies and apply the principle of backward induction.

If, however, Player 1 has other-regarding preferences and is willing to incur a modest cost to greatly increase the payoff to player 2, player 1 may move right at x_1. Her payoff of $7 is only one-eighth smaller than her payoff at the SPE, and yields $14 for player 2. Hence, at a cost of $1 to herself, player 1 can increase her counterpart player 2's payoff by $6. Player 1 need have only a modest preference for an increase in player 2's welfare to induce her to move right since she achieves a six-to-one return for the other player relative to the cost to player 1. Players 1, whose trade-off of other for own payoff is less than six to one, will happily move right at the top of this tree.

At x_2, player 2 may move down, signaling to player 1 that such a move enables both to profit (gain from exchange), provided that at x_3 player 1 cooperates by reciprocating player 2's favor. Alternatively, at x_3 player 1 can defect (D) on the offer to cooperate by choosing her dominant strategy and move down.

The outcome frequencies for the trust game (N = 26 pairs) are entered directly on the tree in Figure 12.4. The first result – overwhelmingly decisive – is that *none* of the twenty-six subjects in the player 1 position chooses the A (altruistic) outcome; all choose to pass to player 2, apparently preferring to seek a higher payoff for themselves and being content to give player 2 a smaller payoff than is achieved at A, depending upon the final outcome of the move sequence. Secondly, 46 percent offer to cooperate (down), and 50 percent of those reciprocate.

Cox (2002) interprets choices in the BDM investment trust game as reflecting other-regarding preferences and reports that many subjects are choosing to respond to a three-to-one private gain from the utility of other versus own payoff. The implication of the Cox finding is that many subjects in Figure 12.4 can be expected to choose low-cost altruism. But none do. Hence, the preference conclusions from BDM games by Cox are not robust to changes in the game structure shown in Figure 12.4. These contradictory results suggest that the dichotomy between reciprocity and social preference explanations of behavior are sensitive to the context, and any resolution may require a better understanding of how contextual features of different trust games account for variations in behavior. The social preference model must somehow deal with this dichotomy. But the results in Figures 12.1 through 12.4 are all consistent with the reciprocity model.

Reciprocity versus Preferences: Does Own Opportunity
Cost Influence Other Choice?

If reciprocity is an exchange in which each player gains relative to the default outcome, SPE, then there must be gains from exchange in which the cooperative outcome yields an increase in the size of the prize to be split between the two players and each does better than at the SPE (see McCabe et al., 2003). Also, player 2 must believe that (1) player 1 made a deliberate choice to make this outcome possible and (2) she incurred an opportunity cost in doing so, that is, she gave up a smaller certain payoff risking a still smaller payoff if C is not attained because of player 2's decision to defect. It then becomes credible to player 2 that player 1 did a favor for player 2 and can reasonably expect reciprocal action in return. Notice that our argument is in the form of a constructivist theory that need not characterize the subjects' self-aware reasoning, even if it has predictive accuracy; that is, constructive rationality may predict emergent ecologically rational outcomes just as CE theory predicts market outcomes that are not part of the intentions of the agents.

McCabe et al. (2001b), however, report fMRI brain-imaging data supporting the hypothesis that subjects who cooperate use the "mind-reading" circuit modules in their brains (see Baron-Cohen, 1995). This circuitry is not activated in subjects who choose not to cooperate (SPE). In a separate study, in responses to post-experiment questions asking them to write their impressions of their decisions, subjects may report that the experiment is all about whether you can trust your counterpart or "partner."[5] They do not refer to returning a favor, reciprocity, an exchange, fairness, and so on, suggesting that if their actions are driven by reciprocity motives, these actions are not part of a self-aware reasoning process that they report when expressing their perceptions. A subject's use of the word "trust," however, does suggest that the first mover wants to do better for himself, and expects the second mover not to defect. (Ashraf et al., 2006, find "trust" strongly associated with expectations of a trustworthy response.) But this does not necessarily rule out "trusting" that player 2 will have other-regarding preferences. Hence the need for continued testing of reciprocity versus preference models of cooperation.

The two game tree forms shown in Figure 12.5 were designed to test reciprocity against the utility interpretation of choice in a cross-subject design. In Figure 12.5a, if player 1 moves down at the top, the potential prize increases from $40 to $50. Player 2 can defect at a cost to player 1, and can

---

[5] Surveys were conducted in the research, but results were not reported, in Gunnthorsdottir et al. (2002b).

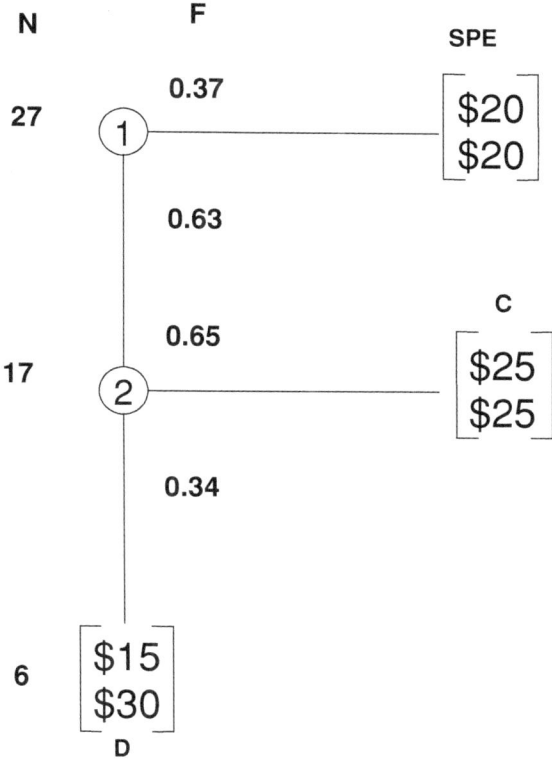

Figure 12.5a. Voluntary Trust Game. *Source:* Reproduced by permission of the Nobel Foundation from Smith (2003).

clearly infer that player 1 deliberately enabled the outcome to increase from ($20, $20) to ($25, $25). But in Figure 12.5b, player 2 can see that player 1 was presented with no voluntary choice to move down. (The experiments were conducted in twelve-person sessions with random assignment to the two decision conditions and to roles.) Consequently, player 1 incurred no opportunity cost to enable player 2 to achieve C. Player 1 did nothing to "work for the welfare" of player 2, and according to reciprocity reasoning, player 2 incurred no implied debt for voluntary repayment. Player 2 can thus move down with impunity. Consequently, in this cross-subject design, reciprocity theory predicts a greater frequency of right moves by players 2 in Figure 12.5a than in Figure 12.5b.

Since only outcomes are supposed to matter, both own and other-regarding utility theories predict no difference in player 2's choices between

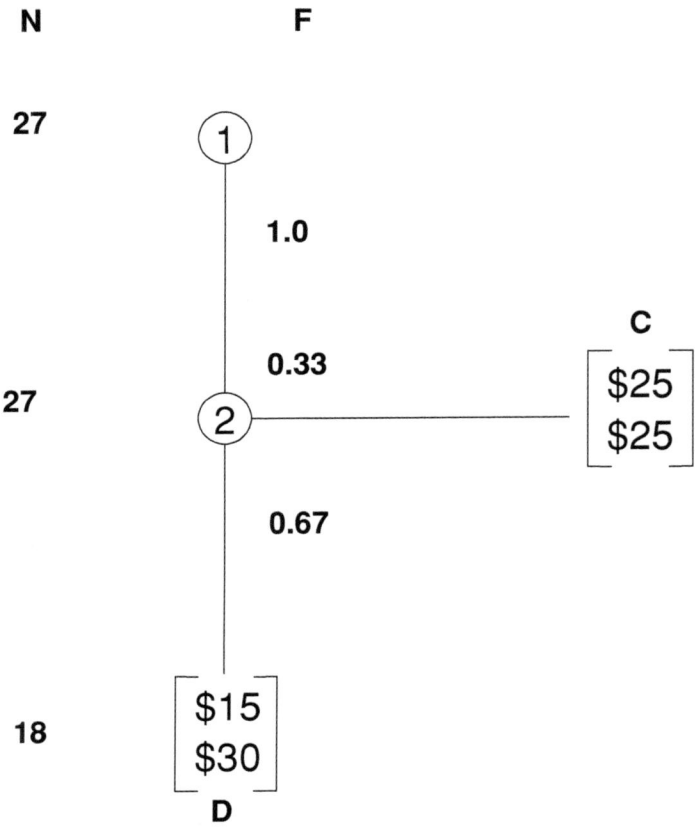

Figure 12.5b. Involuntary Trust Game. *Source:* Reproduced by permission of the Nobel Foundation from Smith (2003).

Figures 12.5a and 12.5b. In fact, as shown by the frequency data on the trees, a third of the players 2 in Figure 12.5b choose C, whereas nearly twice that many choose C in Figure 12.5a.

Opportunity cost and intentions have been found to matter in a study of ultimatum bargaining: " . . . we show that identical offers in an ultimatum game trigger vastly different rejection rates depending on the other offers available to the proposer" (Falk et al., 1999, p. 2). It seems transparent that opportunity cost and intentions acquire their importance because of human sociality and repeat interaction, and not from abstract reasoning or experience obtained through isolated islands of contact between strangers with no history or future, as posited in one-shot games.

Note that the utility program did not generate the experimental design in Figure 12.5a, and could not predict that a larger number of players 2 will cooperate in Figure 12.5a than in Figure 12.5b. The power of reciprocity theory is that in providing a means of falsifying other-regarding utility explanations of interactive decision, it also allows for the distinct possibility that both types may exist.

Although the reciprocity hypothesis receives strong statistical support in this test, there remains a contrary and therefore particularly interesting core of one-third of the players 2 who cooperate when they do not see the outcome forgone by player 1. Prior theoretical hypotheses almost never predict all choices perfectly; in this case, reciprocity outperforms the null hypotheses of no difference by two to one. What can be said about the one-third who cooperate contrary to the predicted outcome? One could arbitrarily assert that this is due to altruistic utility. But if this is so, why did we not observe such behavior in the larger game in Figure 12.4 using a within-subjects design?

In the spirit of reciprocity, here is a testable conjecture as to why people cooperate in Figure 12.5b: As I have argued previously, if most people see life implicitly as an ongoing sequence of games – including many interactions that are just played once – to which they are culturally adapted, and follow a reputation-building life style of reciprocity across sequences, then they are less sensitive to the distinction between a first and second mover, and will "cooperate" even in the game of Figure 12.5b. One implication is that if these subjects return to play a game with the same structure in the role of players 1, they are more likely to offer to cooperate than was observed in Figure 12.5a. An indicator of this style is altruistic egalitarian utility, but the indicator cannot explain itself, and is inconsistent with other choice data reported in this chapter. Such persistent anomalies need ultimately to be unpacked.

I and my coauthors have been skeptical that the social utility program has anywhere to go. In traditional economics, utility is a fixed given property of the individual and there is no next step in the utility research program.[6] Even when it works, it provides no theory showing how or why it is adaptive and better serves individuals in their social context relative to reciprocity norms. The latter are interpreted naturally as exchange, which does not conflict with observed adaptations. The exchange interpretation provides a coherent

---

[6] An exception I should note but cannot here pursue is in evolutionary economic modeling, in which one studies fitness properties of emergent mutant utility functions that improve fitness.

bridge to understand how markets and property rights, which enabled far greater wealth creation, may have evolved in some cultures over historical time as an adaptation and extension of local wealth creation through norms of personal exchange in close-knit societies.

A similar parallel occurs in interpretations of the emergent practice in some cultures of tipping personal service employees such as bellmen, waitresses, and taxi drivers as altruistic actions rather than voluntary payment for services rendered in the context of personal exchange in which people voluntarily reward good service – "giving in gratitude" and receiving in "pride and recognition" (Camerer and Thaler, 1995; Hoffman et al., 1995). The anecdotal "explanation" that tipping is due to "manners" (as in Camerer and Thaler, 1995) is not informative. What explains the good "manners"? Tipping? Its development as a social norm is clearly related to repeat interaction across instances – sequences of one-shot games.

I will return to the topic of repeat play protocols in the following section.

Extensive versus Normal (Strategic) Form Games
A fundamental principle of game theory is that rational behavior is invariant to the form – extensive or normal (strategic)[7] – of the game. Behavior in the extensive and normal forms has been compared by Schotter et al. (1994), Rapoport (1997), and McCabe et al. (2000). All three reject the invariance principle, but in the first study the rationality principles they proposed to explain the invariance either failed to predict the differences, "or they were not what we expected" (Schotter et al., 1994, pp. 446–7). This illustrates the core methodological principle that, when game-theoretic predictions fail, there is little within the corpus of game theory to define a next step in the research program. (This is why, when the dominance axiom fails, the default mode easily degenerates into that of adding parameters to the utility function to preserve that corpus.) In the case of extensive versus normal game forms, the equivalence theorem is based on assumptions about how people think about strategic interactions, and the failure to confirm the equivalence of the two game forms implies that that thinking process needs fundamental reexamination.

Rapoport (1997) provides two versions of the "Battle-of-the-Sexes" game showing how order-of-play information in the extensive form allows players to better coordinate their actions, and making plain the essentially different

---

[7] I prefer to use the term "normal" rather than "strategic" to refer to the simultaneous play alternative to the sequential play extensive form since both forms invite and require strategic choice considerations.

information content of the extensive form. McCabe et al. (2000) argue that the important principle that allows better coordination "derives from the human capacity to read another person's thoughts or intentions by placing themselves in the position and information state of the other person" (p. 4404). Such "mind reading" to detect intentions underlies reciprocity. This interpretation has found support in the brain-imaging study reported by McCabe et al. (2001b). I will summarize here the findings of McCabe et al. (2000) for the trimmed version of the game they study, which is the game we have depicted in Figure 12.4.

These earlier comparisons have been recently extended by Cooper and Van Huyck (2003) in a study of eighteen common games. They confirm the conclusion that behavior in extensive and strategic form games is not the same. In particular, they suggest that players in the first position are motivated to choose moves that include the second player in the decision process, thus inviting the ready development of a relationship. I view this as a direct outgrowth of human sociality that individuals naturally adapt to any new environment, such as the experimental game they are presented with. People are looking to develop a relationship even in single-play two-person interactions. I suggest that this is unsurprising to those not trained in the game-theoretic and economics tradition, which is about modeling people whose relationships are in strict opposition.

In the extensive form of the game in Figure 12.4, player 1 sees the move of player 2 at $x2$ before player 1 has to decide whether to cooperate at node $x3$. In this form of the game, intentions can be clearly communicated along the lines previously described. In the normal form of the same game, each player chooses a move at each node without knowing whether that node will actually be reached in the move sequence. Decisions are thus contingent on the node being reached and may be irrelevant in determining the payoffs. In the normal form, therefore, we can present the game as an $n \times m$ matrix of the $n = 3$ strategies of player 1, right at node $x1$, right or down at node $x3$, and the $m = 2$ strategies of player 2, right or down at node $x2$. Players 1 and 2 each simultaneously choose a row and column among these alternatives not knowing the choice of the other.

McCabe et al. (2000) predict that players 2 will move down at x2, with higher frequency in the extensive than the normal form. They also predict higher rates of cooperation by players 1 (and lower defection rates) in the extensive than normal form. The data are shown in Table 12.1; 46 percent of the players 2 offer to cooperate in the extensive form, and only 29 percent in the normal form. Similarly, they observe a 50 percent cooperative rate

Table 12.1. *Single-play outcome behavior: extensive versus normal form*

Outcome or decision	Extensive form frequency	Normal form frequency
($7, $14)	$\dfrac{0}{26} = 0.0$	$\dfrac{0}{24} = 0.00$
Down at $x2$	$\dfrac{12}{26} = 0.46$	$\dfrac{7}{24} = 0.29$
($10, $10)	$\dfrac{6}{12} = 0.50$	$\dfrac{1}{7} = 0.14$
($12, $6)	$\dfrac{6}{12} = 0.50$	$\dfrac{6}{7} = 0.86$

Source: McCabe et al. (2000). Reproduced by permission.

by players 1 in the extensive form, but only 14 percent in the normal form. Similarly, we observe a larger defection rate (86 percent) in the normal than in the extensive form (50 percent). These comparisons provide little comfort to either the equivalence theorem or to the stability of other-regarding preference interpretations of cooperation, which now becomes contingent on the normal versus extensive form representation.

These one-shot play results and that of others cited previously imply that the extensive and normal forms are not perceived as if they were the same games. (This was insightfully argued long ago by Schelling, 1960.) Players use moves to read intentions that are not the same when actually experienced in sequential move form as when imagined in a mental experiment corresponding to the same sequence, but expressed in the normal form. I would argue that experience and its memory in life are an extensive form process that encodes context along with outcome. The brain is not naturally adapted to solve all sequential move problems by reducing them to a single strategy vector as in a highly structured abstract game. Apparently, we have a built-in tendency to wait, observe, then decide – a process that conserves cognitive resources by applying them only to contingencies that are actually realized, and avoids the need for revision, given the inevitable surprises in the less-structured games of life.[8]

Constructivist modeling glosses over distinctions not part of our cognitive awareness but that nonetheless govern the ecology of choice. This is just the

---

[8] Any such natural process must be deliberately overcome, constructively, in situations where nature serves us poorly. This is a case in which large enough stakes might provide motivation to overcome such a natural process. But we have seen in Chapter 6 that enormous stakes in the FCC auction did not prevent the natural tendency toward "jump bidding" observed much earlier in laboratory experiments. People apparently can rationalize such presumed "irrationalities" even when the stakes are very large.

brain's way of decentralizing function and conserving its scarce resources. Experimental designs conditioned only by constructivist thinking ill prepare us to collect the data that can inform needed revisions in our own thinking. It is both cost-effective and faithful to game-theoretic assumptions in experiments to collect move data from each subject under all contingencies, but it distorts interpretability if game forms are not equivalent.[9] The assumptions of game theory, such as those leading to the logical equivalence of the two game forms, should not be imposed on experimental designs, thereby constraining our understanding of behavior beyond those assumptions and limiting the study of behavior to abstract and arbitrary restrictions on the expression of that behavior.

Trust Games with Punishment Options

Figure 12.6 shows a punishment version of the game in Figure 12.4. Note that if player 1 elects to defect at node $x3$ in Figure 12.4, she chooses directly the defection outcome by moving down. If, however, a player 1 in Figure 12.6 wants to defect at $x3$, she does this by passing play back to player 2, who has the option of not accepting defection in his self-interest and electing the smaller joint outcome ($4, $4). Thus player 2 can punish player 1's failure to reciprocate at node $x3$. Since this defection is costly, the dominance axiom implies that a player will accept the defection by moving right at $x4$; player 2 should see this by backward induction and choose the SPE at node $x2$, exactly as in Figure 12.4. Hence, theoretically, the two forms are equivalent, based on SPE analysis given that both players can be supposed always to choose dominant strategies at every node.

The number of players choosing right or left at each node and their relative frequency are shown in Figure 12.6 for a single play through the game tree. We note first that, as in Figure 12.4, there is zero support for low-cost altruism at node $x1$ in the punishment version of the game – this aspect of decision is indeed the same in the punishment as it is in the nonpunishment version of the tree. Across the two games, we observe the choice behavior of fifty-five subject pairs, no pair having a player 1 who chose the altruism outcome ($7, $14), given the SPE and reciprocity alternatives in the available choice set in each game. Again, in this particular context, we fail to find support for other-regarding utility strong enough to overcome its low $1 opportunity cost

---

[9] Camerer (2003, pp. 48–9) is puzzled as to why economists are reluctant to use the minimum acceptable offer (MAO) method in ultimatum games. I do not know why either, but they *should* be very reluctant because in general the extensive and normal forms of two-person games are not behaviorally equivalent.

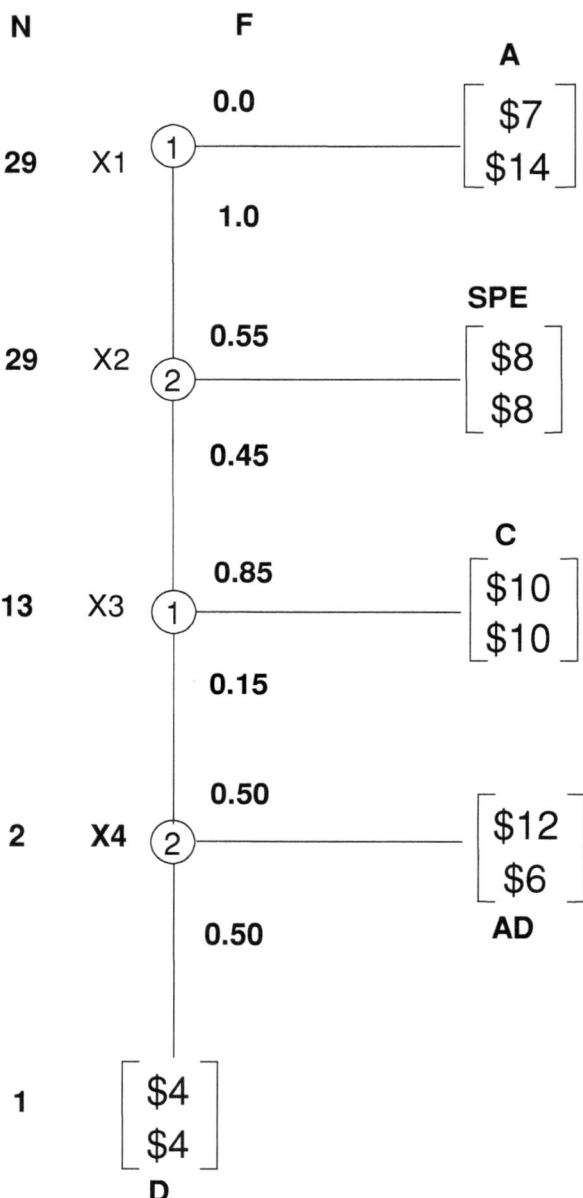

Figure 12.6. Trust/Punishment Game with Altruism: Frequency of Play. *Source:* Trimmed version of tree and data reported in McCabe et al. (1996), with payoffs in dollars.

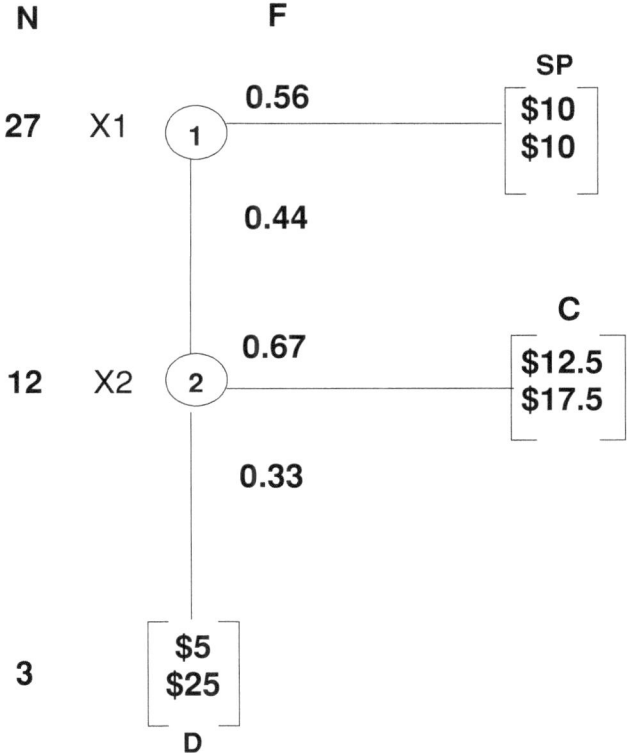

Figure 12.7a. Invest $5 Trust Game: Frequency of Play. *Source:* Reproduced by permission of McCabe et al. (2002, figure 3).

relative to SPE, and making it more generally transparent how restrictions on the choice set in particular contexts – such as ultimatum, dictator, and investment trust games – materially affect decision and the conclusions we draw about behavior.

Comparing the other outcome frequencies in Figures 12.4 and 12.6, it is seen that there is no important difference in the frequency with which players 2 offer to cooperate, but the punishment option brings a nontrivial increase in the reciprocation rate from 50 percent to 85 percent.

These differences can of course be affected by the cost of defection and the rewards to cooperation. This is seen in Figures 12.7a and 12.7b comparing punishment and nonpunishment versions of the invest $5 trust game; that is, think of a move down at x1 as an investment of $5 by player 1, in which the increase is tripled, to $15, then divided equally by player 2, yielding

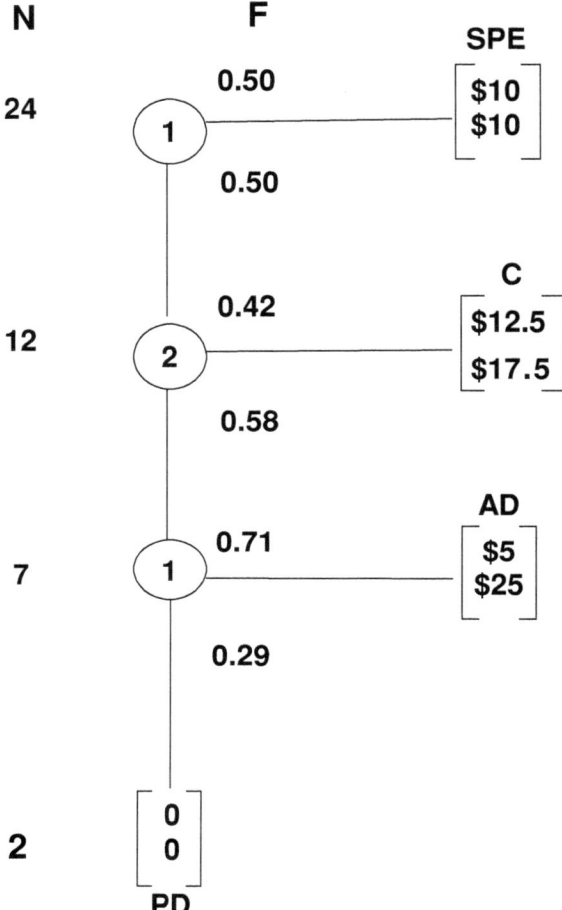

Figure 12.7b. Invest $5 Trust/Punishment Game: Frequency of Play. *Source:* Reproduced by permission of McCabe et al. (2002, figure 4).

($12.50, $17.50), or else player 2 takes all of the $15, yielding the outcome ($5, $25). As in Figure 12.1, we observe not much deviation from an equal split between the SPE and offers to cooperate, with the preponderance of players 2 accepting the offer. But in the punishment version, Figure 12.7b, we observe most players 2 defecting on the cooperative choice, apparently finding not credible the "trust" of player 1, who has the option to punish defection with a costly zero outcome for both players. Nevertheless, two of the nine defectors are punished by players 1 willing to accept a zero payoff to implement the punishment.

Fehr and Rockenbach (2002) have reported a particularly interesting twist on the effect of punishment on trustworthy responses in trust games: The threat of punishment – conveyed in advance of the offer to cooperate – undermines cooperative responses; that is, signaled intentions, and not only incentives, matter in punishment versions of trust games. Why? Here is the argument: If there is no punishment option, I know that when you pass to me you trust me and I am able to show my good offices by rewarding you; if, however, you offer to cooperate with me but reserve the right to punish me "just in case" I fail to accept, that is not conducive to an atmosphere of trust, and for many people, invoking the prospect of punishment backfires and leads to less cooperation.

Houser et al. (2008) study an experimental design for separating the relative importance of intentions and incentives in producing noncooperative behavior in a trust game with punishment. They report that, consistent with Fehr and Rockenbach (2002), credible threats often do fail to result in cooperation, but as it turns out this was not the result of intentions in the following sense: Trustworthiness was not affected by whether punishment was imposed deliberately by the first mover or randomly by nature. Moreover, "trustworthy" responses were substantially increased by the severity of the punishment, showing that negative reciprocity – the endogenous police officer learned in repeat interaction environments – induces more cooperation.

Does the Fehr and Rockenbach (2002) interpretation hold in repeat interaction? If so, it implies that when the same pairs play repeatedly, a direct punishment option in the subgame, as in Figure 12.6 (and 12.7b), may cause cooperation to unravel, with cooperation more likely in the pure trust game; that is, the offer to cooperate on trial t signals less trusting behavior in the punishment version of the game than in the trust version, risking a backlash in the form of less, not more, cooperation. If so, cooperation will be more difficult to achieve over repeat trials in the trust game with the costly punishment option than in the pure trust game.

Differences in behavior between single- and repeat-play experimental games have stimulated a large literature on "learning," but the themes developed in that literature can be discussed briefly, thanks to the ambitious unifying efforts of Camerer (2003) and also Camerer et al. (2004). In the learning research program, equilibrium is the end state of an unspecified, and perhaps unknown, experiential adjustment process; it is about change over time, which can converge to Nash or cooperation. In the extensive form games discussed in the next section, convergence is strongly to cooperation. I believe that much of the "learning" literature begs the question of what is learned and significantly abandons the search for a better understanding of the psychology and ecology of perception and how it changes with experience. But some new insights on this have been provided by Cooper and Kagel (2004; also see 2003), who study the transfer of learning from one game to a second with linear payoff transformations of the first. They find substantial but incomplete transfer compared to a control group as the simple change in the environment disrupts the learning process. Learning models cannot account for the disruptions in learning that occur by changing environments. But in a second, more challenging, experiment, they find encouraging evidence for the development of considerable sophistication – some subjects taking account of the actions of others – over time. The proportion of sophisticated learners increases with experience.

Self-Regarding Cooperation in Repeat Play? Protocols with and without Direct Punishment

Suppose the games in Figures 12.4 and 12.6 are repeated using the same anonymous pairs, the subjects not knowing how many repeat trials will be played but knowing that they will always be matched with the same person.[10] With repeat play, note that even in the pure trust game of Figure 12.4, defection on trial t can be "punished" on the next trial by player 2 choosing SPE. This is the standard Folk Theorem that this credible threat can enable repeat play to support cooperation. But the punishment version of the stage game of Figure 12.4 (shown in Figure 12.6) allows direct, albeit costly, punishment of defection within the subgame of the same trial. The latter option enables a clear, direct, and unmistakable message of intent to be sent by the punisher to the defector. Does it make any difference in the observed behavior? The answer is yes, both in the aggregate and in the dynamics over the whole path of play when the two games are compared.

Table 12.2 lists the frequency (percent) of cooperative outcomes observed by blocks of five trials for the trust games, trimmed versions of which are

---

[10] In the repeated versions, on each of twenty trials, the payoffs are 5 percent of the payoffs in dollars shown in the single-play trees of Figures 12.4 and 12.6. Total payoffs achievable are therefore the same in the single- and repeated-play versions.

Table 12.2. *Percent choosing cooperation by game type, player, and trial block: repeat play, same versus random pairs*

Game type	Punishment				Trust			
	2 Offer cooperation (1)		1 Reciprocate (2)		2 Offer cooperation (3)		1 Reciprocate (4)	
Player Trial Block	Same Pairs	Random Pairs	Same Pairs	Random Pairs	Same Pairs	Random Pairs	Same Pairs	Random Pairs
1–5	71.6	62.4	82.1	64.4	45.0	29.1	64.4	71.9
6–10	81.5	66.7	86.4	64.1	59.1	38.4	84.6	60.5
11–15	88.7	65.5	92.6	65.4	67.6	40.5	89.0	44.7
16–20	84.5	74.6	91.4	76.1	74.3	33.0	91.0	52.6

Source: McCabe et al. (1996).

shown in Figure 12.4 and the punishment game of Figure 12.6 (referred to as "Same 1" and "Same 2" in McCabe et al., 1996). Data columns (1) and (3) compare the frequencies of cooperative play by players 2 (who choose between SPE and offering to cooperate) in the trust and punishment versions of the two games. Table 12.4 also lists the results when the twelve subjects in each session are randomly re-paired on each trial.

Note the following in these results:

- When the "Same" pairs are matched throughout, observe that the propensity of players 2 to initiate cooperation in the trust game, columns (1) and (3), is substantially smaller in every trial block than in the punishment game. These differences narrow as we move from the first to the fourth trial block as cooperation in the trust version gains on the punishment version. Data columns (2) and (4) compare the corresponding cooperativeness of players 1 in the trust and punishment games. The trust game frequencies of cooperation tend to be somewhat smaller than for the punishment game but are almost identical in the final two trial blocks. Consequently, in repeat interaction, it is better to have a direct punishment option in the subgame than to have to rely on punishment in the next period. This is because cooperation takes longer to emerge in the trust game, although it does emerge. This suggests that in long horizon games, there will be little difference between the trust and punishment versions of the game, but the ascension path to cooperation is faster with the direct punishment option. Each case supports the Folk Theorem that repeat interaction favors cooperation.

- With random pairs on each trial, we note reduced cooperation compared with same pairs in every trial block in both the trust and punishment versions of the game (except for trials 1–5 for player 1). Although decay in cooperation occurs across trial blocks, it is not abrupt and substantial.

Effect of Matching Protocol on Frequency of Cooperation in Trust Games with and without Punishment

In trust games like those in Figures 12.4 and 12.6, data have been reported comparing same pairing throughout the session, random pairing (repeat random pairs and roles on each round, twelve subjects per session) and repeat single pairing (each subject plays each other subject exactly once with role alternating between players 1 and 2, sixteen subjects per session) (see McCabe et al., 1996, Result 9). Here are summary results for the relative frequency of play in the cooperative subgame (down move at X2 in Figure 12.4 or 12.6) aggregated across all trials:

| | Matching protocol | | |
Game type	Same pairs	Random pairs	Repeat single
Punishment	0.814	0.672	0.579
Trust	0.616	0.354	NA

Thus, as the probability of meeting the same person is reduced (the poles are 1 with same pairs, 0 with repeat single) across matching protocols, the average frequency of cooperative moves declines. Of course, these aggregate data obscure any trends across trial blocks, but the trend toward increasing cooperation over time when the same pairs interact is indicated in Table 12.2.

Comparison of Behavior in the Repeated Play of Extensive and Normal Form Games

McCabe et al. (2000) also test the equivalence theorem of the extensive and normal game forms in a repeat-play environment with the same matched pairs over twenty trials in which the subjects are not informed as to the number of repeat trials. Both game form and trial number are significant in determining the log odds of cooperation in the following logit regression:

$$\text{Ln}(p(t)/(1 - p(t))) = -0.027 + 0.055\, t + 0.927\, (\text{Game Form})$$
$$(Pr < 0.9)\,(Pr < 0.003)\,(Pr < 0.001)$$

$p(t)$ is the proportion of pairs achieving the cooperative outcome on trial t, and Game Form is an indicator variable with value 1 if Game Form is extensive, 0 if normal. Pr is the significance probability for each of the coefficients in the regression. The results easily reject the null hypothesis that cooperation does not increase in repeat interaction over time by the same pairs. Also strongly rejected is the hypothesis that the two Game Forms are equivalent. Thus, repeat-play experience does not asymptotically rescue the equivalence theorem.

A Matching Protocol Based on Sorting for Cooperative Behavior

In traditional game theory, it is never in one's self-interest to offer to cooperate in a population of individuals who always defect and choose dominant strategies. But a small group of individuals who cooperate if they are matched with another member in that group can invade a population of defectors. That is, if there is a mechanism whereby the probability of being matched with another member of the group exceeds that of being paired with a defector in the host population, such "clustering" can sustain personal exchange within the subpopulation. This may be the important social exchange function of religious and ethnic affiliations – Jew, LDS, Baptist, and Catholic; Irish, Italian, and Chinese; and so on – especially where they are a minority. This argument presumes deliberate engagement and pairing with other identifiable individuals in the subgroup who have a higher probability of being like-minded.

More subtly, if people are acculturated by past success in personal exchange to try cooperation as a means of discovering like-minded individuals or communities of like-minded people who engage in giving and receiving, they can thereby identify social exchangers by trial-and-error sampling. This of course need not generally be a consciously deliberate strategy, but an ecologically fit behavioral life style.

McCabe et al. (2007) implement a qualitative version of this more subtle argument using the game tree of Figure 12.5a. In this version, the key question to be answered is the following: Is cooperation (defection) self-reinforcing over time in an environment in which, unknown to the subjects, people are matched with other like cooperators (defectors) based on their history? That is, does cooperation (defection) beget cooperation (defection) in an environment of "nonstrategic" choice?

This environment is implemented in the following manner.

An agent's history of choices gives her a track record, and the authors track that record and use it to apply a cooperative scoring algorithm to match cooperative types based on their choices over the previous n (not exceeding five) rounds in which the player had an opportunity to move. If player 1 defects, the paired player 2 does not have a play, so the authors compile a score counting only the instances in which players 2 have a move. Therefore, players 2 are tracked to obtain a "trustworthy" opportunity score and players 1 a "trusting" score. The matching algorithm pairs the highest-ranked (trusting) player 1 with the highest-ranked (trustworthy) player 2 for interaction in round n + 1, the next to highest ranked player 1 with the next to highest ranked player 2 for interaction in round n + 1, and so on down the sorted lists.

Consequently, "clustering" in the trust game of Figure 12.5a will be a function of recent cooperative behavior that is hypothesized to emerge endogenously in this bargaining environment and permits the algorithm to sort by type and match individuals by their degree of recent tendencies to cooperate. This emergence may of course be slow or nonexistent, so "clustering" may be empirically problematic, but the experiments we have already reported suggest that "clustering" may have promise even in relatively short sequences.

Each experimental session applying the matching algorithm consists of sixteen people who are rematched in pairs on each round for a total of twenty rounds, but they are not informed of this twenty-round horizon in order to exert some control over potential "end-game" effects. The same environment applies to an equal number of comparison control experiments in which subjects are rematched in pairs randomly on each round. To control for strategic behavior by subjects interacting under the sorting algorithm, subjects are not informed of its use in determining pair composition on each round of play. The relevant instruction is the same for both the sorting and random treatments, namely, "Each period you will be paired with another individual: your counterpart for that period. You will participate for several periods, being re-paired each period."

As I have indicated previously, we already know from other experimental studies (see Table 12.2) that random re-pairing of a finite sample of subjects is less conducive to cooperation than when the same pairs interact repeatedly.

Hence, in this way the clustering argument is implemented by studying the behavior of two contrasting populations: The sorted group in which the probability of selected subjects being paired with another cooperative type is hypothesized to be higher than any similar types emerging in the random control groups. The empirical question that is addressed is whether this way of inducing clustering in the treatment group can sustain cooperative play better than in the control group.

Observe that the prediction of the study – a higher frequency of cooperation in the treatment group than in the control group – could fail completely if people do not alter their tendencies to choose to cooperate when their interaction experience encounters more cooperation than in the control. Any individual inclined to exploit cooperativeness will be more quickly marginalized in the sorted group because on average he will be matched with less cooperative types than in the random group.

Implicit in this design is Adam Smith's (1759; 1982, p. 225) hypothesis that "kindness is the parent of kindness" (and its complement that unkindness begets unkindness), or K. Gibran's (1928, pp. 105–6) person "who gives in gratitude, and . . . receives in pride and recognition." This twist on what might be expected – that you *give* in recognition and *receive* in gratitude, is consistent with the distinction between first and second mover being noncalculative.

Such motives are given maximum opportunity to emerge spontaneously in the sorting treatment because subjects do not have the information needed to maximize in their self-interest consciously by choosing strategically to cooperate in order to manipulate their rank in the sorting. The objective is to ask whether the experience of cooperation encourages "innate" tendencies to cooperate relative to defection behavior, that is, we want potential cooperators to discover that they are in an environment conducive to cooperation and then elect to do whatever comes naturally.

In terms of the theme in this book, we inquire as to whether the sorting treatment induces increased cooperation as a form of ecological rationality. If subjects knew the sorting rule, then increased cooperation could easily follow as a constructively rational response – a distinctly uninteresting result since any strictly self-interested player will be motivated to cooperate if that gives her a higher rank and a sustained increase in payoff. Such a result would be trivial and we would learn nothing about any cultured tendencies for cooperation (or defection) to grow in an environment where people can adapt their choice to their experience.

McCabe et al. (2007) report strong support for the hypothesis that cooperation will increase over time in the sorting treatment relative to the random matching control. The simplest way to see these comparisons is to compute the ratio of the percent cooperative choices in the treatment to those in the random control, by player position in the first five trials and in the last five trials:

	Ratio % cooperation, treatment/control	
Trial Block	Players 1	Players 2
1–5	1.05	1.10
15–20	1.94	1.63

Note first that during the first five trials the ratios are 5 and 10 percent larger, for players 1 and 2 respectively, under the sorting treatment than the random pairing control. But the relative rate of cooperation for players 1 (trust) nearly doubles from the first five to the last five trials, whereas for players 2 trustworthy responses increase by about half, suggesting that trustworthiness lags behind trust as expressed in this environment. This may simply mean that in this environment trustworthiness is necessarily slow to develop because an individual player 2 can express such behavior only when his matched counterpart player 1 offers to cooperate.

More definitively and revealing of the dynamics of cooperation across all the disaggregated rounds of play are the results of the logit regression, shown in Table 12.3. The dependent variable is the natural log odds of cooperation, that is, $\ln (p(t)/(1 - p(t)))$, where $p(t)$ is the proportion of cooperative choices in the sample of players in round t. A cooperative choice by a player 1 is to move down (trust), whereas for a player 2 it is to move right (trustworthy) in Figure 12.5a, allowing the dyad to achieve the cooperative payoff.

Here are some of the interesting behavioral characteristics of choices in this trust game that come out of the marginal analysis shown in rows 2 through 5 in Table 12.3:

- Row 4. What is strong and highly significant (both statistically and behaviorally) is the *interaction* between the treatment and the round of play, shown in row 4, making it clear that sorting relative to random pairing yields increased marginal growth in cooperation for both players 1 and players 2.[11]

---

[11] In terms of the Sobel (2005) model (see Chapter 10), individuals in each treatment group generate distinct endogenous $Vi(H(s))$ functions, and we observe a strong bifurcation in cooperation levels over time.

Table 12.3. *Logit regression of ln $(p(t)/(1 - p(t))$ on round of play, t, treatment, round X treatment and subject's type (based on initial choice)*

	Player 1		Player 2	
Independent variables	Coefficient	t-statistic	Coefficient	t-statistic
1. Constant	−0.43	−2.26*	−2.35	−9.13***
2. Round of Play, t	−0.06	−4.13***	−0.05	−2.68**
3. Treatment: 1, If Sorted; 0 If Random	0.10	0.69	0.10	0.33
4. Interaction: Round X Treatment	0.07	3.22**	0.12	4.62***
5. Subject's iType: Initial Move Choice as Player	1.09	8.97***	2.09	12.9***

Note: * p-value 0.05; ** p-value 0.01; *** p-value 0.001.
$p(t)$ = proportion of cooperative choices in the sample of players in round t.
The independent variables are:
Round of play, t, where $t = 1, 2, \ldots 20$ (row 2).
The treatment is an indicator variable with value 1 if sorted, 0 if random (row 3).
Interaction refers to the joint effect of round and treatment, as a product (row 4).
Individual subjects are typed based on their initial choice as player 1 or 2 (row 4).
Source: McCabe et al. (2002, 2007).

- Row 5. A subject's initial choice in this multiple-round game tree is a strong measure of his or her type and a predictor of subsequent behavior for both players 1 and 2 across both the sorting and the random treatments. I interpret this to reflect the subjects' background evolutionary experience – cultural and biological – which is then modified in specific play experience by the independent variables.
- Row 3. We see that the effect of the sorting treatment *alone*, after adjusting for its dynamic interaction with round and for itype, is small for both players and insignificant.
- Row 2. After correcting for the marginal effects of treatment, interaction, and itype, cooperation decays at significant rates of 6 percent per round for players 1 and 5 percent per round for players 2. I would attribute this decay as a consequence of an inherent decline in coordination across changing pair composition. That decline in cooperation is more than compensated for by the treatment interaction and the itype.

The trust games we report in this chapter provide substantial support for the tendency of people to overcome defection incentives and achieve cooperation even in single-play interactions. Moreover, such cooperation tends to increase under repetition by the same pairs, but not reliably with randomly matched pairs. Overall, the frequency of cooperation increases with the probability of being matched with the same person. Direct tests of the exchange or reciprocity interpretation of cooperation receives more support – much more in some comparisons – than the social preference model, but there are anomalies suggesting the need for further testing. These results apply to extensive form games, and a comparison of behavior in such games with their corresponding normal form rejects the hypotheses that the two forms are behaviorally equivalent both in single and repeat play.

The exchange interpretation of two-person interactions provides an important unifying principle across the range of small group and market interactions treated in Parts II and III of this book. But as emphasized in the quotation from Hayek in Part III, the rules of the two forms of exchange are in perpetual conflict. The fact that in each form of exchange you have to give in order to receive, and that both support wealth creation through the specialization they enable, does not mean that they are perceived in the same way. People are not aware of the benefits resulting from a great range of socioeconomic phenomena, such as the productivity of social exchange systems and the institutional order of markets that underlie the creation of social and economic wealth.

PART FOUR

ORDER AND RATIONALITY IN METHOD AND MIND

Alex's performance was impressive because he was the first bird to demonstrate such abilities, but I was more intrigued by the type labeling errors that he was making than by his success. The reason was that the errors suggested something about how he might be processing information.

Pepperberg (1999, p. 48)

... they were criticized for being unscientific and performing uncontrolled experiments. In science, there's nothing *worse* than an experiment that's uncontrolled.

Grandin and Johnson (2005, p. 249; emphasis added)

THIRTEEN

Rationality in Science

But I believe that there is no philosophical highroad in science, with epistemological signposts. No, we are in a jungle and find our way out by trial and error, building our road behind us as we proceed. We do not find signposts at crossroads, but our scouts erect them, to help the rest.

Born (1943)

... Sir Karl Popper has taught me that natural scientists did not really do what most of them not only told us they did but also urged the representatives of other disciplines to imitate. The difference between the two groups of disciplines (natural and social sciences) has thereby been greatly narrowed. . . .

Hayek (1967, p. viii)

Science is what replaced traditional societies with the modern, technological, industrial world. . . . Science is the body of facts about the natural world that *controlled experiments require us to believe*, together with the logical extrapolations from those facts, and the added things that scientific instruments enable us to see with our own eyes.

H. Smith (2001, pp. 191–2, emphasis added)

Introduction

I have argued that rationality in the economy depends on the behavior of individuals that has been conditioned by cultural norms, and emergent institutions that evolve from human experience and sociality, but that are not ultimately derived from constructivist reason, although clearly constructivist ideas are important sources of variation for the gristmill of ecological selection. Parallel considerations apply to rationality in scientific method, the subject matter of this chapter.

In a nutshell, this is the argument. Scientific methodology reveals a predominately constructivist theme largely guided by the following:

- Falsification criteria for hypotheses derived from theories
- Experimental designs for testing hypotheses
- Statistical tests
- Standard liturgies of reporting style used in scientific papers

But all tests of theory are necessarily joint tests of hypotheses derived from theory and the set of auxiliary hypotheses necessary to implement, construct, and execute the tests; this is the well-known Duhem-Quine (D-Q) problem emphasized by methodologists, but little integrated into the thinking of scientists. Thus, whatever might be the testing rhetoric of scientists, they do not reject hypotheses, and their antecedent theories, on the basis of falsifying outcomes. But this is not cause for despair, let alone retreat into a narrow postmodern sea of denial in which science borders on unintended fraud. D-Q is a property of inquiry, a truth, and as such a source for deepening our understanding of what is, not a clever touché for exposing the ubiquitous rhetorical pretensions of science.

But the failure of all philosophy of science programs to articulate a rational constructivist methodology of science that serves to guide scientists, or explain what they do, as well as what they say about what they do, does not mean that science is devoid of rationality or that scientific communities fail to generate rational programs of scientific inquiry. Thus, scientists engage in commentary, reply, rebuttal, and vigorous discussions over whether the design is appropriate and the tests adequate, whether the procedures and measurements might be flawed, and whether the conclusions and interpretations are correct. One must look to this conversation in the scientific community in asking whether and how science sorts out competing primary and auxiliary hypotheses after each new set of test results is made available.

If emergent method is rational in science, it must be a form of ecological rationality, meaning that it rightly and inevitably grows out of the rule-governed norms, practices, and conversation that characterize meaningful interactions in the scientific community. Listen not only to what scientists say about what they do, ignoring the arrogant tone in their expressions, but also examine what they do. The power behind the throne of accomplishment in the human career is our sociality, and the unintended mansions that are built by that sociality. The long view of that career is in sharp focus: Our accumulation of knowledge and its expression in technology enabled us to survive the Pleistocene age, people the earth, penetrate the heavens, and explore the ultimate particles and forces of matter, energy, and life. That achievement hardly deserves to be described as irrational or nonrational, let alone incompetent.

Rational Constructivism in Method

What does it mean to test a theory or a hypothesis derived from a theory? Scientists everywhere say and believe that the unique feature of science is that theories, if they are to be acceptable, require rigorous support from facts based on replicable observations. But the deeper one examines this belief, the more elusive becomes the task of giving it precise meaning and content in the context of rational constructivist programs of inquiry.

Can We Derive Theory Directly from Observation?

Prominent scientists through history have believed that the answer to this question is "yes," and that this was their modus operandi. Thus, quoting from two of the most influential scientists in history:

I frame no hypotheses; for whatever is not deduced from the phenomena ... have no place in experimental philosophy ... [in which] ... particular propositions are inferred from the phenomena ... (Isaac Newton, *Principia*, 1687; quoted in Segrè, 1984, p. 66).

Nobody who has really gone into the matter will deny that in practice the world of phenomena uniquely determines the theoretical system, in spite of the fact that there is no theoretical bridge between phenomena and their theoretical principles (Einstein, 1934, pp. 22–3).

But these statements are without validity. "One can today easily demonstrate that there can be no valid derivation of a law of nature from any finite number of facts" (Lakatos, vol. 1, 1978, p. 2). Yet in economics, critics of standard theory call for more empirical work on which to base development of the theory, and generally the idea persists that the essence of science is its rigorous observational foundation.

But how are the facts and theories of science to be connected so that each informs the other?

Newton passionately believed that he was not just proffering lowly hypotheses, and that his laws were derived directly, by logic, from Kepler's discovery that the planets moved in ellipses. But Newton showed only that the path was an ellipse if there are n = 2 planets. Kepler was wrong in thinking that the planets followed elliptical paths, and to this day there is no

Newton-style solution for the n (> 2)-body problem, and in fact the paths can be chaotic. Thus, when he published the *Principia*, Newton's model could not account for the motion of our nearest and most accurately observable neighbor, the moon, whose orbit is strongly influenced by both the sun and the earth.

Newton's sense of his scientific procedure is commonplace: One studies an empirical regularity (for example, the "trade-off" between the rate of inflation and the unemployment rate) and proceeds to articulate a model from which a functional form can be derived that yields the regularity. In the preceding confusing quotation, Einstein seems to agree with Newton. At other times, he appears to articulate the more qualified view that theories make predictions, which are then to be tested by observations (see his insightful comment later on Walter Kaufmann's test of special relativity theory), while on other occasions his view is that reported facts are irrelevant compared to theories based on logically complete metatheoretical principles, coherent across a broad spectrum of fundamentals (see Northrup, 1969, pp. 387–408). Thus, upon receiving the telegraphed news that Eddington's 1919 eclipse experiments had "confirmed" the general theory, he showed it to a doctoral student who was jubilant, but he commented, unmoved: "I knew all the time that the theory was correct." But what if it had been refuted? "In that case I'd have to feel sorry for God, because the theory is correct" (quoted in Fölsing, 1997, p. 439).

The main theme I want to develop in this and subsequent sections is captured by the following quotation from a lowbrow source, the mythical character Phaedrus in *Zen and the Art of Motorcycle Maintenance*: "... the number of rational hypotheses that can explain any given phenomena is infinite" (Pirsig, 1981, p. 100).

Proposition 1. Particular hypotheses derived from any testable theory imply certain observational outcomes; the converse is false (Lakatos, 1978, vol. 1, pp. 2, 16, passim).

Theories produce mathematical theorems. Each theorem is a mapping from postulated statements (assumptions) into derived or concluding statements (the theoretical results). Conventionally, the concluding statements are what the experimentalist uses to formulate specific hypotheses (models) that motivate the experimental design that is implemented. The concluding statements are the objects of testing in an economics experiment, insofar as their conditions can be controlled. Since not every assumption can always be reproduced in the experimental design, the problem of the "controlled experiment" is one of trying to minimize the risk that the results will fail to be interpretable as a test of the theory because one or more assumptions were violated. An uncontrolled assumption that is postulated to hold in interpreting test results is one of many possible contingent auxiliary hypotheses to be discussed in this chapter.

We note also that mathematics itself does not have secure (nonempirical) foundations. Bertrand Russell, David Hilbert, and others (including John von Neumann) hoped to rescue mathematics from classical skepticism: that any term (axiom, or so on) must be defined in other terms leading to an infinite regress. Thus, the state of mathematical proofs was found by Russell and Hilbert to be "fallacious," full of "logical absurdities," "paradoxical," and so on, and they set out to remove mathematics from its belief-dependence on inductive or empirical premises, as in science, where, for example, James Clerk Maxwell's equations were considered believable because of the so-called observed truth of their consequences. Their program failed decisively when Kurt Gödel showed (among other things) that any consistent formalization will contain unprovable arithmetical truths. In the end, Russell acknowledged his intellectual sorrow in failing to found mathematics on trivial axioms, consistency, and the rules of logic (Lakatos, 1978, vol. 2, pp. 11–23). Thus, constructivism alone leads nowhere; its roots must find ultimate nourishment outside of reason. "Outside" means knowledge derived from experience, from social interactions and from unconscious sources and processes – the nexus that I have called ecological rationality. Both Gödel and Ludwig von Wittgenstein understood this. As I see it, in the text, the discussion is concerned with applied mathematical science (astronomy, physics, economics), and the questions have to do with all the processes we bring to bear on using evidence and abstract reasoning (theory) to acquire knowledge.

The wellspring of testable hypotheses in economics and game theory is to be found in the marginal conditions defining equilibrium points or strategy functions that constitute a theoretical equilibrium. In games against nature, the subject-agent is assumed to choose among alternatives in the feasible set that which maximizes his or her outcome (reward, utility, or payoff), subject to the technological and other constraints on choice. Strategic games are solved by the device of reducing them to games against nature, as in a noncooperative (Cournot-Nash) equilibrium (pure or mixed), where each agent is assumed to maximize own outcome, given (or subject to the constraints of) the maximizing behavior of all other agents. The equilibrium strategy when used by all but agent i reduces i's problem to an own-maximizing choice of that strategy. Hence, in economics, all testable hypotheses come from the marginal conditions (or their discrete opportunity cost equivalent) for maximization that defines equilibrium for an individual or across individuals interacting through an institution. These conditions are implied by the theory from which they are derived, but given experimental observations consistent with (that is, supporting) these conditions, there is no general way to reverse the steps used to derive the conditions and deduce the theory from a set of observations on subject choice. Behavioral point observations conforming to an equilibrium theory cannot be used to deduce or infer either the equations defining the equilibrium or the logic and assumptions of the theory used to derive the equilibrium conditions.

Suppose that the theory is used to derive noncooperative best-reply functions for each agent that maps one or more characteristics of each individual into that person's equilibrium decision. Suppose next that we perform many repetitions of an experiment varying some controllable characteristic of the individuals, such as their assigned values for an auctioned item, and obtain an observed response for each value of the characteristic. This repetition of course must be assumed always to support equilibrium outcomes. Finally, suppose we use this data to estimate response functions obtained from the original maximization theory. First-order conditions defining an optimum can always be treated formally as differential equations in the original criterion function. Can we solve these equations and "deduce" the original theory?

An example is discussed in Smith et al. (1991) from first-price auction theory. Briefly the idea is this: Each of N individuals in a repeated auction game is assigned value $v_i(t)(i = 1, \ldots, N; t = 1, 2, \ldots, T)$ from a rectangular distribution, and on each trial bids $b_i(t)$. Each i is assumed to

$$\max_{0 \leq b_i \leq v_i} (v_i - b_i)^{r_i} G_i(b_i) \tag{1}$$

where $r_i(0 < r_i \leq 1)$ is i's measure of constant relative risk aversion, and $G_i(b_i)$ is i's probability belief that a bid of b_i will win. This leads to the first-order condition,

$$(v_i - b_i)G'_i(b_i) - r_i G_i(b_i) = 0. \tag{2}$$

If all *i* have common rational probability expectations

$$G_i(b_i) = G(b_i). \tag{3}$$

this leads to a closed-form equilibrium bid function (see Cox et al., 1988; reprinted in Smith, 1991, Chapter 29):

$$b_i = (N-1)v_i/(N-1+r_i), \; b_i \lesseqgtr \bar{b} = 1 - G(\bar{b})/G'(b), \;\; \text{for all } i. \quad (4)$$

The data from experimental auctions strongly support linear subject bid rules of the form

$$b_i = \alpha_i v_i, \quad (5)$$

obtained by linear regression of b_i on v_i using the T observations on (b_i, v_i), for given N, with $\alpha_i = (N-1)v_i/(N-1+r_i)$. Can we reverse the preceding steps, then integrate (2) to get (1)? The answer is no. Equation (2) can be derived from maximizing either (1) or the criterion

$$(v_i - b_i)G(b_i)^{1/r_i}, \quad (1')$$

in which subjective probabilities rather than profit are "discounted." That is, without all the assumptions used to get (4), we cannot uniquely conclude (1). In (1') we have $G_i(b_i) = G(b_i)^{1/r_i}$ instead of (3), and this is not ruled out by the data.

In fact, instead of (1) or (1') we could have maximized

$$(v_i - b_i)^{\beta_i} G(b_i)^{\beta_i/r_i}, \text{ with } r_i \le \beta_i \le 1, \quad (1'')$$

which implies an infinite mixture of subjective utility and subjective probability models of bidding.

Thus, in general, we cannot backward induct from empirical equilibrium conditions, even when we have a large number of experimental observations, to arrive at the original parameterized model within the general theory. The purpose of theory is precisely one of imposing much more structure on the problem than can be inferred from the data. This is because the assumptions used to deduce the theoretical model contain more information, such as (3), than the data – the theory is underdetermined. In this example, decisions are postulated to depend upon utility and expectations, and these two elements are confounded in the observations on an individual's decisions. Moreover, there is nothing in the data that need rule out a completely different theoretical approach, as in the hypothesis of Radner (1997) and his coworkers, who model decision under uncertainty as a problem in the economics of survival. Hence, the preceding derivation based on empirical bid functions imposes the constraints of a particular theory that may represent an incomplete picture, or subspace, of larger issues in the spectrum of decision.

Economics: Is It an Experimental Science?

All editions of Paul Samuelson's *Principles of Economics* refer to the inability of economists to perform experiments. This continued for a short time after William Nordhaus joined Samuelson as a coauthor. Thus, "Economists...cannot perform the controlled experiments of chemists and biologists because they cannot easily control other important factors" (Samuelson and Nordhaus, 1985, p. 8).

My favorite quotation, however, is supplied by one of the twentieth century's foremost Marxian economists, Joan Robinson. To wit, "Economists cannot make use of controlled experiments to settle their differences" (Robinson, 1979, p. 1319). Like Samuelson, she was not accurate – economists do perform controlled experiments – but how often have they, or their counterparts in any science, used them to "settle their differences"? Here she was expressing the popular image of science, which is perceived incorrectly as one in which "objective" facts are the arbiters of truth that in turn "settle" differences. The caricatured image is that of two scientists, who, disagreeing on a fundamental principle, go to the lab, do a "crucial experiment," and learn which view is assuredly right. They do not argue about the result; their question is answered and they move on to a new topic that is not yet "settled."

Although these quotations provide telling commentaries on the state of the profession's knowledge of the development of experimental methods in economics during the past half-century, there is a deeper question of whether there are more than a very small number of nonexperimentalists in economics who understand key features of our methodology. These features are twofold: (1) the use of a reward scheme to motivate individual behavior in the laboratory within an economic environment defining gains from trade that are controlled by the experimenter, for example, the supply and demand for an abstract item in an isolated market or an auction; and (2) the use of the observations to test predictive hypotheses derived from one or more models (formal or informal) of behavior in these environments using the rules of a particular trading institution, for example, the equilibrium clearing price and corresponding exchange volume when subjects trade under some version of the double auction. This differs from the way that economics is commonly researched, taught, and practiced, which implies that it is largely an a priori science in which economic problems come to be understood by thinking about them. This generates logically correct, internally consistent theories and models. The data of econometrics are then used for "testing" between alternative model specifications within basic equilibrium theories that are not subject to challenge, or to estimate the supply and/or demand parameters assumed to generate data representing equilibrium outcomes by an unspecified process.[1] It is constructivism all the way down.

I want to report two examples indicating how counterintuitive it has been for prominent economists to see the function of laboratory experiments in economics. The first example is contained in a quotation from Hayek, whose Nobel citation was for his theoretical conception of the price system as an information system for coordinating agents with dispersed information in a world where no single mind or control center possesses, or can ever have knowledge of, this information. His critique and rejection of mainstream quantitative methods, "scientism," in economics are well known (see, e.g., Hayek 1942/1979, 1945). But in his brilliant paper interpreting competition as a discovery process, rather than a model of equilibrium price determination, he argues:

> ... wherever the use of competition can be rationally justified, it is on the ground that we do not know in advance the facts that determine the actions of competitors.... [C]ompetition is valuable only because, and so far as, its results are unpredictable and on the whole different from those which anyone has, or could have, deliberately aimed at .... The necessary consequence of the reason why we use competition is that, in those cases in which it is interesting, the validity of the theory can never be tested empirically. We can test it on conceptual models, and we might conceivably test it in artificially created real situations, where the facts that competition is intended to discover are already known to the observer. But in such cases it is of no practical value, so that to carry out the experiment would hardly be worth the expense (Hayek, 1978/1984, p. 255).

He describes with clarity an important use (unknown to him) that has been made of experiments – testing competitive theory "in artificially created real situations, where the facts which competition is intended to discover are already known to the observer" – then proceeds to fail completely to see how such an experiment could be used to test his own proposition that competition is a discovery procedure, under the condition that neither agents as a whole nor any one mind needs to know what each agent knows. Rather, his concern for dramatizing what is one of the most important socioeconomic ideas of the twentieth century seems to have caused him to interpret his suggested hypothetical experiment as "of no practical value," since it would (if successful) merely reveal what the observer already knew.

---

[1] Leamer (1978) and others have challenged the interpretation of this standard econometric methodology as a scientific "testing" program as distinct from a program for specification searches of data.

I find it astounding that one of the most profound thinkers in the twentieth century did not see the demonstration potential and testing power of the experiment that he suggests for testing the proposition that with competition no one in the market need know in advance the actions of competitors, and that competition is valuable only because, and so far as, its results are unpredictable by anyone in the market and on the whole different from those at which anyone in the market has, or could have, deliberately aimed. Yet, unknown to me at the time, this is precisely what my first experiment, conducted in January 1956 and published later as "Test 1," was about (Smith, 1962).

I assembled a considerable number of experiments for the paper "Markets as Economizers of Information: Experimental Examination of the 'Hayek Hypothesis,'" presented at the Fiftieth Jubilee Congress of the Australian and New Zealand Association for the Advancement of Science in Adelaide, Australia, May 12–16, 1980. A substantially shortened version of the paper was published in Smith (1982b) and reprinted in Smith (1991, pp. 221–35). Here is what I called the Hayek Hypothesis: Strict privacy together with the trading rules of a market institution (the oral double auction in this case) is sufficient to produce efficient competitive market outcomes. The alternative was called the Complete Knowledge Hypothesis: Competitive outcomes require perfectly foreseen conditions of supply and demand, a statement attributable to many economists, such as Samuelson, that can be traced back to W. S. Jevons in 1871.[2] In this empirical comparison, the Hayek Hypothesis was strongly supported. This theme had been visited earlier (without the deserved attribution to Hayek) in Smith (1976; reprinted in Smith, 1991, pp. 100–5; 1980, pp. 357–60), wherein eight experiments comparing private information with complete information showed that complete information was neither necessary nor sufficient for convergence to a competitive equilibrium: Complete information, compared to private information, interfered with and slowed convergence. Shubik (1959, pp. 169–71) had noted earlier the confusion inherent in ad hoc claims that perfect knowledge is a requirement of pure (or sometimes perfect) competition. The experimental proposition that private information increases support for the noncooperative equilibrium, under repeat interaction, applies not only to markets, but also to the two-person extensive-form repeated games reported by McCabe et al. (1998). Hence it is clear that without knowledge of the other's payoff, it is not possible for players to identify and consciously coordinate on a cooperative outcome, such conscious coordination being essential in two-person interactions but

---

[2] Stigler (1957) provides a historical treatment of the concept of perfect competition.

irrelevant, if not pernicious, in impersonal market exchange. We note in passing that the large number of experiments demonstrating the Hayek Hypothesis in no sense implies that there may not be exceptions (Holt, 1989). It's the other way around: This large set of experiments demonstrates clearly that there are exceptions everywhere to the Complete Knowledge Hypothesis and that these exceptions were not part of a prior design created to provide a counterexample.

Holt and his coauthors have asked, "Are there any conditions under which double-auction markets do not generate competitive outcomes? The only exception seems to be an experiment with a 'market power' design reported by Holt et al. (1986) and replicated by Davis and Williams (1991)" (Davis and Holt, 1993, p. 154). The example reported in this exception was a market in which there was a constant excess supply of only one unit – a market with inherently weak equilibrating properties. Actually, there were two earlier reported exceptions, neither of which required market power: (1) one in which information about private circumstances is known by all traders (the alternative to the Hayek Hypothesis as stated previously), and (2) an example in which the excess supply in the market was only two units. Exception (1) was reported in Smith (1976, reprinted 1991, pp. 104– 5; 1980) and (2) in Smith (1965, reprinted 1991, p. 67). The preceding cited exceptions, attributed to market power, would need to be supplemented with comparisons in which there was just one unit of excess demand, but no market power, in order to show whether or not the exception was driven by market power, and not the fact that there is only one unit of excess supply.

My second example involves the same principle as the first; it derives from a personal conversation in the early 1980s with one of my favorite Nobel laureates in economics, a prominent theorist. In response to a question, I described the experimental public goods research I had been doing in the late 1970s and early 1980s comparing the efficacy of various incentive compatible public good mechanisms. (See the public goods papers reprinted in Smith, 1991). He wondered how I had achieved control over the efficient allocation as the benchmark used in these comparisons. So I explained what I had thought was commonly understood by then: I give each subject a payoff function (table) in monetary rewards defined jointly over variable units of a public (common outcome) good, and variable units of a private good. This allows the experimenter to solve for the social optimum and then use the experimental data to judge the comparative performance of alternative public good incentive mechanisms. Incredibly, he objected that if, as the experimenter, I have sufficient information to know what constitutes the socially optimal allocation, then I did not need a mechanism. I could just impose the optimal allocation!

So there I was, essentially an anthropologist on Mars, trying to convey to one of the best and brightest that the whole idea of laboratory experiments was to evaluate mechanisms in an environment where the Pareto optimal outcome was known by the experimental designer but not by the agents so that performance comparisons could be made; that in the field such knowledge was never possible, and we had no criteria, other than internal theoretical properties such as incentive compatibility to judge the efficacy of the mechanism. He didn't get it; psychologically, this testing procedure is not comprehensible if somehow your thinking has accustomed you to believe that allocation mechanisms require agents to have complete information, but not mechanism designers. In fact, with that world view, what is there to test in mechanism theory?[3]

For me, it is not natural to think about economics as it is manifest in the preceding quotations; you come to live in a different way of thinking as you live in your own skin.

The issue of whether economics is an experimental science is moot among experimental economists who are, and should be, too busy having fun doing their work to reflect on the methodological implications of what they do. But when we do, as in comprehensive introductions to the field, what do we say? Quotations from impeccable sources will serve to introduce the concepts to be developed next. The first emphasizes that an important

---

[3] Why do we tend to believe or assert that economic theories assume complete (or probabilistic) information by each agent on the circumstances (tastes, costs) of all other agents? As noted in Chapter 5, this interpretation cannot be because there is a theorem about behavior dealing with information in a constructive process of equilibrium formation. Rather, we confuse the cognitive requirements of analysis used to deduce the properties of end states, with interactive decision processes that are as consciously inaccessible to the agent as to the scientist. Modeling mechanisms is not the same thing as modeling interactive minds. In the former, it is unimaginable that a noncooperative equilibrium obtains except under a conceptual regime in which each agent calculates his or her strategy given the information on all other agents needed to make the calculation. From this perspective, it isn't that Hayek (1945) is wrong; he is irrelevant to game theory in which the concept of a strategy requires one to have information on adversaries – enough information (perfect or probabilistic) to calculate mutually compatible optimal strategies. It is that, or it is nothing. Hence, observations are inadmissible outside this conceptual framework, and therefore are not separable from that framework. In my view, Hayek is talking about unconscious mental processes that serve the interests of the person, conserve working memory, and achieve unintended ends that are not comprehensible to the individual precisely because the welfare of all *requires* him to focus on his local circumstances. If he focuses on ends, he is all too likely to want to impose inappropriate policies.

category of experimental work ". . . includes experiments designed to test the predictions of well articulated formal theories and to observe unpredicted regularities, in a controlled environment that allows these observations to be unambiguously interpreted in relation to the theory" (Kagel and Roth, 1995, p. 22). In a similar vein, it is often suggested that because game theory predictions are sharp, it is not hard to spot likely deviations and counterexamples. Experimental economists strongly believe, I think, that this is our most powerful scientific defense of experimental methods: We ground our experimental inquiry in the firm bedrock of economic or game theory. A second crucial advantage, recognizing that field tests involve hazardous joint tests of multiple hypotheses, is that "Laboratory methods allow a dramatic reduction in the number of auxiliary hypotheses involved in examining a primary hypothesis" (Davis and Holt, 1993, p. 16).

Hence, the strongly held belief that, in the laboratory, we can test well-articulated theories, interpret the results unambiguously in terms of the theory, and do so with minimal, or at least greatly reduced, dependence on auxiliary hypothesis. These views are not unique to experimental economics, but they are illusions.

Both field and laboratory experiments are powerful sources of learning in economics largely because experimental controls enable a better identification of the effect of treatments on measured performance, but "control" can also be a prison for exploring decision making that is inevitably based on human socialization. I have emphasized this in earlier discussions of the effect of context on behavior and the illusion of thinking that there are clear criteria for conveying to subjects "entirely abstract explanations of the game, and experimental context" (Henrich et al., 2005, p. 805). I want simply to note here that there are similar illusions that control is a panacea for ensuring the quality of the information we gather in experiments. The study of cognition in parrots took on whole new dimensions using "social modeling theory" in which the subject bird watches two trainers who interact as one trainer "teaches" the other trainer to label objects with color, shape, and type of material. Alex, a gray parrot, upon occasion made queries to his trainers after he had learned the labels for various objects (squares, triangles) and colors (red, green, and blue). One day, observing himself in his mirror, he inquired, "What color?" This is how he learned that he was a gray parrot and learned the color gray. Similar inquiries, "What's that?" led to "rock" for his lava stone beak conditioner, "bock" ("box") for an object container, "carrot," and "orange" (Pepperberg, 1999, p. 244).

None of the Alex experiments used classical operant conditioning, or food rewards. Training was based on observation and on proactive interaction with his trainers; the "reward" was to be given the correctly described object – green wood, green hide – to play with. Basically, the experiments were "uncontrolled," although systematic procedures were developed and followed carefully. By eschewing classical operant conditioning, Pepperberg (1999) succeeded for the first time in teaching a bird abstract concepts of color, shape, and material.

What Is the Scientists' qua Experimentalists' Image of What They Do?

The standard experimental paper within and without economics uses the following format in outline form: (1) state the theory; (2) implement it in a particular context (with "suitable" motivation in economics); (3) summarize the implications in one or more testable hypotheses; (4) describe the experimental design; (5) present the data and results of the hypothesis tests; and (6) conclude that the experiments either reject or fail to reject the theoretical hypotheses. This format is shown in Figure 13.1. In the case in which we have two or more competing theories and corresponding hypotheses, the researcher offers a conclusion as to which one is supported by the data using some measure of statistical distance between the data and each of the predictive hypotheses, reporting which distance is statistically the shortest.

Suppes (1969; also see Mayo, 1996, chapter 5) has observed that there exists a hierarchy of models behind the process in Figure 13.1. The primary model using the theory is contained in (1) and (2), which generate particular topical hypotheses that address primary questions. Experimental models are contained in (3) and (4). These serve to link the primary theory with data. Finally, we have data models, steps (5) and (6), which link operations on raw data (not the raw data themselves) to the testing of experimental hypotheses.

This process describes much of the rhetoric of science and reflects the self-image of scientists, but it does not adequately articulate what scientists actually do. Furthermore, the rhetoric does not constitute a viable, defensible, and coherent methodology. But what we actually do, I believe, is highly defensible and, on the whole, positively impacts what we think we know from experiment. Implicitly, as experimentalists, we understand that every step, (1) through (6), in the preceding process is subject to judgments, to learning from past experiments, to our knowledge of protocols and technique, and

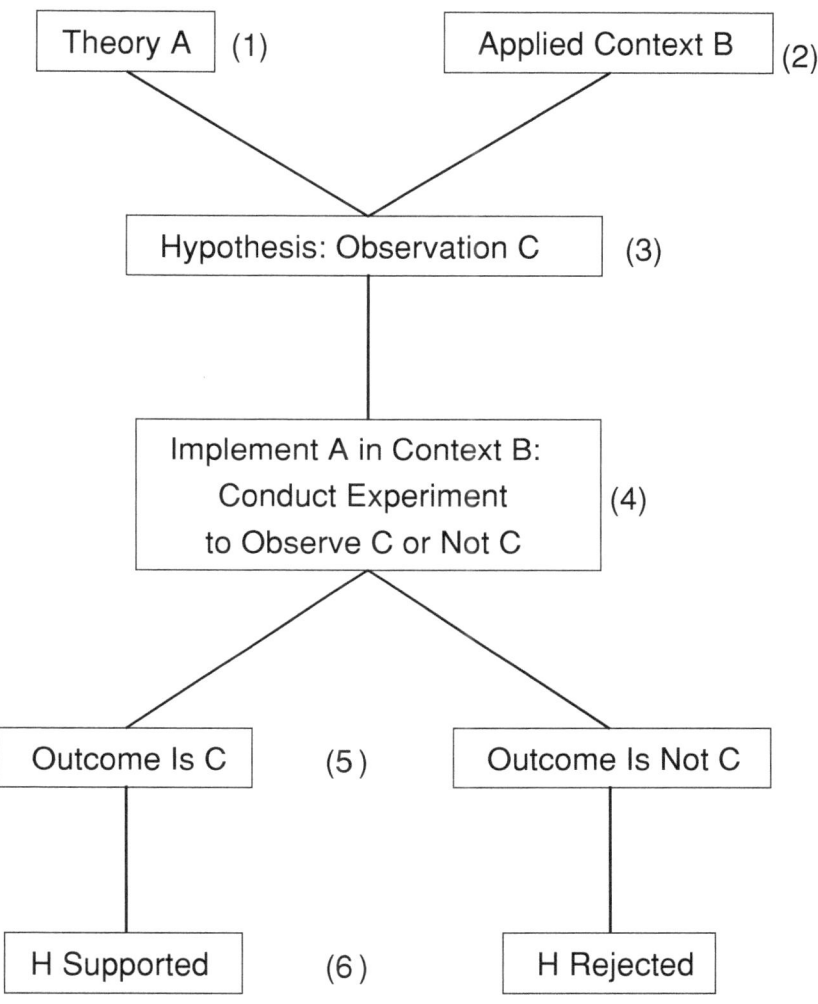

Figure 13.1. What Is the Scientist's Image of What He Does?

to error. This is reflected in what we do as a professional community, if not in what we say about what we do in the standard scientific paper.

Auxiliaries and the Ambiguity of Rejecting the "Test" Hypothesis

The problem with the preceding image is known as the "Duhem-Quine (D-Q) problem": Experimental results always present a joint test of the

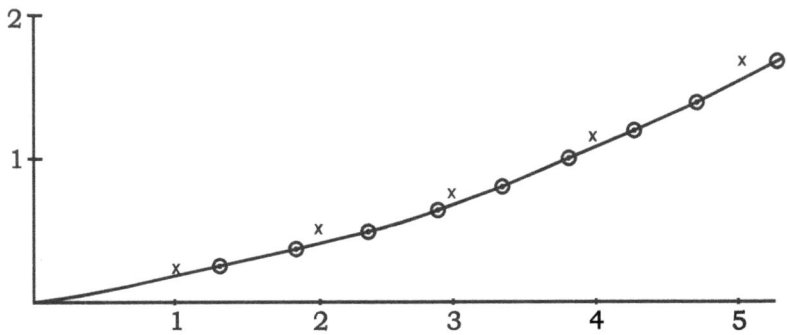

Figure 13.2. Einstein's Reconstruction of Kaufmann's Figure Using Data from Beck (1989).

theory (however well articulated, formally) that motivated the test, and all the things you had to do to implement the test.[4] Thus, if theoretical hypothesis H is implemented with context-specific auxiliary hypotheses required to make the test operational, $A_1, A_2, \ldots, A_n$; then it is (H$|A_1, A_2, \ldots, A_n$) that implies observation C. If you observe not-C, this can be because any of the antecedents (H; $A_1, \ldots, A_n$) can represent what is falsified. Thus, the interpretation of observations in relation to a theoretical hypothesis is inherently and inescapably ambiguous, contrary to our accustomed rhetoric.

The reality of what we do, and indeed must do, is implied by the truth that "No theory is or can be killed by an observation. Theories can always be rescued by auxiliary hypotheses" (Lakatos, 1978, vol. 1, p. 34).

A D-Q Example from Physics

Here is a famous historical example from physics. In 1905, Kaufmann (cited in Fölsing, 1997, p. 205), a very accomplished experimentalist,[5] published a paper "falsifying" Einstein's special theory of relativity the same year in which the latter was published (Einstein, 1905; 1989). Subsequently, Einstein (1907; 1989) in a review paper reproduced Kaufmann's Figure 2, shown here as Figure 13.2, commenting that

The little crosses above the curve (Kaufmann's) indicate the curve calculated according to the theory of relativity. In view of the difficulties involved in the experiment

---

[4] Good discussions of the D-Q thesis and its relevance for experimental economics can be found in Soberg (2005) and Guala (2005, pp. 54–61 and passim). Soberg provides interesting theoretical results showing how the process of replication can be used, in the limit, to eliminate inductively clusters of alternative hypotheses and lend increasing weight to the conclusion that the theory itself is in doubt.

[5] In 1902, he showed that the mass of an electron increases with its velocity, a property that led to Einstein's famous equation expressing the energy equivalence of mass.

one would be inclined to consider the agreement as satisfactory. However, the deviations are systematic and considerably beyond the limits of error of Kaufmann's experiment. That the calculations of Mr. Kaufmann are error-free is shown by the fact that, using another method of calculation, Mr. Planck arrived at results that are in full agreement with those of Mr. Kaufmann. Only after a more diverse body of observations becomes available will it be possible to decide with confidence whether the systematic deviations are due to a not yet recognized source of errors or to the circumstance that the foundations of the theory of relativity do not correspond to the facts (Einstein, 1907; 1989, p. 283).

Kaufmann was testing the hypothesis that (H|A) implies C, where H was an implication of special relativity theory, A was the loose auxiliary hypothesis that his context-specific test and enabling experimental apparatus was without "a not yet recognized source of errors," C was the theoretical curve indicated by the "little crosses," and not-C was Kaufmann's empirical curve in Figure 13.2 (Einstein's Figure 2).

In effect, Einstein's comment states that either of the antecedents (H|A) represents what is falsified by Kaufmann's results. Others, such as Planck and Kaufmann himself, however, recognized that the observation might conceivably be in error (Fölsing, 1997, p. 205). In less than a year, Hans Bucherer (see Fölsing, 1997, p. 207) showed that there indeed had been a "problem" with Kaufmann's experiments and obtained new results supporting Einstein's theory.

There is an important lesson in this example if we develop it a little more fully. Suppose Bucherer's experiments had not changed Kaufmann's results enough to change the conclusion – a common outcome when any experiment is challenged and a new test is constructed.[6] Einstein could still have argued that there may be "a not yet recognized source of errors." Hence, the implication is that H is either not falsifiable or predictive, for the same argument can be made after each new test in which the new results are outside the range of error for the apparatus. Recall that the deviations were alleged to be "considerably beyond the limits of error of Kaufmann's experiment." But here "error" is used in the sense of internal variations arising from the apparatus and procedure, not in the sense that there is a problem with the apparatus/procedure itself. We can go still further in explicating the problem of testing H conditional on any A. The key is to note in this example the strong

---

[6] Thus, many retests of ultimatum game behavior have been conducted using increased payoffs (Hoffman et al., 1996a, and List and Cherry, 2002, are just two examples; see Camerer 2003, pp. 60–2, for a summary). Although payoffs affect certain minor features of the results, the overall conclusions have been impervious to stakes: Responders reject "low" offers in defiance of game-theoretic predictions based on the abstract conception of a single-play game.

dependence of any test outcome on the state of experimental knowledge: Bucherer found a way to "improve" Kaufmann's experimental technique so as to rescue Einstein's "prediction." But the predictive content of H (and therefore of the special theory) was inextricably bound up with A. Einstein's theory did not embrace any of the elements of Kaufmann's (or Bucherer's) apparatus: A is based on experimental knowledge of testing procedures and operations in the physics laboratory and has nothing to do with the theory of relativity, a separate and distinct body of knowledge on which all tests are implicitly conditioned. In this example, ecological rationality is at work in the community of physicists who discuss results, new tests, and problems with the theory, and end up with a local topically satisfactory resolution of the discrepancy between predictions and observations. Other topical tests may or may not fail to reject the theory.

There is nothing about the results of any particular "controlled experiment" that require us to believe them (as in the previous quote from Houston Smith). Specific experimental results are only part of a conversation that must be ongoing and open to new experimental knowledge and technique. Moreover, it is commonplace for initial path-breaking and informative experiments such as Kaufmann's, in retrospect, to have been "uncontrolled" in the light of new experimental knowledge.

A Proposition and Some Economics Examples

All of us have encountered a common objection to economics experiments: The payoffs are too small. This objection is one of several principal issues in a target article by Hertwig and Ortmann (2001), with comments by thirty-four experimental psychologists and economists. This objection sometimes is packaged with an elaboration to the effect that economic or game theory is about situations that involve large stakes, and you "can't" study these in the laboratory. (Actually, of course, you can, but funding is not readily available.)

A theme of discussion in experimental economics that is sometimes linked with the adequacy of stakes is whether or not propositions supported or not supported in laboratory experiments have the same implications in the field. The "field" here refers to situations in the economy (sometimes referred to as "the real world" in pejorative contrast to laboratory) that may serve to motivate theoretical models that have been tested in laboratory experiments. Thus, as we have seen in Chapter 6, Vickrey (1961) proposed equilibrium models of auctions the implications of which were subsequently tested in Coppinger et al. (1980) and Cox et al. (1988), and by many others, both in laboratory and field environments. I want to emphasize yet again (see, for example, Smith, 1982a, pp. 935–7) that the transferability of results from one environment (lab or field) to any other is a subject for comparative empirical studies. Laboratory or field observations obtained in one locality (university, business group, country, culture, and so on) may or may not be statistically distinguishable from results obtained in another locality. Thus, the rhetorical question, "What do laboratory experiments tell us about the real world?" (Levitt and List, 2007), after meaningful translation,[7] has a simple and transparent answer: nothing. As indicated in this chapter, hypotheses derived from theory have contextual, topical implications for observations (data), not the reverse. You can't go from data to generalizations (you can go from data to summaries of results from different localities, which is not the same thing) and thence to other data sets. It is of course very informative to conduct tests of theory in a range of circumstances and environments, as we have all done, to see which predictive results hold or fail to hold across different environments. But all the action is in the empirical findings, as empiricists probe what might be different about two specific test environments that accounts for any specific difference in what is observed.

Suppose, therefore, that we have the following:

H (from theory): Subjects will choose the equilibrium outcome (for example, Nash or subgame perfect).
A (auxiliary hypothesis): Payoffs are adequately motivating to test H.

Proposition 2. Suppose a specific rigorous test rejects (H|A), and someone (for example, T) protests that what must be rejected is A, not H. Let E replicate the original experiment with an n-fold increase in payoffs. There are only two outcomes and corresponding interpretations, neither of which is comforting to the preceding rhetorical image of science:

- The test outcome is negative. Then T can imagine a still larger payoff multiple, $N > n$, and argue for rejecting A, not H. But this implies that H cannot be falsified.
- After repeated increases in payoffs, the test outcome is positive for some $N \geqq n^*$. Then H has no predictive content. E, with no guidelines from the theory, has searched for and discovered an empirical payoff multiple, n^*, that "confirms" the theory, but n^* is an extratheoretical property of considerations outside H and the theory that was being

---

[7] Translated, the coherent scientific question posed is, "What does one set of data tell us about a different set of data, whatever their context, lab or field?"

tested. Finding this multiple is not something for T or E to crow about, but rather an event that should send T or E back to his desk. The theory is inadequately specified to embrace the observations from all the experiments.

Proposition 2 holds independent of any of the following considerations:

- How well articulated, rigorous, or formal is the theory? Game theory in no sense constitutes an exception.
- How effective is the experimental design in reducing the number of auxiliary hypotheses? It only takes one to create the problem.
- What is the character or nature of the auxiliary hypothesis? A can be any feature of the test environment that affects the outcome.

In experimental economics, reward adequacy is just one of a standard litany of objections to experiments in general, and to many experiments in particular. Here are four additional categories in this litany:

- *Subjects must be sophisticated.* The standard claim is that undergraduates are not sophisticated enough. They are not out there participating in the "real world." In the "real world," where the stakes are large, such as in the FCC spectrum rights auctions, bidders use game theorists as consultants. This is a very convenient and self-serving argument for anyone who does not want to come to terms with the problem of auxiliary hypotheses in interpreting test results. Of course, those who deny the relevance of all experiments and work only with field observations face this problem in spades.
- *Subjects need an opportunity to learn.* This is a common response from both experimentalists and theorists when you report the results of single-play games in which "too many" people cooperate. The usual proposed "fix" is to do repeat single protocols in which distinct pairs play on each trial and apply a model of learning to play noncooperatively. But there are many unanswered questions implicit in this auxiliary hypothesis: Since repeat single protocols require large groups of subjects (twenty subjects to produce a ten-trial sequence), have any of these games been run long enough to measure adequately the extent of learning? In single-play two-person anonymous trust games, data have been reported showing that group size matters; that is, it makes a difference whether you are running twelve subjects (six pairs) or twenty-four subjects (twelve pairs) simultaneously in a session (Burnham et al., 2000). In the larger groups, pairs are less trusting than in the small groups – which perhaps is not too surprising. But in repeat

single, larger groups are needed for longer trial sequences. Hence, learning and group size as auxiliary hypotheses lose independence, and we have knotty new problems of complex joint hypothesis testing. The techniques, procedures, and protocol tests we fashion for solving such problems are the sources of our experimental knowledge. All testing depends on, and is limited by, the state of experimental knowledge at any given time.

- *Instructions are adequate* (or decisions are robust with respect to instructions, and so on). What does it mean for instruction to be adequate? Clear? If so, clear about what? What does it mean to say that subjects understand the instructions? (See the Appendix to Chapter 10.) Usually this is interpreted to mean that subjects can perform the task, which is judged by how they answer questions about what they are to do. In two-person interactions, instructions often matter so much that they must be considered a (powerful) treatment.[8] As indicated in Chapter 10, instructions are important because they define context, and context can and often matters. Ultimatum and dictator game experiments yield statistically and economically significant differences in results due to differing instructions and protocols.

- *Cooperation in single-play games is due to social preferences.* As we have seen in the discussion of other-regarding preferences in Chapters 10 and 11, when we get a prediction failure in two-person single-play games, based on the dominant choice hypothesis, H, there is ambiguity about whether we reject H or A, where A is the hypothesis that the subjects "think about" (treat) single-play games in the same way as the standard game-theoretic model uncontaminated by their experience in repeat interaction in society. If we reject H in favor of not-H – preferences are other-regarding, as discussed in Chapters 10 and 11 – this implies that we must accept A, but if the problem is with A, then we cannot reject H in favor of not-H.[9] I have already noted in Chapter 10 that tests of the effect of social context (earned property rights, exchange, social distance, earned endowments) provide an independent challenge to social preference models.

---

[8] Thus, Hertwig and Ortmann (2001, section 2) argue that scripts (instructions) are important for replication, and that "ad-libbing" should be avoided.

[9] In the ultimatum game, if you accept A, then you would feel comfortable saying, "The responder thus faces a conflict in case of low offers between his economic self-interest, which encourages him to accept the offer, and his fairness goals, which drive him toward rejecting it" (Knoch et al., 2006, p. 829). But if you do not accept A, you might say it in this way: The responder thus faces a conflict in case of low offers between choosing according to dominance, which encourages him to accept the offer, and his social reputation (reciprocity) goals, which drive him toward rejecting it.

The Methodology of Positive Economics

There is a methodological perspective associated with Milton Friedman (1953) that fails to provide an adequate foundation for experimental (field or laboratory) science, but which influenced economists for decades and still has some currency. Friedman's proposition is that the truth value of a theory should be judged by its testable and tested predictions, not by its assumptions. This proposition is deficient for at least three reasons:

- If a theory fails a test, we should ask why, not always just move on to seek funding for a different new project; obviously, one or more of its assumptions may be wrong, and it behooves us to design experiments that will probe conjectures about which assumptions failed. Thus, if first-price auction theory fails a test, is it a consequence of the failure of one of the axioms of expected utility theory, for example, the compound lottery axiom? If a subgame perfect equilibrium prediction fails, does the theory fail because the subjects do not satisfy the assumption that the agents choose dominant strategies? Or did the subjects fail to use backward induction? Or was it none of the above because the theory was irrelevant to how some motivated agents solve such problems in a world of bilateral reciprocity in social exchange? When a theory fails, there is no more important question than to ask what it is about the theory that has failed.
- Theories may have the if-and-only-if property that one (or more) of its "assumptions" can be derived as implication(s) from one (or more) of its "results."
- If a theory passes one or more tests, this need not be because its assumptions are correct. A subject may choose the subgame perfect equilibrium because she believes she is paired with a person who is not trustworthy, and not because she always chooses dominant strategies, assumes that others always so choose, or that this is common knowledge. This is why you are not done when a theory is corroborated by a test. You have only examined one point in the parameter space of payoffs, subjects, tree structure, and such. Your results may be a freak accident of nature, due to a complex of suitabilities, or in any case may have other explanations.

In View of Proposition 2, What Are Experimentalists and Theorists to Do?

Consider first the example in which we have a possible failure of A: Rewards are adequate to motivate subjects. Experimentalists should do what comes naturally, namely, do new experiments that increase rewards and/or lower decision cost by simplifying experiment procedures. The literature is full of examples (for surveys and interpretations, see Smith and Walker, 1993a, 1993b; Camerer and Hogarth, 1999; Holt and Laury, 2002; Harrison et al., 2005).

Theorists should ask whether the theory is extendable to include A, so that the effect of payoffs is explicitly modeled. As I have emphasized, it is something of a minor scandal when economists – whose models predict the optimal outcome independently of payoff levels however gently rounded is the payoff in the neighborhood of the optimum – object to an experiment because the payoffs are inadequate. What is adequate? Why are payoff inadequacies a complaint rather than a spur to better and more relevant modeling of payoffs? (See Chapter 8; Smith and Szidarovzsky, 2004.)

Generally, both groups must be aware that for any H, any A, and any experiment, one can say that if the outcome of the experiment rejects (H|A), then both should assume that either H or A may be false, which is an obvious corollary to Proposition 2. This was Einstein's position concerning the Kaufmann results, and was the correct position, whatever later tests might show. After every test, either the theory (leading to H) is in error or the circumstances of the test, A, are in error.

Experimental Knowledge Drives Experimental Method

Philosophers have written eloquently and argued intently over the implications of D-Q and related issues for the interpretation of science. Karl Popper tried to demarcate science from pseudoscience with the basic rule requiring the scientist to specify in advance the conditions under which he will give up his theory. This is a variation on the idea of a so-called crucial experiment, a concept that cannot withstand examination (Lakatos, 1978, vol. 1, pp. 3–4, 146–8), as is evident from our Proposition 2.

The failure of all programs attempting to articulate a defensible rational science of scientific method has bred postmodern negative reactions to science and scientists. From that failure, none of these alternatives follow: the excesses of postmodernism, the irrationality of science, or the nonrationality of science. Postmodernism misses much of what is most important in the working lives of all practitioners. Yes, Popper was wrong in thinking he could demarcate science from pseudoscience by an exercise in logic, but that does not imply that the Popperian falsification rule failed as a milestone contribution to the scientific conversation; nor does it mean that "anything goes" (Feyeraband, 1975). Rather, what one can say is much less open-ended: Anything goes only insofar as what can be decisively concluded about constructive rationality in science. For the scientific enterprise is also about ecological rationality in science, which is about discovery, about probes into Max Born's "jungle," about thinking outside the box, and, as I shall argue later, about the technology of observation in science that renders obsolete long-standing D-Q problems while introducing new ones for a time. That dynamism is what is missing in the static logic of hypothesis testing.

You do not have to know anything about D-Q and statements like Proposition 2 to appreciate that the results of an experiment nearly always suggest new questions precisely because the interpretation of results in terms of the theory is commonly ambiguous. This ambiguity is reflected in the discussion whenever new results are presented at seminars and conferences. Without ambiguity, there would be nothing to discuss. What is the response to new results? Invariably, if it is a matter of consequence, experimentalists design new experiments with the intention of confronting the issues in the controversy and in the conflicting views that have arisen in interpreting the previous results. This leads to new experimental knowledge of how results are influenced in economics, or not, by changes in procedures, context, instructions, and control protocols. The new knowledge may include new techniques that have application to areas other than the initiating circumstance. This ecological process is driven by the D-Q problem, but practitioners need have no knowledge of the philosophy of science literature to take the right next steps in the laboratory. Parallel social processes apply to other sciences.

This is because the theory or primary model that motivates the questions tells you nothing definitive, or even very useful, about precisely how to construct tests. Tests are based on extratheoretical intuition, conjectures, and experiential knowledge of procedures. In economics, the context, subjects, instructions, parameterization, and such are determined largely outside the theory, and their evolution constitutes the experimental knowledge that defines our methodology. The forms taken by the totality of individual research testing programs cannot be accurately described in terms of the rhetoric of falsification, no matter how much we speak of the need for theories to be falsifiable, the tests discriminating or "crucial," and the results robust.

Whenever negative experimental results threaten perceived important new theory, the resulting controversies galvanize experimentalists everywhere into a search for different or better tests – tests that examine the robustness of the original results. Hence, Kaufmann's experimental apparatus was greatly improved by Bucherer, although there was no question about Kaufmann's competence in laboratory technique. The point is that with escalated personal incentives, and a fresh perspective, scientists found improved techniques. This scenario is common as new challenges bring forth renewed effort. This process generates the constantly changing body of experimental knowledge whose insights in solving one problem often carry over to applications in many others.

Just as often, experimental knowledge is generated from curiosity about the properties of phenomena that we observe long before a body of theory exists that deals specifically with the phenomenon at issue. An example in experimental economics is the continuous double auction trading mechanism discussed in Part II.

An example from physics is Brownian motion, discovered in 1827 by the botanist Robert Brown, who first observed the unpredictable motion of small particles suspended in a fluid. This motion is what keeps them from sinking under gravity. This was seventy-eight years before Einstein's famous paper (one of three) in 1905 developed the molecular kinetic theory that enabled subsequent research to account for the phenomenon, although he was not motivated by those long-familiar observations (see Mayo, 1996, chapter 7, for references and details). The long "inquiry into the cause of Brownian motion has been a story of hundreds of experiments... [testing hypotheses attributing the motion]... either to the nature of the particle studied or to the various factors external to the liquid medium..." (ibid., pp. 217–18). The essential point is "that these early experiments on the possible cause of Brownian motion were not testing any full-fledged theories. Indeed it was not yet known whether Brownian motion would turn out to be a problem in chemistry, biology, physics, or something else. Nevertheless, a lot of information was turned up and put to good use by those later researchers who studied their Brownian motion experimental kits" (ibid., p. 240). The problem was finally solved by drawing on the extensive bag of experimental tricks, tools, and past mistakes that constitute "a log of the extant experimental knowledge of the phenomena in question" (ibid., p. 240).

Again, concerning the maze of auxiliary hypotheses, "... the infinitely many alternatives really fall into a few categories. Experimental methods (for answering new questions) coupled with experimental knowledge (for using techniques and information already learned) enable local questions to be split off and answered" (ibid., p. 242).

The bottom line is that good-enough solutions emerge to the baffling infinity of possibilities, as new measuring systems emerge, experimental tool kits are updated, and understanding is sharpened.

This conclusion also goes far to writing the history of experimental economics and its many detailed encounters with data and the inevitable ambiguity of subsequent interpretation. And in most cases, the jury continues in session on whether we are dealing with a problem in psychology (perception), economics (opportunity cost and strategy), social psychology (equality, equity, or reciprocity), or neuroscience (functional imaging and brain modeling). So be it.

The Machine Builders

Mayo's (1996) discussion and examples of experimental knowledge leaves unexamined the question of how technology impacts the experimentalist's tool kit. Important among the heroes of science are not only the theorists and the experimentalists, but the unsung tinkerers, mechanics, inventors, and engineers who create the new generation of machines that make obsolete yesterday's observations and heated arguments over whether it is T or A that has been falsified. Scientists, of course, are sometimes a part of this creative destruction, but what is remembered in academic recognition is the new scientific knowledge they enabled, not the instruments they invented that made possible the new knowledge. Michael Faraday, "one of the greatest physicists of all time" (Segrè, 1984, p. 134), had no formal scientific education. He was a bookbinder who had the habit of reading the books that he bound. He was preeminently a tinkerer for whom "some pieces of wood, some wire and some pieces of iron seemed to suffice him for making the greatest discoveries" (quoted from a letter by Helmholz, in Segrè, 1984, p. 140). Yet he revolutionized how physicists thought about electromagnetic phenomena, invented the concept of lines of force (fields), and inspired Maxwell's theoretical contributions. "He writes many times that he must experience new phenomena by repeating the experiments, and that reading is totally insufficient for him" (ibid., p. 141). This is what I mean, herein, when I use the term "experimental knowledge." It is what Mayo (1996) is talking about.

Technology and Science

With the first moon landing, theories of the origin and composition of our lunar satellite, contingent on the state of existing indirect evidence, were upstaged and rendered obsolete by direct observation; the first Saturn probe sent theorists back to their desks and computers to reevaluate the planet's mysterious rings, whose layered richness had not been anticipated.

Similar experiences have followed the results of ice core sampling in Greenland and instrumentation for mapping the genome of any species. Galileo's primitive telescope opened a startling window on the solar system, as does Roger Angel's multiple-mirror light-gathering machines (created under the Arizona football stadium) that opened the mind to a view of the structure of the universe.[10] The technique of tree ring dating, invented by an Arizona astronomer, has revolutionized the interpretation of archeological data from the last five thousand years.

Yesterday's reductionisms, shunned by mainstream "practical" scientists, create the demand for new subsurface observations, and hence for the machines that can deliver answers to entirely new questions. Each new machine – microscope, telescope, Teletype, computer, the Internet, fMRI – changes the way teams of users think about their science. The host of auxiliary hypotheses needed to give meaning to theory in the context of remote and indirect observations (inferring the structure of Saturn's ring from earth-based telescopes) are suddenly made irrelevant by deep and more direct observations of the underlying phenomena (computer-enhanced photos). It's the machines that drive the new theory, hypotheses, and testing programs that take you from atoms, to protons, to quarks. Yet with each new advance comes a blizzard of auxiliary hypotheses, all handcuffed to new theory, giving expression to new controversies seeking to rescue T and reject A, or to accept A and reject T.

Technology and Experimental Economics

When Arlington Williams (1980) created the first electronic double auction software program for market experiments in 1976, all of us thought we were simply making it easier to run experiments, collect more-accurate data, observe longer time series, facilitate data analysis, and so on. What were computerized were the procedures, recording, and accounting that heretofore had been done manually. No one was anticipating how this tool might impact and change the way we thought about the nature of doing experiments. But with each new electronic experiment, we were "learning" (impacted by, but without conscious awareness) the fact that traders could be matched at essentially zero cost, that the set of feasible rules that could be considered was no longer restricted by costly forms of implementation and monitoring, that vastly larger message spaces could be accommodated,

[10] For a brief summary of the impact of past, current, and likely future effects of rapid change in optical and infrared (terrestrial and space) telescopes on astronomy, see Angel (2001).

and that optimization algorithms could now be applied to the messages to define new electronic market forms for trading energy, air emission permits, water, and the services of other network supply chains. In short, the transactions cost of running experimental markets became minuscule in comparison with the preelectronic days, and this opened up new directions that previously had been unthinkable.

The concept of smart computer-assisted markets appeared in the early 1980s (Rassenti, 1981; Rassenti et al., 1982), extended conceptually to electric power and gas pipelines in the late 1980s (Rassenti and Smith, 1986; McCabe et al., 1989), with practical applications to electric power networks and the trading of emission permits across time and regions in the 1990s. Laboratory experiments followed by field testing led to implementation in the form of new markets. These developments continue as a major new effort in which the laboratory is used as a test bed for measuring, modifying, and further testing the performance characteristics of new institutional forms in the field.

What is called e-commerce has spawned a rush to reproduce on the Internet the auction, retailing, and other trading systems people know about from past experience. But the new experience of being able to match traders at practically zero cost is sure to change how people think about trade and commerce, and ultimately this will change the very nature of trading institutions. In the short run, of course, efforts to protect existing institutions and stakeholders will spawn efforts to shield them from entry by deliberately introducing cost barriers, but in the long run, these efforts will be increasingly uneconomical and ultimately will fail.

Neuroscience carries the vision of changing the experimental study of individual two-person interactive, and indeed all other, decision making. The neural correlates of decision making, how it is affected by rewards, cognitive constraints, working memory, repeat experience, and a host of factors that in the past we could neither control nor observe can in the future be expected to become an integral part of the way we think about and model decision making. Models of decision, now driven by game and utility theory and based on trivial, patently false models of mind must take account of new models of cognitive, calculation, and memory properties of mental function that are accessible to more direct observational interpretation. Game-theoretic models assume consciously calculating, rational mental processes, but models of mind include non-self-aware processes just as accessible to neural brain imaging as the conscious. For the first time, we may be able to give some observational content to the idea of "bounded rationality" (see Chapter 14).

In Conclusion

In principle, the D-Q problem is a barrier to any defensible notion of a rational science that selects theories by a logical process of confrontation with scientific evidence. There is no such thing as a science of the methodology of science, but this principle had to be discovered and the many efforts to rationalize the scientific process were important contributions to understanding. Moreover, the failed objective of this constructivist adventure is cause for joy, not despair. Failure is an important part of learning and error reduction; think how dull would be a life of science if, once we were trained, all we had to do was to turn on the computer of science, feed it the facts, and send its output to the printer.

My personal experience as an experimental economist since 1956 resonates well with Mayo's (1996) critique of Lakatos:

Lakatos, recall, gives up on justifying control; at best we decide – by appeal to convention – that the experiment is controlled. . . . I reject Lakatos and others' apprehension about experimental control. Happily, the image of experimental testing that gives these philosophers cold feet bears little resemblance to actual experimental learning. Literal control is not needed to correctly attribute experimental results (whether to affirm or deny a hypothesis). Enough experimental knowledge will do. Nor need it be assured that the various factors in the experimental context have no influence on the result in question – far from it. A more typical strategy is to learn enough about the type and extent of their influences and then estimate their likely effects in the given experiment (p. 240).

And, as in Polanyi (1962):

. . . personal knowledge in science is not made but discovered, and as such it claims to establish contact with reality beyond the clues on which it relies . . . [and] commits us . . . to a vision of reality. Of this responsibility we cannot divest ourselves by setting up criteria of . . . falsifiability, or testability. . . . For we live in it as in the garment of our own skin (p. 65).

I would add only that this vision also does not permit us to divest ourselves of the sociality of science. Controlled experiments, logic, and machines alone require us to believe nothing; what we believe must always pass through the filter of human sociality in science if it is ultimately to be credible.

FOURTEEN

Neuroeconomics

The Internal Order of the Mind

Psychology will be based on a new foundation, that of the necessary acquirement of each mental power and capacity by gradation.
Darwin (1859; 1979, p. 458)

Introduction

Neuroeconomics is concerned with the study of the connections between how the mind/brain works – the internal order of the mind – and behavior in (1) individual decision making, (2) social exchange, and (3) institutions such as markets – in short, the neurocorrelates of decision. The way I would define the working hypothesis of neuroeconomics is that the brain has evolved interdependent, adaptive mechanisms for each of these tasks involving experience, memory, perception, and personal tacit knowledge. This theme has been prominent for many years in evolutionary psychology, but new tools are now available. The tools include brain-imaging technologies, the existence of patients with localized brain lesions associated with specific loss of certain mental functions, and inherited variations in how brains work.

For example, in the spectrum of autism "disorders," the emergent ability of children around age four to infer what others are thinking from their words and actions is compromised (Baron-Cohen, 1995). The interpretation is that there is a development failure in the child's natural awareness of mental phenomena in other people that severely limits the normal development of their capacity for social exchange. The do's and don'ts in personal exchange behavior have to be learned the way you learn to read a language – by explicit instruction. Otherwise, high-functioning autistics may be unusually accomplished. The best-known such case is probably Grandin (1996), a fascinating report on how differently another mind can work. As you might expect from autistic children, compared with non-autistic children, in the ultimatum game there is a much higher proportion of offers of one or zero units. This is less the case for autistic adults, who have memorized some of the principles of human sociality (Hill and Sally, 2003).

Neuroeconomics is driven primarily by the machines for brain imaging, a technology sure to become ever more sophisticated that has much promise for changing the way we think about brain function, decision, and human sociality. It is likely to chart entirely new directions once the research community gets beyond trying to get better answers to the old questions. My discussion will be very brief, largely a commentary, not a survey, and will ask: What are the exciting questions that are open for specialized examination by neuroeconomic methods? (See Camerer et al., 2005, for an excellent survey of the state of development and learning in neuroeconomics.)

Neuroeconomics, like mathematical economics and econometrics decades before it, is controversial and its worth is often questioned. For many, it is too reductionist and oblique to the main questions in economics to be of lasting significance. In some interpretations, neuroeconomics is said to be important in proportion to its challenges to the validity of many of the propositions thought to have been central to economic theory. But those propositions have long been subjected to qualifications stemming from experimental and empirical studies quite apart from neuroeconomic developments. Neuroeconomcs will not achieve distinction in a focus confined to correcting the "errors" believed to pervade professional economics of the past, an exercise of interest to a narrow few. Nor will notoriety likely stem from better answers to the traditional questions. Rather, neuroeconomic achievement more likely will be determined by its ability to bring a new perspective and understanding to the examination of important economic questions that have been intractable for, or beyond the reach of, traditional economics. Initially new tools tend naturally to be applied to the old questions, but as I indicated in Chapter 13, their ultimate importance emerges when those tools change how people think about their subject matter, ask new questions, and pursue answers that would not have been feasible before the innovation. Neuroeconomics has this enormous nonstandard potential, but it is far too soon to judge how effective it will be in creating new pathways of comprehension.

Individual Decision Making

Decision making has drawn the attention of neuroscientists who study the deviant behavior of neurological patients with specific brain lesions, such as front lobe (ventromedial prefrontal) damage. It has long been known that such patients are challenged by tasks involving planning and coordination over time, although they score quite normally on battery after battery of standard psychological tests (Damasio, 1994). Such tests measure what is on the surface of human memory and perception, but tacit knowledge appears to be beyond their reach. A landmark experimental study of such patients, compared with normal (nonpatient) controls in individual decision making under uncertainty, is that of Bechara et al. (1997). The "Iowa Gambling Task" starts with (fictional) endowments of $2,000; each subject on each trial draws a card from one of four freely chosen decks (A, B, C, D). In decks (A, B), each card has a payoff value of $50, whereas in decks (C, D), each is worth $100. Also, the $100 decks contain occasional large negative payoff cards, while the $50 decks have much smaller negative payoff cards. All this must be learned from single-card draws in a sequence of trials, with a running tally of cumulative payoff value. A subject performs much better by learning to avoid the $100 decks in favor of the $50 decks. By period 60, normal control subjects have learned to draw only from the $50 (C, D) decks, while the brain-damaged subjects continue to sample disadvantageously from the $100 (A, B) decks. Moreover, the control subjects shift to the (C, D) decks before they are able to articulate why, in periodic questioning. The brain responds autonomically with the self-conscious mind reporting out later. Also, the normal subjects preregister emotional reactions to the (A, B) decks as measured by real-time skin conductivity test (SCT) readings before they shift to the (C, D) decks. But the brain-damaged patients tend verbally to rationalize continued sampling of the (A, B) decks, and some types (with damaged amygdala) register no SCT response. Results consistent with those of Bechara et al. (1997) have been reported by Goel et al. (1997), who study patient performance in a complex financial planning task.

This one study neatly demonstrates the human decision-making process involving subconscious action-based decision, integrated with emotional support – what Gazzaniga (1998) calls the mind's past – and finally, with the work done, the mind arriving with a verbal articulation and rationalization of what has occurred, but believing all the while that it has been in full control.

Over fifty years ago, experiments with animal behavior demonstrated that the motivation driving choice was based on relative or foregone reward – opportunity cost – and not on an absolute scale of values generated by the brain. Thus Zeaman (1949) reported experiments in which rats were trained to run for a large reward-motivated goal. When shifted to a small reward, the rats responded by running more slowly than they would to the small reward only. A control group began with a small reward and shifted to a large one, and these rats immediately ran faster than if the large reward alone was applied. Monkeys similarly respond to comparisons of differential rewards. It is now established that orbital frontal cortex (just above the eyes) neuron activity in monkeys enables them to discriminate between rewards that are directly related to the animals' relative, as distinct from absolute, preference among rewards such as cereal, apples, and raisins (in order of increasing preference in monkeys) (Tremblay and Schultz, 1999).

Thus, suppose A is preferred to B is preferred to C based on choice responses. Then neuronal activity is greater for A than for B when the subject is viewing A and B, and similarly for B and C when comparing B and C. But the activity associated with B is much greater when compared to C than when it is compared to A. This is contrary to what one would expect to observe if A, B, and C are encoded on a fixed properties scale rather than a relative scale (ibid., p. 706). These studies have a parallel significance for humans. As already noted, prospect theory proposes that the evaluation of a gamble depends not on the total asset position but focuses myopically on the opportunity cost, gain or loss, relative to one's current asset position. There is also asymmetry – the effect of a loss looms larger than the effect of gain of the same magnitude (Kahneman and Tversky, 1979). Mellers et al. (1999) have shown that the emotional response to the outcome of a gamble depends on the perceived value and likelihood of the outcome and on the foregone outcome. It feels better (less bad) to receive $0 from a gamble when you forgo +$10 than when you forgo +$90. (They use the equivalent term "counterfactual" rather than "opportunity cost" to refer to the alternative that might have prevailed.) Thus, our ability to form opportunity cost comparisons for decision receives important neurophysiologic support from our emotional circuitry. This is just one of the many mechanisms in the brain that provide autonomic decision skills that are not mindfully constructivist and that underscore ecologically rational personal knowledge.

Breiter et al. (2001) use these same principles in the design of an fMRI study of human hemodynamic responses to both the expectation and experience of monetary gains and losses under uncertainty. They observed significant activation responses in the amygdala and orbital gyrus, with both activations increasing with the expected value of the gamble. There was also some evidence that the right hemisphere is predominantly active for gains and the left for losses – a particularly interesting possibility inviting deeper examination, perhaps by imaging split-brain subjects in the same task (also see K. Smith et al., 2002).

Rewards and the Brain

The effect of paying subjects is informed by Thut et al. (1997), who compare brain activation under actual monetary rewards with the feedback of an "OK" reinforcement in a dichotomous choice task. The monetary rewards yielded a significantly higher activation of the orbit frontal cortex and other related brain areas. (See Schultz, 2000, 2002. Siegel, 1959, 1961, documents early comparisons showing the significant effect of monetary rewards in predicting Bernoulli trials choice.) The human brain acquired its reward reinforcement system for food, drink, ornaments, and other items of cultural value long before money was introduced in exchange. Our brains appear to have adapted to money, as an object of value, or "pleasure," by simply treating it like another "commodity," latching on to the older receptors and reinforcement circuitry. To the brain, money is like food, ornaments, and recreational drugs, lending some credibility to those economists who included money in the utility function.

Strategic Interaction: Moves, Intentions, and Mind Reading

The neural correlates of individual decision making were extended by McCabe et al. (2001b) to an fMRI study of behavior in two-person strategic interactions in extensive form trust games like those in Chapter 12. Many of the same game trees were first pretested behaviorally to establish time lines and typical behavioral responses, as a prelude to the fMRI study, and reported in Corelli et al. (2000). The prior hypothesis, derived from reciprocity theory, the theory-of-mind literature, and supported by imaging results from individual studies of cued thought processes (Fletcher et al., 1995), was that cooperators would show greater activation in the prefrontal cortex (specifically BA-8) and supporting circuitry than did noncooperators. The control, for comparison with the mental processes used when a subject is playing against a human, is for the subject to play against a computer knowing the programmed response probabilities and therefore in principle having no need to interpret moves as intentions. This, of course, assumes that subjects think about the task as we do in constructivist modeling, a topic I visited in the Appendix to Chapter 10; it is of interest that McCabe et al. (2001b) imaged one subject who cooperated when matched with the computer, as well as when he was matched with a human. The predicted activations were significantly greater, relative to controls, for cooperators than for noncooperators and are consistent with the reciprocity interpretation of behavior discussed in Chapters 11 and 12, in which reading intentions is integral to predicting cooperative responses.

What Are the Neuroeconomics Questions?

The essential feature of markets is that knowledge in the sense of knowing how to function is fundamentally private, dispersed, and asymmetric.

As we have seen in Chapter 4, experiments corroborate that under these conditions and with repeat interaction in trade, prices and exchange volumes converge to equilibrium predictions. This was a stunning discovery, but no victory for economic theory, which has never solved the dynamic information and convergence-to-equilibrium problem, although the Hurwicz program made important contributions, as I have indicated in Chapter 5. Moreover, an important attempt to solve it with a complete information game-theoretic analysis of the double auction process encountered an end-game problem that prevented convergence. And the deficiencies were at the heart of the game-theoretic formulation. We seem to be up against the ultimate limits of the current game-theoretic tool kit (Wilson, 1987, p. 411).

How can it be that naïve cash-motivated subjects solve this problem so effortlessly, and have no idea of what they have wrought, while theorists have found it too hard and are all too cognizant of their bounded rationality in attempting to solve it?

Here is a summary of what I think we have learned:

Market institutions define the language of trade and state the rules providing algorithms that carry agent messages into outcomes.

Based on these rules, the social brains of subject agents provide algorithms that process incoming market-trading information, and respond with messages that over time converge to equilibrium outcomes best for the individual and the group. What we observe is the outcome of an interaction between the institutional and mental algorithms.

The mental algorithms are not accessible to our explanatory and articulation skills, nor are the rule change dynamics of institutions as they have evolved over time.

Two of the key questions for understanding the demonstrated high performance of market institutions are:

- What are the neuroeconomic foundations of these algorithmic mental processes?
- Do they, for example, resemble language computational processes?

Utility was an imaginary construct about subjective value; whatever it was supposed to measure, you maximized it, but the attempt to identify consistent and reliable measures of utility have persistently failed. Conceptually, utility was about proximate, not ultimate, cause. Moreover, the ordinal revolution showed that in consumer choice, we did not need it – we thought that we needed utility only to deal with uncertainty, risk aversion, and optimal diversification. But what we observe in the brain are centers that respond hemodynamically to reward, and behavior responds to reward. Although hypothetical and "OK" rewards activate these brain centers, the activation is much stronger with actual rewards. The existence of reward centers is clearly functional. We want to understand function and how rewards serve the individual.

A very useful neuroscience perspective is that of homeostasis, which "... involves detectors that monitor when a system departs from a 'set point,' and mechanisms that restore equilibrium when such departures are detected." Thus: "The traditional economic account of behavior, which assumes that humans act so as to maximally satisfy their preferences, starts in the middle ... of the neuroscience account ... [which views] ... pleasure as a homeostatic cue – an informational signal" (Camerer et al., 2005, p. 27). Of course, in relating this perspective to the issue of reciprocity in extensive form games, we do not know what determines an individual's "set point," which I think of as emerging from cultural experience across time, but also having biological coevolutionary origins. This "set point" is reflected in the empirical finding in experiments that an individual's initial decision in a sequence serves to "type" the cooperativeness of that person in subsequent decisions, whatever the matching treatment. (See Table 12.3 and the Houser references in note 5, Chapter 10.) It is further reflected in the many studies of the effect of context, game form, stakes, and so on, constituting different cues that distill a provisional "set point" from the individual's imported sociality characteristics depending on her autobiographical experience, and then further modified (or not) by her specific new experience in the experiment. These are mechanisms through which individuals seek to implement a life-style "type" across the situations encountered.

- Can we identify how people encode other as well as own reward, and how the brain processes reward to other in interactive decision?
- Can we represent the heterogeneity of individuals in terms of homeostatic processes that better track the effects of own and other payoffs, and of context in predicting and understanding behavior?

A recent study uses rTMS (a new technology for low-frequency repetitive transcranial magnetic stimulation) to show that disruption of the right dorsolateral prefrontal cortex greatly reduces normal responder rejection rates (also response times) of low ultimatum offers ($4 out of a $20 stake) (Knoch et al., 2006). But disruption of the same left cortex region and a sham treatment has no such effect. They used a repeat single protocol: "Each responder played the Ultimatum Game 20 times with 20 different anonymous partners" (ibid., p. 830).

The key new observation is that right TMS disruption significantly raises the acceptance rates of low offers; the left TMS and sham treatments do not. Although the authors' perspective seems to be that of "reciprocity" ("reciprocal fairness"), not social preference, their descriptive rhetoric uses the language of social preference concepts – for example, "fairness goals," "inequitable," "insultingly unfair" – that are extraneous to their basic findings. They conduct a survey of all subjects, asking them to rate the "fairness" of different offers in the available set, showing that the subjects rate low offers as "unfair" to "very unfair" for all three treatments groups. What the authors want to be able to say is that *hypothetical* reported acceptance rates are the same as the *actual* rates in the experiment with the sole exception of the right TMS disruption treatment. So why ask subjects to rate the "fairness" of offers on a seven-point scale (1 = "very unfair," to 7 = "very fair")? Just ask the subjects to rate the offers according to their acceptability (1 = "very unacceptable," to 7 = "very acceptable"). The unique English word "fairness" (Wierzbicka, 2006) is a diversion from the authors' important findings and there should be no need for either the subjects or the researchers to have to speculate about the equivalence of "acceptability" and "fairness" judgments.

Hence, whatever might be the process whereby different brain regions jointly determine decision, this right cortex area appears to be implicated in enabling the brain to resist the immediate game-theoretic domination motive to accept low offers. In personal exchange systems, dominance defection in the present jeopardizes longer-term benefit, gains from exchange, and reputations valuable both to the individual and to the social group. Perhaps the right cortex is what keeps active the second term in Sobel's (2005) equation, enabling an individual to stay focused on long-term benefit across interactions: $(1 - d) u_i(s) + d V_i(H(s))$ (see Chapter 10). These are also critical elements in social cohesion that enable groups to resist internal (and external) intrusion by dominance motives that free-ride on the community, building norms of positive and negative reciprocity. Thus, in trust games, cooperation grows, relative to controls, when pairs are (unknowingly) formed based on the history of their cooperation (see Chapter 12).

List and Cherry (2002) report a reduction in ultimatum responder rejection rates over time under the condition of repeat single play, suggesting that people are gradually approaching equilibrium play over time. In both their low and high earned stakes treatment, the acceptance rates for low offers are higher than those reported in Knoch et al. (2006) *without* right TMS disruption; and the acceptance rates are essentially the same *with* disruption.

- Does the right dorsolateral prefrontal cortex have a general role in mediating dominance behavior across different contexts?
- Or is the cortex more specialized in interacting with other brain regions that combine to produce different outcomes in different contexts?

Regarding loss avoidance and survival economics, according to Adam Smith, as quoted in Chapter 7, "We suffer more ... when we fall from a better to a worse situation, than we ever enjoy when we rise from a worse to a better." Kahneman and Tversky (1979) have powerfully demonstrated the ubiquity of this maxim in explaining choice. It is interpreted as contrary to expected utility maximization, but it need not be irrational if you want to maximize the probability of never going bankrupt. Does the evidence mean that people are "irrational," or does it mean merely that they have different implicit objectives? In Chapter 8, we have seen a model (Radner, 1997) showing that asymmetry of gains and losses can be a characteristic of optimality in the following precise sense: Below a critical level of wealth, a "survivalist," one who is out to minimize the probability of going bankrupt, will behave as if he were risk preferring, and above this level, his behavior will appear to be risk-averse. But the behavior has nothing to do with "risk aversion" as we traditionally have modeled it; rather, it is simply a characteristic of a survivalist's decision behavior. Hence, the expected utility model of risk aversion is confounded with other possible motives, and the behavior, although inconsistent with the utility model, may be consistent with a survival model.

- Is the survival model, which allows the derivation of an asymmetric gain/loss representation of decision behavior, better than expected utility for organizing individual decision data?
- Can neuroeconomics help resolve this question?

In developing such a research program, "wealth" is not likely to be the right state variable for signaling the critical state below which behavior changes from minimizing to maximizing the variance of return if the chances of survival are to be increased.

Humans do not seem to be good processors of sample evidence. But the Bayesian model postulates that a person can a priori list all possible states of nature – information not naturally given to an adapted brain accustomed to learning about states from observations. Life experiences are analogous to sampling an urn that at a given time is perhaps thought to contain only red and black balls, but a new event experience yields a new outcome – an orange ball. I find it very implausible that human brains are not well adapted to the prospect of surprise events. Just because the phenomenon is not part of our standard models of rational choice does not mean that it is not ecologically rational.

How can it be fit, as in Bayesian modeling, for any well-adapted organism to assign a zero posterior probability to any event surprise merely because there was no prior experiential basis for a positive assignment? The implication is that no new category learning is possible. I would hypothesize that the adapted brain's "sampling model" will always update on "surprise" events, initially foreign to experience, but once discovered requiring one to be open to event reoccurrence and the encoding of the frequency of such states.

From the perspective of the formal Bayesian model, this may create an apparent "representative bias"; far from being irrational, such openness may be an essential part of adapting one's information state to any environment of uncertainty. Nor are event occurrences and their relative rate of reoccurrence likely to be encoded independent of their value/cost consequences for the individual.

- How does the brain encode surprises, and how does it learn from them?

As in Fuster (1999) and developed originally in Hayek (1952), contemporary neuroscience would say that the brain automatically creates a category for the surprise, commits it to memory, perhaps weights it according to its value (cost), and draws on it as needed. The brain creates categories out of experience as it goes along, not out of abstract analysis motivated by generalist concerns for optimality applicable to all possible cases that are conjectured to arise. The brain does not naturally see principles common across multiple applications. Such principles are constructivist and must necessarily ignore variation in the fine structure and clutter of details inimical to extracting theorems. But when we leave our closet to relate those principles to observations, replicable contradictions must always precipitate a willingness to reexamine the provisional assumptions we needed to derive testable and tested implications.

FIFTEEN

A Summary

The great question ... is whether man's mind will be allowed to continue to grow as part of this process [of creating undesigned achievements] or whether human reason is to place itself in chains of its own making.

Hayek (1948, p. 32)

Cartesian constructivism applies reason to individual action and for the design of rules for institutions that are to yield socially optimal outcomes, and constitutes the standard socioeconomic science model. But most of our operating knowledge and ability to decide and perform is nondeliberative. Because such processing capacities are scarce, our brains conserve attention and conceptual and symbolic thought resources, and proceed to delegate most decision making to autonomic processes (including the emotions) that do not require conscious attention.

Emergent arrangements and behaviors, even if initially constructivist, must have fitness properties that incorporate opportunity costs and social environmental challenges invisible to constructivist modeling, but that are an integral part of experience and selection processes. This leads to an alternative, ecological concept of rationality: an emergent order based on trial-and-error cultural and biological coevolutionary change. These processes yield home- and socially-grown rules of action, traditions, and moral principles that underlie emergent (property) rights to act that create social cohesion in personal exchange. In personal exchange, rights to take action emerge by mutual consent:

- Positive reciprocity serves to generate gains from specialization and social exchange that create social wealth; we do things for each other that are based on our respective heterogeneous knowledge and skills. Negative reciprocity is the endogenous police officer serving to remind

those using dominance choice behavior of their larger cultural respon-
sibilities in repeat interaction.
- In market economies, these principles are further developed and modified for impersonal exchange supported by codified (property) rights
to act whose force still must derive from and not contradict tradition
and the daily practice of norms.

To study ecological rationality, we use rational reconstruction – for example, reciprocity or other-regarding preference models – to examine individual behavior, emergent order in human culture and institutions, and their
persistence, diversity, and development over time. Experiments enable us to
test propositions derived from these rational reconstructions.

The study of both kinds of rationality has been prominent in the work
of experimental economists. This is made plain in the many direct tests
of the observable implications of propositions derived from economic and
game theory. It is also evident in the great variety of experiments that have
reached far beyond the theory to ask why the tests have succeeded, failed,
or performed better (under weaker conditions) than was expected. Here is
a partial summary of the state of our knowledge that I have tried to convey
in this book.

Markets constitute an engine of productivity by supporting resource specialization through trade and creating a diverse wealth of goods and services. They are rule-governed institutions providing algorithms that select,
process, and order the exploratory messages of agents who are each uniquely
informed as to their personal circumstances, experience, and "can do" (tacit,
personal) knowledge. Simultaneously, agents generate these messages once
they become practiced in the institutional rules that convert those messages into realizations. As precautionary probes by agents yield to contracts,
each becomes more certain of what must be given in order to receive. Out
of this interaction between minds through the intermediary of rules, the
process aggregates the dispersed asymmetric information, converging more
or less rapidly to a competitive efficient equilibrium if it exists. Each experimental market carries its own unique path-dependent mark with different
dynamics.

All this information is captured in the static or time variable supply-
and-demand environment and must be aggregated to yield efficient clearing prices. We can never fully understand how this process works in the
world because the required information is not given, or available, to any one
mind. Thus, for many, the arguments of the Scottish philosophers and of
Hayek are obscure and mystical. But we can design experiments in which the

information is not given to any participant, then compare market outcomes with efficient competitive outcomes and gauge a market institution's performance conditional on the givens in this economic environment.

The resulting order is invisible to the participants, unlike the visible individual gains they reap. Agents discover what they need to know to achieve outcomes judged provisionally to be best against the constraining limits imposed by others.

More generally, in the larger society the rules themselves also emerge as institutions in a spontaneous order – they are found, not deliberately designed by only one calculating mind. Initially constructivist institutions undergo evolutionary change adapting beyond the circumstances that gave them birth. What emerges is a form of "social mind" that solves complex organization problems without conscious cognition. This social mind is born of the interaction among all individuals through the rules of institutions that have to date survived cultural selection processes.

This process of institutional change and discovery accommodates trade-offs between the cost of transacting, attending, and monitoring, and the efficiency of the allocations so that the institution itself generates an order of economy that fits the problem for which it evolved to solve. Hence, the hundreds of variations on the fine structure of institutions, each designed without a designer to accommodate disparate conditions but all of them subservient to the reality of dispersed agent knowledge and skill. The negative effects of certain cases of asymmetric information tend to beget compensatory responses – warranties, reputations, signaling of quality, entry, arbitrage, and interminable gossip allowing unique experiences to be shared more widely – to convert efficiency loss into individual gains. However, these effects do not necessarily eliminate efficiency loss, as not all problems have efficient solutions based on the technologies available in a particular epoch.

We understand little about how rule systems for social interaction and markets emerge, but it is possible in the laboratory to do variations on the rules, and thus to study that which is *not*. In this way, we can access the opportunity cost of alternative institutional arrangements and better understand what *is*.

Markets require enforcement – voluntary or involuntary – of the rules of exchange. These are the right of possession, its transference by consent, and the performance of promises (the "three laws of human nature" articulated by Hume, 1739; 1985). Voluntary enforcement occurs when in some cultures people in the market reward good services with gratuities or tips, an example, perhaps, of an emergent cultural norm in which people recognize that such practices are part of an informal exchange. If self- or community-enforcement conditions are not present in markets, the result tends to yield unintended consequences for the bad, as markets are compromised or may fail. The game of "trade" yields to the game of "steal."

Hume's three laws constitute an eighteenth century rearticulation of human universal norms forged from ancient cultural and religious experience, the great "shalt nots." These prohibitions allow wide-ranging freedom within their boundaries for more specific definitions of "property" rights to act under local cultural variations.

Using the laboratory as a test bed to assist in the design of new or modified market, management, and social institutions has opened exciting prospects for exploring the use of reason to provide constructivist models of new arrangements, then subjecting the models to ecological tests of viability and modifying them in the light of the test results before further testing and implementation in the field. This "try it before you fly it" program seeks to combine the advantages of reason in providing variation, with the advantages of selection through trial-and-error experimental test bedding to minimize the human propensity to err in exploring unfamiliar territory. Examples of those errors are prominent in electricity liberalization, fisheries and water management systems, and the FCC spectrum auctions, but they can be greatly diminished by more thorough testing in advance of implementation in the field. There will be time enough to test them further in the field, where errors are much more costly.

Reciprocity, trust, and trustworthiness are universal mechanisms of personal exchange, where formal markets are not worth their cost, yet there are endless opportunities for small-scale local gains from exchange to be captured to reinforce the value of diversity, and to grow wealth beyond the vision of the individual. They also may be of crucial importance as supplements to formal contracting, as not every margin of gain at the expense of another can be anticipated and formalized in written contracts.

People are not required to be narrowly self-interested (domination); rather, the point of the Scottish philosophers and Mandeville was that people did not have to be good to produce good outcomes. Markets economize on information, understanding, rationality, numbers of agents, and virtue.

Markets in no way need destroy the foundation upon which they probably emerged – social exchange between family, friends, and associates. This statement is supported in the studies reported by Henrich et al. (2005). Thus, individuals can be habitual social exchangers and vigorous traders as well, but as in Hayek's "two worlds" text, the ecologically rational coexistence of personal and impersonal exchange is not a self-aware Cartesian construct. Consequently, there is the ever-present danger that the rules of "personal exchange" will be applied inappropriately and destructively to govern or modify the extended order of markets. Equally dangerous, the rules of impersonal market exchange may be applied insensitively to our cohesive social networks and crush viable interpersonal exchange systems based on mutual trust.

Experiments have helped us to see that there is no reason to suppose that our pencil and paper models of rationality are adequate. This is revealed in impersonal exchange, where complete information is neither necessary nor sufficient for people to discover through replication complex efficient equilibria by processes that defy our modeling skills.

In personal exchange, people can do better for themselves by cooperating and not following dominant choice strategies. This is demonstrated in anonymous interaction between the same players in two-person extensive form games with complete information. (Private information disables this coordination and we observe convergence to equilibria based on dominance.) Cooperative behavior persists, but reduced in frequency, when people are paired randomly, in single-play interactions and in repetition with distinct pairs.

Cooperation in single-play games has recently been explained by recourse to social preferences, a utility for other as well as own payoff that accepts without challenge all the other paraphernalia of game theory. Social preference theory constitutes a research end game that stops short the examination of social exchange as dynamic interaction over time. It is an explanation that does not accord well with the discovery that cooperation is sensitive to social context: double blind protocols, a buyer/seller exchange context; the strategic versus extensive form of game representation; and earned rights to a privileged role and earned endowments.

The power of earned endowments to alter behavior in two-person games falsifies the game-theoretic auxiliary hypothesis that money is money whatever its source. It suggests the need for reappraisal of behavioral experiments that have been trapped in a loop that has thoroughly explored only the social condition in which decisions are made using OPM (other people's money). We have learned a lot about what game players do with money imported free from the external world, but we do not know how far those results can be generalized. These new results call for the study of variations on the protocols for defining rights in the sense of "mine" and "yours." It is these new baselines from which we must launch new studies of personal exchange systems to examine whether gains from exchange, governed by trust and trustworthiness, can be leveraged through investment of resources that are "mine" and "yours." Will those gains be enhanced, made more credible, and not diminished by earned resources brought to the table? Or will we discover otherwise? What is the reach of entanglement between stake origins, rights, and strategic interactive decision?

Are we poor at expected utility maximization because we are adapted for survival? What is the right model with which to understand choice under uncertainty? Are people poor Bayesian learners because we are adapted to environments in which the states of the world include surprise outcomes and therefore – contrary to the Bayesian model – samples are a source of knowledge of the states as well as the relative frequency with which states occur?

Just as the socioeconomic order evolves out of human sociality in response to random as well as constructed variations that must survive ecological tests, so science does not evolve only through logical falsification paths of selection. Regardless of whether hypotheses fail or pass particular tests, scientific communities debate both the circumstances of the test and the conceptual framework generating the test hypotheses. Personal scientific knowledge, including that of experimental operations, is created out of these design, testing, and discussion processes.

Machines are the coequal partners of theory and experiment in creating new scientific knowledge that allows old controversies, about which hypothesis – research or auxiliary – is falsified, to be finessed by deeper, previously unavailable observational learning. Neuroeconomics is the latest such addition to the study of economic behavior made possible by brain-imaging machines and their associated technology. But its main promise is still in the future beyond the myopic vision of previous narrowly drawn questions.

References

Acheson, J. 1975. "The Lobster Fiefs: Economic and Ecological Effects of Territoriality in the Maine Lobster Industry." *Human Ecology* 3:183–207.

Akerlof, G. A. 2002. "Behavioral Macroeconomics and Macroeconomic Behavior." *American Economic Review* 92:411–33.

Alchian, A. 1950. "Uncertainty, Evolution and Economic Theory." *Journal of Political Economy* 21:39–53.

Anderson, T., and P. J. Hill. 1975. "The Evolution of Property Rights: A Study of the American West." *Journal of Law and Economics* 18:163–79.

Andrioni, J. 1995. "Cooperation in Public Good Experiments: Kindness or Confusion?" *American Economic Review* 85:891–904.

Angel, R. 2001. "Future Optical and Infrared Telescopes." *Nature Insights* 409:427–30.

Arrow, K. 1987. "Rationality of Self and Others in an Economic System." In R. Hogarth and M. Reder, eds., *Rational Choice*. Chicago: University of Chicago Press, pp. 201–15.

Arthur, W. B. 1989. "Competing Technologies, Increasing Returns, and Lock-in by Historical Events." *Economic Journal* 99(395):116–31.

Ashraf, N., I. Bohnet, and N. N. Piankov. 2006. "Decomposing Trust and Trustworthiness." *Experimental Economics* 9:193–208.

Augier, M., and J. March. 2004. *Models of a Man: Essays in Memory of Herbert A. Simon.* Cambridge: MIT Press.

Aumann, R. 1990. Foreword to A. Roth and M. Sotomayer. *Two-Sided Matching: A Study of Game Theoretic Modeling and Analysis.* Cambridge: Cambridge University Press, pp. ix–xi.

Backerman, S., M. Denton, S. Rassenti, and V. L. Smith. 2001. "Market Power in a Deregulated Electrical Industry." *Journal of Decision Support Systems* 30:357–81.

Backerman, S., S. Rassenti, and V. L. Smith. 2000. "Efficiency and Income Shares in High Demand Energy Network: Who Receives the Congestion Rents When a Line Is Constrained?" *Pacific Economic Review* 5:331–47.

Banks, J., J. Ledyard, and D. Porter. 1989. "Allocating Uncertain and Unresponsive Resources: An Experimental Approach." *Rand Journal of Economics* 20:1–25.

Banks, J., M. Olson, D. Porter, S. Rassenti, and V. Smith. 2003. "Theory, Experiment and the Federal Communications Commission Spectrum Auctions." *Journal of Economic Behavior and Organization* 51:303–50.

Barkow, J., L. Cosmides and J. Tooby. 1992. *The Adapted Mind*. New York: Oxford University Press.

Baron-Cohen, S. 1995. *Mindblindness: An Essay on Autism and Theory of Mind*. Cambridge: MIT Press.

Baumol, W. 1979. "Quasi-Permanence of Price Reductions: A Policy for Prevention of Predatory Pricing." *Yale Law Journal*, November:1–26.

Baumol, W., J. Panzer, and R. Willig. 1982. *Contestable Markets and the Theory of Industrial Structure*. New York: Harcourt-Brace-Jovanovich.

Bechara, A., H. Damasio, D. Tranel, and A. R. Damasio. 1997. "Deciding Advantageously before Knowing the Advantageous Strategy." *Science* 275(5304):1293–5.

Beck, A. (translator). 1989. *The Collected Papers of Albert Einstein*, vol. 2. Princeton: Princeton University Press.

Berg, J., J. Dickhaut, and K. McCabe. 1995. "Trust, Reciprocity, and Social History." *Games and Economic Behavior* 10:122–42.

———. 2005. "Risk Preference Instability across Institutions: A Dilemma." *Proceedings National Academy of Science* 102:4209–14.

Berg, J., R. Forsythe, F. Nelson, and T. Rietz. 2000. "Results from a Dozen Years of Election Futures Market Research." *Iowa Electronic Market*, available online at http://www.biz.uiowa.edu/iem/archive/BFNR_2000.pdf.

Binmore, K. 1994. *Game Theory and the Social Contract*, vol. 1: *Playing Fair*. Cambridge: MIT Press.

———. 1997. *Game Theory and the Social Contract*, vol. 2: *Just Playing (Economic Learning and Social Evolution)*. Cambridge: MIT Press.

Binswanger, H. P. 1980. "Attitudes toward Risk: Experimental Measurement in Rural India." *American Journal of Agricultural Economics* 63:395–407.

———. 1981. "Attitudes toward Risk: Theoretical Implications of an Experiment in Rural India." *Economic Journal* 91:867–90.

Bisson, T. 1995. *Bears Discover Fire and Other Stories*. New York: St. Martin's Press.

Bloom, P. 2004. "Can a Dog Learn a Word?" *Science*, June 11, 304:1605–6.

Bohnet, I., B. Frey, and S. Huck. 2001. "More Order with Less Law: On Contract Enforcement, Trust, and Crowding." *American Political Science Review* 95:131–51.

Bolton, G. E. 1991. "A Comparative Model of Bargaining: Theory and Evidence." *American Economic Review* 81:1096–136.

Bolton, G., E. Katok, and R. Zwick. 1998. "Dictator Game Giving: Rules of Fairness Versus Acts of Kindness." *International Journal of Game Theory* 27:269–99.

Bolton, G., and A. Ockenfels. 2000. "ERC: A Theory of Equity, Reciprocity, and Competition." *American Economic Review* 90:166–93.

Borenstein, S., and J. Bushnell. 1999. "An Empirical Analysis of the Potential for Market Power in California's Electricity Industry." *Journal of Industrial Economics* 47:285–324.

Born, M. 1943. *Experiment and Theory in Physics*. New York: Dover.

Bortoft, H. 1996. *The Wholeness of Nature: Goethe's Way of Science*. Edinburgh: Floris.

Boudoukh, J., M. Richardson, Y. Shen, and R. Whitelaw. 2007. "Do Asset Markets Reflect Fundamentals? Freshly Squeezed Evidence from the FCOJ Market." *Journal of Financial Economics* 83: 397–412.

Breiter, H. C., I. Aharon, D. Kahneman, A. Dale, and P. Shizgal. 2001. "Functional Imaging of Neural Responses to Expectancy and Experience of Monetary Gains and Losses." *Neuron* 30:619–39.

Bronfman, C., K. McCabe, D. Porter, S. Rassenti, and V. L. Smith. 1996. "An Experimental Examination of the Walrasian Tatonnement Mechanism." *Rand Journal* 27:681–99.

Brown, D. E. 1991. *Human Universals.* New York: McGraw-Hill.

Buchan, J. 2004. *Crowded with Genius the Scottish Enlightenment: Edinburgh's Moment of the Mind.* New York: Perennial.

Buchanan, J., and G. Tullock. 1962. *The Calculus of Consent: Logical Foundations of Constitutional Democracy.* Ann Arbor: University of Michigan Press.

Burnham, T. 2003. "Engineering Altruism: An Experimental Investigation of Anonymity and Gift Giving." *Journal of Economic Behavior and Organization* 50(1):133–44.

Burnham, T., K. McCabe, and V. Smith. 2000. "Friend-or-Foe Intentionality Priming in an Extensive Form Game." *Journal of Economic Behavior and Organization* 44:1–17.

Caldwell, B. 2003. *Hayek's Challenge: An Intellectual Biography of F. A. Hayek.* Chicago: University of Chicago Press.

Camera, G., C. Noussair, and S. Tucker. 2001. "Rate-of-Return Dominance and Efficiency in an Experimental Economy." Department of Economics, Krannert School of Management, Purdue University.

Camerer, C. 2003. *Behavioral Game Theory: Experiments on Strategic Interaction.* Princeton: Princeton University Press.

Camerer, C., T-H. Ho, and J. Chong. 2004. "Behavioral Game Theory: Thinking, Learning, and Teaching." In S. Huck, ed., *Advances in Understanding Strategic Behavior.* New York: Palgrave.

Camerer, C., and R. M. Hogarth. 1999. "The Effects of Financial Incentives in Experiments: A Review and Capital-Labor-Production Framework." *Journal of Risk and Uncertainty* 19:7–42.

Camerer, C., G. Loewenstein, and D. Prelec. 2005. "Neuroeconomics: How Neuroscience Can Inform Economics." *Journal of Economic Literature* 43:9–64.

Camerer, C., and R. H. Thaler. 1995. "Anomalies: Ultimatums, Dictators, and Manners." *Journal of Economic Perspectives* 9(2):209–19.

Camerer, C., and K. Weigelt. 1988. "Experimental Tests of a Sequential Equilibrium Reputation Model." *Econometrica* 56(1):1–36.

Cason, T., and D. Friedman. 1993. "An Empirical Analysis of Price Formation in Double Auction Markets." In Friedman and J. Rust, eds., *The Double Auction.* Redwood City: Addison-Wesley.

Cassady, R. 1967. *Auctions and Auctioneering.* Berkeley: University of California Press.

Cesari, M., and C. Plott. 2003. "Decentralized Management of Common Property Resources: Experiments with a Centuries-Old Institution." *Journal of Economic Behavior and Organization* 51:217–47.

Charles River and Associates, Inc., and Market Design, Inc. 1998a. Report 1A: *Auction Design Enhancements for Non-Combinatorial Auctions.* Charles River and Associates No. 1351–00.

———. 1998b. Report 1B: *Package Bidding for Spectrum Licenses.* Charles River and Associates No. 1351–00.

———. 1998c. Report 2: *Simultaneous Ascending Auctions with Package Bidding.* Charles River and Associates No. 1351–00.

Charness, G., and M. Rabin. 2003. "Understanding Social Preferences with Simple Tests." *Quarterly Journal of Economics* 117:817–69.

Charness, G., G. Frechette, and J. Kagel. 2004. "How Robust Is Gift Exchange?" *Experimental Economics* 7:189–205.

Cherry, T., P. Frykblom, and J. Shogren. 2002. "Hardnose the Dictator." *American Economic Review* 92:1218–21.

Cherry, T., S. Kroll, and J. Shogren. 2005. "The Impact of Endowment Heterogeneity and Origin on Public Good Contributions." *Journal of Economic Behavior and Organization* 57:357–65.

Chew, S. H., and N. Nishimura. 1999. "Revenue Non-Equivalence between the English and the Second Price Auctions: Experimental Evidence." Working paper, Hong Kong University of Science and Technology.

Christie, R., F. Colgin, and L. Geis. 1970. *Studies in Machiavellianism.* New York: Academic Press.

Chu, Y.-P., and R.-L. Chu. 1990. "The Subsidence of Preference Reversals in Simplified and Market-Life Experimental Settings: A Note." *American Economic Review* 80:902–11.

Coase, R. 1960. "The Problem of Social Cost." *Journal of Law and Economics* 3:1–44.

———. 1974. "The Lighthouse in Economics." *Journal of Law and Economics* 17:357–76.

Colgin, L., and E. Moser. 2006. "Neuroscience: Rewinding the Memory Record." *Nature* 440: 615.

Cooper, D., and J. Kagel. 2003. "Lessons Learned: Generalizing Learning across Games." *American Economic Review* 93(2):202–7.

———. 2004. "Learning and Transfer in Signaling Games." Working paper, Ohio State University, Department of Economics.

Cooper, D., and J. B. Van Huyck. 2003. "Evidence on the Equivalence of the Strategic and Extensive Form Representation of Games." *Journal of Economic Theory* 110(2):290–308.

Coppinger, V., V. Smith, and J. Titus. 1980. "Incentives and Behavior in English, Dutch and Sealed-Bid Auctions." *Economic Inquiry* 18(1):1–22.

Corelli, K. McCabe, and V. L. Smith. 2000. "Theory-of-Mind Mechanism in Personal Exchange." Proceedings, 13th Annual Toyota Conference on Affective Minds. In G. Hatano, N. Okada, and H. Tanabe, eds., *Affective Minds.* Amsterdam: Elsevier, chapter 26.

Cosmides, L., and J. Tooby. 1992. "Cognitive Adaptations for Social Exchange." In J. Barkow, L. Cosmides, and J. Tooby, eds., *The Adapted Mind: Evolutionary Psychology and the Generation of Culture.* New York: Oxford University Press, pp. 163–228.

———. 1996. "Are Humans Good Intuitive Statisticians after All? Rethinking Some Conclusions from the Literature on Judgment under Uncertainty." *Cognition* 58:1–73.

Coursey, D., R. M. Isaac, M. Luke, and V. Smith. 1984a. "Market Contestability in the Presence of Sunk (Entry) Cost." *Rand Journal of Economics* Spring: 69–84.

Coursey, D., R. M. Isaac, and V. Smith. 1984b. "Natural Monopoly and Contested Markets: Some Experimental Results." *Journal of Law and Economics* April:91–113.

Cox, J. 2000. "Implications of Game Triads for Observations of Trust and Reciprocity." Working paper, University of Arizona, Department of Economics.

———. 2002. "Trust, Reciprocity and Other-Regarding Preferences: Groups vs. Individuals and Males vs. Females." In R. Zwick and A. Rapoport, eds., *Advances in Experimental Business Research.* Boston: Kluwer, pp. 33–350.

————. 2004. "How to Identify Trust and Reciprocity." *Games and Economic Behavior* 46:260–81.

Cox, J., and C. Deck. 2005. "On the Nature of Reciprocal Motives." *Economic Inquiry* 43:623–35.

Cox, J., and D. Grether. 1996. "The Preference Reversal Phenomenon: Response Mode, Markets and Incentives." *Economic Theory* 7:381–405.

Cox, J., R. M. Isaac, and V. Smith. 1983. "OCS Leasing and Auctions: Incentives and the Performance of Alternative Bidding Institutions." *Supreme Court Economic Review* 2:43–87.

Cox, J., B. Roberson, and V. Smith. 1982. "Theory and Behavior of Single Object Auctions." In V. Smith, ed., *Research in Experimental Economics*, vol. 2. Greenwich: JAI Press.

Cox, J., and V. Smith. 1992. "Endogenous Entry and Exit in Common Value Auction." Mimeograph, Economic Science Laboratory, University of Arizona.

Cox, J., V. Smith, and J. Walker. 1988. "Theory and Individual Behavior of First Price Auctions." *Journal of Risk and Uncertainty* 1:61–99.

Cramton, P., Y. Shoham, and R. Steinberg, eds. 2006. *Combinatorial Auctions.* Cambridge: MIT Press.

Crockett, S., V. Smith, and B. Wilson. 2006. "Specialization and Exchange as a Discovery Process." Working paper, Interdisciplinary Center for Economic Science. George Mason University. Available online at http://gunston.gmu.edu/bwilson3/papers/CrockettSmithWilson.pdf.

Dalton, G. 1979. "Aboriginal Economies in Stateless Societies." In T. Earle and J. Ericson, eds., *Exchange Systems in Prehistory*. New York: Academic Press.

Damasio, A. R. 1994. *Descartes Error: Emotion, Reason, and the Human Brain*. New York: Putnam.

Darwin, C. 1859; 1979. *The Origin of Species*. Darby: Arden Library.

————. 1872; 1998. *The Expression of the Emotions in Man and Animals*, 3rd edition. Oxford: Oxford University Press.

Davis, D., and C. Holt. 1993. *Experimental Economics*. Princeton: Princeton University Press.

Davis, D., and A. W. Williams. 1991. "The Hayek Hypothesis in Experimental Auctions: Institutional Effects and Market Power." *Economic Inquiry* 29:261–74.

Dawes, R. 1988. *Rational Choice in an Uncertain World*. New York: Harcourt-Brace-Jovanovich.

Deck, C., K. McCabe, and D. Porter. 2006. "Why Stable Fiat Money Hyperinflates: Evidence from an Experimental Economy." *Journal of Economic Behavior and Organization* 61:471–86.

Deck, C., and B. Wilson. 2004. "Economics at the Pump." *Regulation* 1:22–9.

————. 2006. "Tracking Customer Search to Price Discriminate." *Economic Inquiry* 44:280–95.

————. 2007. "Experimental Gasoline Markets." FTC Bureau of Economics Working Paper No. 253. *Journal of Economic Behavior and Organization*.

Dehaene, S. 1997. *The Number Sense*. New York: Oxford University Press.

Demmert, H., and D. B. Klein. 2003. "Experiment on Entrepreneurial Discovery: An Attempt to Demonstrate the Conjecture of Hayek and Kirzner." *Journal of Economics and Behavioral Organization* 50(3):295–310.

Dentin, M., S. Rassenti, and V. L. Smith. 2001. "Spot Market Mechanism Design and Competitivity Issues in Electric Power." *Journal of Economic Behavior and Organization* 44:435–53.

de Waal, F. B. M. 1989. "Food Sharing and Reciprocal Obligations among Chimpanzees." *Journal of Human Evolution* 18(5):433–59.

———. 1997. "Food-Transfers through Mesh in Brown Capuchins." *Journal of Comparative Psychology* 111:370–8.

———. 2005. "How Animals Do Business." *Scientific American*, April: 73–9.

Donahue, G. 2002. "The US Air Transportation System: A Bold Vision for Change." Paper prepared for the Commission on the Future of the U.S. Airspace Industry, September 19.

Dufwenberg, M., T. Lindqvist, and E. Moore. 2002. "Bubbles and Experience: An Experiment." *American Economic Review* 95:1731–7.

Durham, Y., K. McCabe, M. Olson, and S. Rassenti. 2004. "Oligopoly Competition in Fixed Cost Environments." *International Journal of Industrial Organization* 22(2):147–62.

Dutta, P. K., and R. Radner. 1999. "Profit Maximization and the Market Selection Hypothesis." *Review of Economic Studies* 66:769–98.

Dyer, D., and J. H. Kagel. 1996. "Bidding in Common Value Auctions: How the Commercial Construction Industry Corrects for the Winners' Curse." *Management Science* 42:1463–75.

Eckel, C., and P. Grossman. 1996. "Altruism in Anonymous Dictator Games." *Games and Economic Behavior* 16:181–91.

Eckel, C. C., and R. K. Wilson. 2004. "Is Trust a Risky Decision?" *Journal of Economic Behavior and Organization* 55(4):447–65.

Economic Science Laboratory Research Group. 1985. "Alternatives to Rate of Return Regulation." *Final Report for the Arizona Corporation Commission*, February 15.

Einstein, A. 1905; 1989. "On the Electrodynamics of Moving Bodies." In A. Beck, tr., *The Collected Papers of Albert Einstein*, vol. 2. Princeton: Princeton University Press.

———. 1907; 1989. "On the Relativity Principle and the Conclusions Drawn from It." In A. Beck, tr., *The Collected Papers of Albert Einstein*, vol. 2, pp. 252–311. Princeton: Princeton University Press.

———. 1934. *The World as I See It.* New York: Covici, Friede.

El-Gamal, M., and D. Grether. 1995. "Are People Bayesian? Uncovering Behavioral Strategies." *Journal of the American Statistical Association* 90:1137–45.

Ellickson, R. C. 1991. *Order without Law: How Neighbors Settle Disputes.* Cambridge: Harvard University Press.

Engelmann, D., and M. Strobel. 2004. "Inequality Aversion, Efficiency and Maximum Preferences in Simple Distribution Experiments." *American Economic Review* 94:857–69.

Erev, I., and A. Roth. 1998. "Predicting How People Play Games: Reinforcement Learning in Experimental Games with Unique, Mixed-Strategy Equilibria." *American Economic Review* 88:848–81.

Falk, A., E. Fehr, and U. Fischbacher. 1999. "On the Nature of Fair Behavior." Working Paper No. 17, Institute for Empirical Research in Economics, 1424–0459, University of Zurich.

Fehr, E., and U. Fischbacher. 2002. "Why Social Preferences Matter: The Impact of Nonselfish Motives on Competition, Cooperation, and Incentives." *Economic Journal* 112:C1–C33.

Fehr, E., and S. Gachter. 2000. "Do Incentive Contracts Crowd Out Voluntary Cooperation?" Working paper, Institute for Empirical Research in Economics, University of Zurich.

Fehr, E., G. Kirchsteiger, and A. Riedl. 1993. "Does Fairness Prevent Market Clearing? An Experimental Investigation." *Quarterly Journal of Economics* 108(2):437–59.

Fehr, E., and B. Rockenbach. 2002. "Detrimental Effects of Sanctions on Human Altruism." *Nature* 422:137–40.

Fehr, E., and K. M. Schmidt. 1999. "A Theory of Fairness, Competition, and Cooperation." *Quarterly Journal of Economics* 114(3):817–68.

Feyerabend, P. 1975. *Against Method: Outline of an Anarchistic Theory of Knowledge.* London: NLB.

Fiske, A. P. 1991. *The Structures of Social Life: The Four Elementary Forms of Human Relations.* New York: Free Press.

Fletcher P, F. Happe, U. Frith, S. Baker, R. Dolan, R. Frakowiak, and C. Frith. 1995. "Other Minds in the Brain: A Functional Imaging Study of 'Theory of Mind' in Story Comprehension." *Cognition* 57:109–28.

Fölsing, A. 1997. *Albert Einstein.* New York: Viking.

Forsythe, R., J. L. Horowitz, N. E. Savin, and M. Sefton. 1994. "Fairness in Simple Bargaining Experiments." *Games and Economic Behavior* 6:347–69.

Forsythe, R., F. Nelson, G. R. Neumann, and J. Wright. 1992. "Anatomy of an Experimental Political Stock Market." *American Economic Review* 82:1142–61.

Forsythe, R., T. A. Rietz, and T. W. Ross. 1999. "Wishes, Expectations, and Actions: Price Formation in Election Stock Markets." *Journal of Economic Behavior and Organization* 39:83–110.

Foster, D., and M. Wilson. 2006. "Reverse Replay of Behavioral Sequences in Hippocampal Place Cells during the Awake State." *Nature,* 440:680–3.

Fouraker, L., and S. Siegel. 1963. *Bargaining Behavior.* New York: McGraw-Hill.

Franciosi, R., P. Kujal, R. Michelitsch, V. L. Smith, and G. Deng. 1995. "Fairness: Effect on Temporary and Equilibrium Prices in Posted Offer Markets." *Economic Journal* 105:938–50.

Frank, R. H. 1988. *Passions within Reason: The Strategic Role of the Emotions.* New York: W. W. Norton.

Freuchen, P. 1960. *Book of the Eskimos.* Cleveland: World Publishing.

Friedman, D. 1984. "On the Efficiency of Double Auction Markets." *American Economic Review* 74:60–72.

————. 1993. "The Double Auction Market Institution: A Survey." In J. Friedman and J. Rust, eds., *The Double Auction Market: Institutions, Theories, and Evidence: Proceedings of the Workshop on Double Auction Markets Held June 1991 in Santa Fe, NM.* Santa Fe: Perseus, pp. 3–26.

Friedman, M. 1953. *Essays in Positive Economics.* Chicago: University of Chicago Press.

————. 1960. *A Program for Monetary Stability.* New York: Fordham University Press.

Friedman, M., and L. J. Savage. 1948. "The Utility Analysis of Choices Involving Risk." *Journal of Political Economy* 56:279–304.

Fuster, J. 1999. *Memory in the Cerebral Cortex.* Cambridge: MIT Press.

Garvin, S., and J. Kagel. 1994. "Learning in Common Value Auctions." *Journal of Economic Behavior and Organization* 25:351–72.

Gazzaniga, M. S. 1998. *The Mind's Past*. Berkeley: University of California Press.

Gazzaniga, M., R. Ivry, and G. Mangun. 1998. *Cognitive Neuroscience*. New York: W. W. Norton.

Gibran, K. 1918; 2002. *The Madman*. Mineola: Dover.

———. 1928. *Jesus the Son of Man*. New York: Knopf.

Gigerenzer, G. 1991. "How to Make Cognitive Illusions Disappear: Beyond Heuristics and Biases." In W. Stroebe and M. Hewstone, eds., *European Review of Social Psychology*, vol. 2. New York: Wiley, pp. 83–115.

———. 1993. "The Bounded Rationality of Probabilistic Models." In K. Manktelow and D. Over, eds., *Rationality*. London: Routledge, pp. 284–313.

———. 1996. "On Narrow Norms and Vague Heuristics: A Reply to Kahneman and Tversky." *Psychological Review* 103:592–6.

Gigerenzer, G., and R. Selten, eds. 2002. *Bounded Rationality: The Adaptive Toolbox*. Cambridge, MA: MIT Press.

Gigerenzer, G., P. M. Todd, and the ABC Research Group. 1999. *Simple Heuristics That Make Us Smart*. New York: Oxford University Press.

Giocoli, N. 2003. "Fixing the Point: The Contribution of Early Game Theory to the Tool-Box of Modern Economics." *Journal of Economic Methodology* 10 (1):1–39.

Gode, D., and S. Sunder. 1993. "Allocative Efficiency of Markets with Zero Intelligence Traders: Market as a Partial Substitute for Individual Rationality." *Journal of Political Economy*, 101:119–37.

———. 1997. "What Makes Markets Allocationally Efficient?" *Quarterly Journal of Economics* 57: 603–30.

Goel, V., J. Grafman, J. Tajik, S. Gana, and D. Danto. 1997. "A Study of the Performance of Patients with Frontal Lobe Lesions in a Financial Planning Task." *Brain* 120(10):1805–22.

Goeree, J., and C. Holt. 2001. "Ten Little Treasures of Game Theory and Ten Intuitive Contradictions." *American Economic Review* 91:1402–22.

Goldberg, I., M. Harel, and R. Malach. 2006. "When the Brain Loses Its Self: Prefrontal Inactivation during Sensorimotor Processing." *Neuron* 50:329–39.

Goldberger, A. 1979. "Heritability." *Economica* 46(184):327–47.

Gomez, R., J. K. Goeree, and C. A. Holt. 2008. "Predatory Pricing: Rare Like a Unicorn?" In C. A. Plott and V. L. Smith, eds., *Handbook of Experimental Economics Results*. New York: Elsevier.

Gouldner, A. W. 1960. "The Norm of Reciprocity: A Preliminary Statement." *American Sociological Review* 25:161–78.

Grandin, T. 1996. *Thinking in Pictures*. New York: Vintage.

Grandin, T., and C. Johnson. 2005. *Animals in Translation*. New York: Scribner.

Grether, D., and C. Plott. 1984. "The Effects of Market Practices in Oligopolistic Markets: The Ethyl Case." *Economic Inquiry* 22(4):479–507.

Guala, F. 2005. *The Methodology of Experimental Economics*. Cambridge: Cambridge University Press.

Gunnthorsdottir, A., D. Houser, K. McCabe, and H. Ameden. 2002a. *Disposition, History, and Contributions in a Public Good Experiment*. Working Paper, George Mason University.

Gunnthorsdottir, A., K. McCabe, and V. Smith. 2002b. "Using the Machiavellianism Instrument to Predict Trustworthiness in a Bargaining Game. *Journal of Economic Psychology* 23:49–66.

Güth, W., S. Huck, and P. Ockenfels. 1996. "Two-Level Ultimatum Bargaining with Incomplete Information: An Experimental Study." *Economic Journal* 106(436):593–604.

Güth, W., R. Schmittberger, and B. Schwarze. 1982. "An Experimental Analysis of Ultimatum Bargaining." *Journal of Economic Behavior and Organization* 3:367–88.

Gwartney, J., and R. Lawson. 2003. "The Concept and Measurement of Economic Freedom." *European Journal of Political Economy* 19(3):405–30.

Hamowy, H. 2003. "F. A. Hayek and the Common Law." *Cato Journal* 23(2):241–64.

Harrison, G. 1988. "Predatory Pricing in a Multiple Market Experiment: A Note." *Journal of Economic Behavior and Organization* 9:405–17.

———. 1989. "Theory and Misbehavior of First-Price Auctions." *American Economic Review* 79(4):749–62.

Harrison, G., E. Johnson, M. McInnes, and E. Rutström. 2005. "Risk Aversion and Incentive Effects: Comment." *American Economic Review* 95(3):897–901.

Harrison, G., and J. List. 2004. "Field Experiments." *Journal of Economic Literature* 52:1005–55.

Harrison, G., and M. McKee. 1985a. "Experimental Evaluation of the Coase Theorem." *Journal of Law and Economics* 28:653–70.

———. 1985b. "Monopoly Behavior, Decentralized Regulation, and Contestable Markets: An Experimental Evaluation." *Rand Journal of Economics* 16(1):51–69.

Hawkes, K. 1990. "Showing Off: Tests of an Hypothesis about Men's Foraging Goals." *Ethnology and Evolutionary Biology* 12:29–54.

Hayek, F. A. 1937. "Economics and Knowledge." *Economica* 4:33–54.

———. 1942; 1979. "Scientism and the Study of Society." *Economica* 9(35):267–91. Reprinted in *The Counter-Revolution of Science*, pp. 19–40. Indianapolis: Liberty Press.

———. 1945. "The Use of Knowledge in Society." *American Economic Review* 35(4):519–30.

———. 1948. *Individualism and Economic Order (Essays)*. Chicago: University of Chicago Press.

———. 1952. *The Sensory Order*. Chicago: University of Chicago Press.

———. 1956. "The Dilemma of Specialization." In L. White, ed., *State of the Social Sciences*. Chicago: University of Chicago Press.

———. 1960. *The Constitution of Liberty*. Chicago: University of Chicago Press.

———. 1967. *Studies in Philosophy Politics and Economics*. Chicago: University of Chicago Press.

———. 1973. "Law, Legislation, and Liberty." In *Rules and Order*, vol. 1. Chicago: University of Chicago Press.

———. 1974. "Banquet Speech." December 10. Available online at http://nobelprize.org/nobel_prizes/economics/laureates/1974/hayek_speech.html.

———. 1976; 1991. "Adam Smith (1723–1790): His Message in Today's Language." In W. Bartley and S. Kresge, eds., *The Trend of Economic Thinking*. Chicago: University of Chicago Press, pp. 119–24.

———. 1978; 1984. "Competition as a Discovery Procedure." In *The Essence of Hayek*. Stanford: Hoover Institution Press, chapter 13.

———. 1988. *The Fatal Conceit*. Chicago: University of Chicago Press.

Henrich, J. 2000. "Does Culture Matter in Economic Behavior? Ultimatum Game Bargaining among the Machiguenga of the Peruvian Amazon." *American Economic Review* 90(4):973–9.

———. 2004. "Cultural Group Selection, Co-Evolutionary Processes and Large-Scale Cooperation." *Journal of Economic Behavior and Organization* 53:3–35.

Henrich, J., R. Boyd, S. Bowles, C. Camerer, E. Fehr, H. Gintis, R. McElreath, M. Alvard, A. Barr, J. Ensminger, N. Henrich, K. Hill, F. Gil-White, M. Gurven, F. Marlowe, J. Patton, and D. Tracer. 2005. "'Economic Man' in Cross-Cultural Perspective: Behavioral Experiments in 15 Small-Scale Societies." *Behavioral and Brain Sciences* 28(6):795–815.

Hertwig, R., and A. Ortmann. 2001. "Experimental Practices in Economics: A Methodological Challenge for Psychologists?" *Behavioral and Brain Sciences* 24(3):383–403.

———. 2003. "Economists' and Psychologists' Experimental Practices: How They Differ, Why They Differ, and How They Could Converge." In I. Brocas and J. D. Carrillo, eds., *The Psychology of Economic Decisions*. Oxford: Oxford University Press, pp. 253–72.

Hill, E., and D. Sally. 2003. "Dilemmas and Bargains: Autism, Theory of Mind, Cooperation and Fairness." Working paper, University College, London.

Hoffman E., K. McCabe, K. Shachat, and V. Smith. 1994. "Preferences, Property Rights, and Anonymity in Bargaining Games." *Games and Economic Behavior* 7(3):346–80.

Hoffman, E., K. McCabe, and V. Smith. 1995. "Ultimatum and Dictator Games," *Journal of Economic Perspectives* 9:236–9.

———. 1996a. "On Expectations and the Monetary Stakes in Ultimatum Games." *International Journal of Game Theory* 25(3):289–301.

———. 1996b. "Social Distance and Other-Regarding Behavior in Dictator Games." *American Economic Review* 86(3):653–60.

———. 1998. "Behavioral Foundations of Reciprocity: Experimental Economics and Evolutionary Psychology." *Economic Inquiry* 36(3):335–53.

———. 2000. "The Impact of Exchange Context on the Activation of Equity in Ultimatum Games." *Experimental Economics* 3(1):5–9.

———. 2008. "Reciprocity in Ultimatum and Dictator Games." In C. A. Plott and V. L. Smith, eds., *Handbook of Experimental Economics Results*. New York: Elsevier, chapter 4.

Hoffman, E., and M. L. Spitzer. 1985. "Entitlements, Right and Fairness: An Experimental Examination of Subject's Concepts of Distributive Justice." *Journal of Legal Studies* 14(2):259–97.

Hoffrage, U., G. Gigerenzer, S. Krauss, and L. Martignon. 2002. "Representation Facilitates Reasoning: What Natural Frequencies Are and What They Are Not." *Cognition* 84(3):343–52.

Hogarth, R., and M. Reder. 1987. *Rational Choice*. Chicago: University of Chicago Press.

Holt, C. 1989. "The Exercise of Market Power in Experiments." *Journal of Law and Economics* 32(2):S107–30.

———. 1995. "Industrial Organization." In J. Kagel and A. Roth, eds., *The Handbook of Experimental Economics*. Princeton: Princeton University Press, chapter 5.

Holt, C., L. Langan, and A. Villamil. 1986. "Market Power in Oral Double Auction Experiments." *Economic Inquiry* 24:107–23.

Holt, C., and S. Laury. 2002. "Risk Aversion and Incentive Effects." *American Economic Review* 92(5):1644–55.

Holt, D. 1999. "An Empirical Model of Strategic Choice with an Application to Coordination Games." *Games and Economic Behavior* 27:86–105.

Homans, G. C. 1967. *The Nature of Social Science*. New York: Harcourt, Brace and World.

Houser, D. 2003. "Classification of Types of Dynamic Decision Makers." In L. Nadel, ed., *Encyclopedia of Cognitive Science*, vol. 1. London: Nature Publishing Group, pp. 1020–6.

Houser, D., A. Bechera, M. Keane, K. McCabe, and V. Smith. 2005. "Identifying Individual Differences: An Algorithm with Application to Phileas Gage." *Games and Economic Behavior* 52:373–85.

Houser, D., A. Gunnthorsdottir, and K. McCabe. 2006. "Disposition, History and Contributions in a Public Goods Experiment." *Journal of Economic Behavior and Organization.*

Houser, D., M. Keane, and K. McCabe. 2004. "Behavior in a Dynamic Decision Problem." *Econometrica* 72:781–822.

Houser, D., and R. Kurzban. 2002. "Revisiting Kindness and Confusion in Public Good Experiments." *American Economic Review* 92:1062–9.

Houser, D., K. McCabe, and V. L. Smith. 2004. "Cultural Group Selection, Co-Evolutionary Processes and Large-Scale Cooperation: Discussion." *Journal of Economic Behavior and Organization* 53(1):85–8.

Houser, D., E. Xiao, K. McCabe, and V. Smith. 2008. "When Punishment Fails: Research on Sanctions, Intentions and Non-Cooperation." *Games and Economic Behavior.*

Huck, S., H. Norman, and J. Oechssler. 2004. "Two Are Few and Four Are Many: Number Effects in Experimental Oligopolies." *Journal of Economic Behavior and Organization* 53:435–46.

Humboldt, A., and A. Bonpland. 1815; 1907. *Personal Narrative of Travels to the Equinoctial Regions of the New Continent, during the Years 1799–1804*. T. Ross, ed. 3 volumes. London: George Bell and Sons.

Hume, D. 1739; 1985. *A Treatise of Human Nature*. London: Penguin.

Hurwicz, L. 1960. "Optimality and Informational Efficiency in Resource Allocation Processes." In K. Arrow, S. Karlin, and P. Suppes, eds., *Mathematical Methods in the Social Sciences*. Stanford: Stanford University Press.

———. 1973 "The Design of Mechanism for Resource Allocation." *American Economic Review* 63(2):1–30.

Hurwicz, L., R. Radner, and S. Reiter. 1975. "A Stochastic Decentralized Resource Allocation Process: Part II." *Econometrica* 43(3):363–93.

Hussam, R., D. Porter, and V. L. Smith. 2006. "Thar She Blows: Rekindling Bubbles with Inexperienced Subjects." Working paper, Interdisciplinary Center for Economic Science, George Mason University.

Insul, S. 1915. *Central Station Power Systems*. Chicago: Privately published.

Isaac, R., and D. James. 2000a. "Just Who Are You Calling Risk Averse?" *Journal of Risk and Uncertainty* 20:177–87.

———. 2000b. "Robustness of the Incentive Compatible Combinatorial Auction." *Experimental Economics* 3(1):31–53.

Isaac, R., T. Salmon, and A. Zillante. 2007. "A Theory of Jump Bidding in Ascending Auctions." *Journal of Economic Behavior and Organization* 62(1):144–64.

Isaac, R. M., and V. L. Smith. 1985. "In Search of Predatory Pricing." *Journal of Political Economy* 93:320–45.

Ishikida, T., J. Ledyard, M. Olson, and D. Porter. 2001. "The Design of a Pollution Trading System for Southern California's RECLAIM Emission Trading Program." In R. M. Isaac, ed., *Research in Experimental Economics*, vol. 8. Greenwich, CT: JAI Press.

Jackson, M. O. 2005. "Non-Existence of Equilibrium in Vickrey, Second-Price, and English Auctions." Division of Humanities and Social Sciences Working Paper 1241, California Institute of Technology.

Jamal, K., and S. Sunder. 1991. "Money vs. Gaming: Effects of Salient Monetary Payments in Double Oral Auctions." *Organizational Behavior and Human Decision Processes* 49(1):151–66.

Jevons, W. S. 1871; 1888. *The Theory of Political Economy*, 3rd edition. London: Macmillan. Available online at http://www.econlib.org/LIBRARY/YPDBooks/Jevons/jvnPE.html.

Jung, Y., J. Kagel, and D. Levin. 1994. "On the Existence of Predatory Pricing: An Experimental Study of Reputation and Entry Deterrence in the Chain-Store Game." *Rand Journal of Economics* 25:253–79.

Kachelmeier, S. J., and M. Shehata. 1992. "Examining Risk Preferences under High Monetary Incentives: Evidence from the People's Republic of China." *American Economic Review* 82:1120–41.

Kagan, J. 1994. *Galen's Prophecy*. New York: Basic Books.

Kagan, J., and S. Lamb. 1987. *The Emergence of Morality in Young Children*. Chicago: University of Chicago Press.

Kagel, J. H., and D. Levin. 1986. "The Winner's Curse and Public Information in Common Value Auctions." *American Economic Review* 76:894–920.

Kagel, J., and A. Roth. 1995. *Handbook of Experimental Economics*. Princeton: Princeton University Press.

Kahneman, D. 2002. Nobel Interview. Available online at http://www.nobel.se/economics/laureates/2002/kahneman-interview.html.

Kahneman, D., J. Knetsch, and R. Thaler. 1986. "Fairness as a Constraint on Profit Seeking: Entitlements in the Market." *American Economic Review* 76(4):728–41.

———. 1991. "Fairness and the Assumptions of Economics." In R. Thaler, ed., *Quasi Rational Economics*. New York: Russell Sage Foundation.

Kahneman, D., and A. Tversky. 1979. "Prospect Theory: An Analysis of Decision under Risk." *Econometrica* 47:263–91.

———. 1996. "On the Reality of Cognitive Illusions: A Reply to Gigerenzer." *Psychological Review* 103:582–91.

Kalai, E., and E. Lehrer. 1993. "Rational Learning Leads to Nash Equilibrium." *Econometrica* 61(5):1019–45.

Kaminski, J., J. Call, and J. Fischer. 2004. "Word Learning in a Domestic Dog: Evidence for 'Fast Mapping.'" *Science* 304:16832–3.

Kaplan, H., and K. Hill. 1985. "Food Sharing among Ache Foragers: Tests of Explanatory Hypotheses." *Current Anthropology* 26:223–46.

Ketcham, J., V. L. Smith, and A. Williams. 1984. "A Comparison of Posted-Offer and Double-Auction Pricing Institutions." *Review of Economic Studies* 51:595–614.

Klemperer, P. 1998. "Auctions with Almost Common Values: The 'Wallet Game' and Its Applications." *European Economic Review* 42:757–69.

Klemperer, P., and M. Meyer. 1989. "Supply Function Equilibria in Oligopoly under Uncertainty." *Econometrica* 57:1243–77.

Knoch, D., A. Pascual-Leone, K. Meyer, V. Treyer, and E. Fehr. 2006. "Diminishing Reciprocal Fairness by Disrupting the Right Prefrontal Cortex." *Science* 314:829–32.

Knowlton, B., J. Mangels, and L. Squire. 1996. "A Neostriatal Habit Learning System in Humans." *Science* 273(5280):1399–1402.

Koehler, J. J. 1996. "The Base-Rate Fallacy Reconsidered: Descriptive, Normative, and Methodological Challenges." *Behavioral and Brain Sciences* 19(1):1–53.

Kroll, Y., H. Levy, and A. Rapoport. 1988. "Experimental Tests of the Separation Theorem and the Capital Asset Pricing Model." *American Economic Review* 78(3):500–19.

Krueger, J. I., and D. C. Funder. 2004. "Towards a Balanced Social Psychology: Causes, Consequences, and Cures for the Problem-Seeking Approach to Social Behavior and Cognition." *Behavioral and Brain Sciences* 27:313–76.

Krutch, J. 1954. *The Voice of the Desert.* New York: William Sloane Associates.

Kurzban, R., and D. Houser. 2005. "An Experimental Investigation of Cooperative Types in Human Groups: A Complement to Evolutionary Theory and Simulation." *Proceedings of the National Academy of Sciences* 102:1803–7.

Kwasnica, A. M., J. Ledyard, D. Porter, and C. DeMartini. 2003. "A New and Improved Design for Multiobject Iterative Auctions." *Management Science* 51(3):419–35.

Lakatos, I. 1978. *The Methodology of Scientific Research Programs,* vol. 1, 2. Cambridge: Cambridge University Press.

Lazzarini, S., G. Miller, and T. Zenger. 2002. "Order with Some Law: Complementarity vs. Substitution of Formal and Informal Arrangements." Paper presented at the Interdisciplinary Conference on Trust and Reciprocity in Experimental Economics, October 11–12. Washington University, St. Louis.

Leamer, E. 1978. *Specification Searches.* New York: Wiley.

Ledyard, J. 1986. "The Scope of the Hypothesis of Bayesian Equilibrium." *Journal of Economic Theory* 39:59–82.

Ledyard, J., M. Olson, D. Porter, J. A. Swanson, and D. P. Torma. 2002. "The First Use of a Combined-Value Auction for Transportation Services." *Interfaces* 32(5):4–12.

Ledyard, J., D. Porter, and A. Rangle. 1997. "Experiment Testing Multi Object Allocation Mechanism." *Journal of Economics and Management Strategy* 6:639–75.

Lei, V., C. Noussair, and C. Plott. 2001. "Nonspeculative Bubbles in Experimental Asset Markets: Lack of Common Knowledge of Rationality vs. Actual Irrationality." *Econometrica* 69:831–59.

Levitt, S., and J. List. 2007. "What Do Laboratory Experiments Tell Us about the Real World?" University of Chicago prepublication paper, June 15. *Journal of Economic Perspectives.*

Libet, B. 2004. *Mind Time.* Cambridge: Harvard University Press.

List, J. 2004. "Neoclassical Theory versus Prospect Theory: Evidence from the Marketplace," *Econometrica* 72(2):615–25.

List, J., and T. Cherry. 2002. "Learning to Accept in Ultimatum Games: Evidence from an Experimental Design that Generates Low Offers." *Experimental Economics* 3:11–29.

Lohr, S. 1992. "Lessons from a Hurricane: It Pays Not to Gouge." *New York Times*, September 22, p. C2.

Lopes, L. 1991. "The Rhetoric of Irrationality." *Psychology and Theory* 1:65–82.

Lucas, R. E. 1986. "Adaptive Behavior and Economic Theory." *Journal of Business* 56:S401–26.

Lynch, M., R. M. Miller, C. R. Plott, and R. Porter. 1986. "Product Quality, Consumer Information and 'Lemons' in Experimental Markets." In P. Ippolito and D. Scheffman, eds., *Empirical Approaches to Consumer Protection Economics.* Washington: FTC Bureau of Economics, pp. 251–306.

Lyyken, D. T. 1995. *The Antisocial Personalities.* Hillsdsale, NJ: Erlbaum.

Majumdar, M. K., and R. Radner. 1991. "Linear Models of Economic Survival under Production Uncertainty." *Economic Theory* 1:13–30.

———. 1992. "Survival under Production Uncertainty." In M. K. Majumdar, ed., *Equilibrium and Dynamics.* London: Macmillan, pp. 179–200.

Mandaville, A. 2006. "Look and Learn." *Nature,* 441:271–2.

Mandeville, B. 1705; 1924; 2005. "The Grumbling Hive: Or, Knaves Turned Honest." In F. B. Kaye, ed., *The Fable of the Bees: Or, Private Vices, Public Benefits,* vol. 1. London: Oxford University Press, Amen House, pp. 17–37. Available online at http://oll.libertyfund.org/Home3/Set.php?recordID=0014.

Maquet, P. 2001. "The Role of Sleep in Learning and Memory." *Science* 294(5544):1048–52.

Markowitz, H. 1952. "The Utility of Wealth." *Journal of Political Economy* 60:151–8.

Mayo, D. 1996. *Error and the Growth of Experimental Knowledge.* Chicago: University of Chicago Press.

Mazzarello, P. 2000. "What Dreams May Come?" *Nature* 408:523.

McAfee, R. P., and J. McMillan. 1996. "Analyzing the Airways Auction." *Journal of Economic Perspectives* 10:159–75.

McCabe, K., J. Henrich, W. Albers, R. Boyd, G. Gigerenzer, A. Ockenfels, P. E. Tetlock, and H. P. Young. 2001a. "Group Report: What Is the Role of Culture in Bounded Rationality?" In G. Gigerenzer and R. Selten, eds., *Bounded Rationality: The Adaptive Toolbox.* Cambridge: MIT Press, pp. 343–59.

McCabe, K., D. Houser, L. Ryan, V. Smith, and T. Trouard. 2001b. "A Functional Imaging Study of Cooperation in Two-Person Reciprocal Exchange." *Proceedings of the National Academy of the Sciences* 98:11832–5.

McCabe, K., S. J. Rassenti, and V. L. Smith. 1989. "Designing 'Smart' Computer Assisted Markets." *European Journal of Political Economy* 5(2–3):259–83.

———. 1990. "Auction Institutional Design: Theory and Behavior of Simultaneous Multiple-Unit Generalizations of the Dutch and English Auctions." *American Economic Review* 80(5):1276–83.

———. 1991a. "Smart Computer-Assisted Markets." *Science* 254:534–8.

———. 1991b. "Testing Vickrey's and Other Simultaneous Multiple Unit Versions of the English Auction." In R. M. Isaac, ed., *Research in Experimental Economics,* vol. 4. Greenwich: JAI Press, pp. 45–79.

———. 1993. "Designing a Uniform-Price Double Auction: An Experimental Evaluation." In J. Friedman and J. Rust, eds., *The Double Auction Market: Institutions,*

Theories, and Evidence: Proceedings of the Workshop on Double Auction Markets Held June 1991 in Santa Fe, NM. Santa Fe: Perseus, pp. 307–32.

———. 1996. "Game Theory and Reciprocity in Some Extensive Form Experimental Games." *Proceedings of the National Academy of Sciences* 93:13421–8.

———. 1998. "Reciprocity, Trust and Payoff Privacy in Extensive Form Bargaining." *Games and Economic Behavior* 24(1–2):10–24.

McCabe, K., M. Rigdon, and V. L. Smith. 2002. "Cooperation in Single Play, Two-Person Extensive Form Games between Anonymously Matched Players." In R. Zwick and A. Rapoport, eds., *Experimental Business Research.* Boston: Kluwer, chapter 3.

———. 2003. "Positive Reciprocity and Intentions in Trust Games." *Journal of Economic Behavior and Organization* 52(2):267–75.

———. 2007. "Sustaining Cooperation in Trust Games," Revised working paper, *Economic Journal.*

McCabe, K., and V. L. Smith. 2000. "A Comparison of Naïve and Sophisticated Subject Behavior with Game Theoretic Predictions." *Proceedings of the National Academy of Sciences* 97:3777–81.

———. 2001. "Goodwill Accounting and the Process of Exchange." In G. Gigerenzer and R. Selten, eds., *Bounded Rationality: The Adaptive Toolbox.* Cambridge: MIT Press, pp. 319–40.

McCabe, K., V. Smith, and M. LePore. 2000. "Intentionality Detection and 'Mindreading': Why Does Game Form Matter?" *Proceedings of the National Academy of the Sciences* 97:4404–9.

Mealy, L. 1995. "The Sociobiology of Sociopathy." *Behavioral and Brain Sciences* 18:523–99.

Meardon, S., and A. Ortmann. 1996. "Self-Command in Adam Smith's Theory of Moral Sentiments: A Game-Theoretic Re-Interpretation." *Rationality and Society* 8:57–80.

Mehler, J., and T. G. Bever. 1967. "Cognitive Capacity of Very Young Children." *Science New Series* 158 (3797):141–2.

Mellers, B., A. Schwartz, and I. Ritor. 1999. "Emotion-Based Choice." *Journal of Experimental Psychology: General* 128:1–14.

Metzger, B., and R. Murphy, eds. 1994. *The New Oxford Annotated Bible.* New York: Oxford University Press.

Milgrom, P. 2000. "Putting Auction Theory to Work: The Simultaneous Ascending Auction." *Journal of Political Economy* 108(2):245–72.

———. 2004. *Putting Auction Theory to Work.* New York: Cambridge University Press.

Milgrom, P. R., and R. J. Weber. 1982. "A Theory of Auctions and Competitive Bidding." *Econometrica* 50:1089–1122.

Mill, J. S. 1848; 1900. *Principles of Political Economy,* vols. 1–2. London: Colonial Press.

Miller, R. 2002. "Don't Let Your Robots Grow Up to Be Traders." Working paper, Miller Risk Advisors.

Miller, R., and C. Plott. 1985. "Product Quality Signaling in Experimental Markets." *Econometrica* 53:837–72.

Miller, R., C. Plott, and V. Smith. 1977. "Intertemporal Competitive Equilibrium: An Empirical Study of Speculation." *Quarterly Journal of Economics* 91:599–624.

Montes, L. 2003. "Das Adam Smith Problem: Its Origins, the Stages of the Current Debate, and One Implication for Our Understanding of Sympathy." *Journal of the History of Economic Thought* 25:63–90.

Moser, E. 2006. "Rewinding the Memory Record." *Nature*, 440:615–17.

Mullainathan, S., and R. Thaler. 2001. "Behavioral Economics." In N. J. Smelser and P. B. Baltes, eds., *International Encyclopedia of the Social and Behavioral Sciences*. Oxford: Elsevier.

Myerson, R. 1991. *Game Theory*. Cambridge: Harvard University Press.

Nash. J. F. 1996. *Essays in Game Theory*. Cheltenham: Edward Elgar.

Nelson, K., and R. Nelson. 2002. "On the Nature and Evolution of Human Know-How." *Research Policy* 31:719–33.

Nelson, L. 2004. "While You Were Sleeping." *Nature* 430 (7003):962–4.

Netting, R. 1976. "What Alpine Peasants Have in Common: Observations on Communal Tenure in a Swiss Village." *Human Ecology* 4:135–46.

Norman, J. 2002. "Two Visual Systems and Two Theories of Perception: An Attempt to Reconcile the Constructivist and Ecological Approaches." *Behavioral and Brain Sciences* 25(1):73–96.

North, D. C. 1990. *Institutions, Institutional Change and Economic Performance*. Cambridge: Cambridge University Press.

———. 2005. *Understanding the Process of Economic Change*. Princeton: Princeton University Press.

Northrup, F. S. C. 1969. "Einstein's Conception of Science." In P. A. Schilpp, ed., *Albert Einstein Philosopher – Scientist*. LaSalle: Open Court.

O'Keefe, J., and L. Nadel. 1978. *The Hippocampus as a Cognitive Map*. London: Clarendon.

Olson, M., S. Rassenti, V. Smith, and M. Rigdon. 2003. "Market Design and Motivated Human Trading Behavior in Electricity Markets." *Institute of Industrial Engineering Transactions* 35(9):833–49.

Ortmann, A. 2005. "Field Experiments: Some Methodological Caveats." In J. Carpenter, G. W. Harrison, and J. A. List, eds., *Research in Experimental Economics*, vol. 10: *Field Experiments in Economics*. Greenwich: JAI Press.

Ortmann, A., J. Fitzgerald, and C. Boening. 2000. "Trust, Reciprocity, and Social History." *Experimental Economics* 3(1):81–100.

Ortmann, A., and O. Rydval. 2004. "How Financial Incentives and Cognitive Abilities Affect Task Performance in Laboratory Settings: An Illustration." *Economic Letters* 85:315–20.

Ostrom, E. 1982. *Strategies of Political Inquiry*. Beverly Hills: Sage Publications.

———. 1990. *Governing the Commons*. Cambridge: Cambridge University Press.

Ostrom, E., R. Gardner, and J. Walker. 1994. *Rules, Games and Common Pool Resources*. Ann Arbor: University of Michigan Press.

Oxoby, R., and J. Spraggon. 2007. "Mine and Yours: Property Rights in Dictator Games." *Journal of Economic Behavior and Organization*, forthcoming.

Pepperberg, I. M. 1999. *The Alex Studies: Cognitive and Communicative Abilities of Grey Parrots*. Cambridge: Harvard University Press.

Pillutla, M., D. Malhotra, and J. K. Murnighan. 2003. "Attributions of Trust and the Calculus of Reciprocity." *Journal of Experimental Social Psychology* 39(5):448–55.

Pinker, S. 1994. *The Language Instinct*. New York: William Morrow.

———. 2002. *The Blank Slate*. New York: Viking.

Pirsig, R. M. 1981. *Zen and the Art of Motorcycle Maintenance*. New York: Bantam.

Pitchik, C., and A. Schotter. 1984. "Regulating Markets and Asymmetric Information: An Experimental Study." Working Paper No. 84–12. C. V. Starr Center for Applied Economics, New York University.

Plott, C. 1988. "Research on Pricing in a Gas Transportation Network." Technical Report No. 88–2. Washington: Federal Energy Regulatory Commission, Office of Economic Policy.

———. 2001. "Equilibrium, Equilibration, Information and Multiple Markets: From Basic Science to Institutional Design." Paper presented to the Nobel Symposium, Behavioral and Experimental Economics, Stockholm, December 4–6.

Plott, C. R., and G. Agha. 1983. "Intertemporal Speculation with a Random Demand in an Experimental Market." In R. Tietz, ed., *Aspiration Levels in Bargaining and Economic Decision Making*. Berlin: Springer-Verlag.

Plott, C., A. B. Sugiyama, and G. Elbaz. 1994. "Economies of Scale, Natural Monopoly, and Imperfect Competition in an Experimental Market." *Southern Economic Journal* 61(2):261–87.

Plott, C., and S. Sundar. 1982. "Efficiency of Experimental Security Markets with Insider Information: An Application of Rational-Expectations Models." *Journal of Political Economy* 90:663–98.

———. 1988. "Rational Expectations and the Aggregation of Diverse Information in Laboratory Securities Markets." *Econometrica* 56:1085–118.

Plott, C., and T. Turocy III. 1996. "Intertemporal Speculation under Uncertain Future Demand." In W. Albers, W. Güth, P. Hammerstein, B. Moldovanu, and E. van Damme, eds., *Understanding Strategic Interaction: Essays in Honor of Reinhard Selten*. Berlin: Springer-Verlag, pp. 475–93.

Plott, C., and J. Uhl. 1981. "Competitive Equilibrium with Middlemen." *Southern Economic Journal* 47:1063–71.

Plott, C., and L. Wilde. 1982. "Professional Diagnosis vs. Self-Diagnosis: An Experimental Examination of Some Special Features of Markets with Uncertainty." In V. L. Smith, ed., *Research in Experimental Economics*, vol. 2. Greenwich: JAI Press, pp. 63–112.

Plott, C., and K. Zeiler. 2005. "The Willingness to Pay/Willingness to Accept Gap, the 'Endowment Effect,' Subject Misconceptions, and Experimental Procedures for Eliciting Valuations." *American Economic Review* 95(3):530–45.

Poincaré, H. 1913. *Foundations of Science*. New York: Science Press.

Polanyi, M. 1962. *Personal Knowledge*. Chicago: University of Chicago Press.

———. 1969. *Knowing and Being*. Chicago: University of Chicago Press.

Porter, D. 1999. "The Effect of Bid Withdrawal in a Multi-Object Auction." *Review of Economic Design* 4(1):73–97.

Porter, D., S. Rassenti, A. Roopnarine, and V. Smith. 2003. "Combinatorial Auction Design." *Proceedings of the National Academy of Sciences* 100(19):11153–7.

Porter, D., S. Rassenti, V. Smith, A. Winn, and W. Shobe. 2005. "The Design, Testing and Implementation of Virginia's NOx Allowance Auction." Working Paper, Interdisciplinary Center for Economic Science, George Mason University.

Porter, D., and V. L. Smith. 1994. "Stock Market Bubbles in the Laboratory." *Applied Mathematical Finance* 1:111–27.

Rabin, M. 1993. "Incorporating Fairness into Game Theory and Economics." *American Economics Review* 83(5):1281–302.

Radner, R. 1997. "Economic Survival." In D. P. Jacobs, E. Kalai, and M. I. Kamien, eds., *Frontiers of Research in Economic Theory: The Nancy L. Schwartz Memorial Lectures, 1983–1997.* Cambridge: Cambridge University Press.

Radner, R., and A. Schotter. 1989. "The Sealed-Bid Mechanism: An Experimental Study." *Journal of Economic Theory* 48:179–220.

Rapoport, A. 1997. "Order of Play in Strategically Equivalent Games in Extensive Form." *International Journal of Game Theory* 26:113–36.

Rapoport, A., and J. A. Sundali. 1996. "Ultimatums in Two Persons Bargaining with One-Sided Uncertainty: Offer Games." *International Journal of Game Theory* 25(4):475–94.

Rassenti, S. 1981. "O-1 Decision Problems with Multiple Resource Constraints: Algorithms and Applications." Ph.D. Thesis, University of Arizona.

Rassenti, S., and V. Smith. 1986. "Electric Utility Deregulation." In *Pricing Electric Gas and Telecommunication Services.* Institute for the Study of Regulation, Proceedings of a Conference, December.

Rassenti, S., V. L. Smith, and R. Bulfin. 1982. "A Combinatorial Auction Mechanism for Airport Time Slot Allocation." *Bell Journal of Economics* 13:402–17.

Rassenti, S., V. L. Smith, and B. Wilson. 2002a. "Controlling Market Power and Price Spikes in Electricity Networks: Demand Side Bidding." Working paper, Interdisciplinary Center for Economic Science, George Mason University.

————. 2002b. "Using Experiments to Inform the Privatization/Deregulation Movement in Electricity." *Cato Journal* 21:515–44.

————. 2003. "Controlling Market Power and Price Spikes in Electricity Networks: Demand-Side Bidding." *Proceedings of the National Academy of Sciences* 100(5):2998–3003.

Rassenti, S., and B. Wilson. 2004. "How Applicable Is the Dominant Firm Model of Price Leadership?" *Experimental Economics* 7(3):271–88.

Reiley, D.. 2000. "Auctions on the Internet: What's Being Auctioned, and How?" *Journal of Industrial Economics* 48(3):227–52.

Reiter, S. 1959; 1981. "A Dynamic Process of Exchange." In G. Horwich and J. Quirk, eds., *Essays in Contemporary Fields of Economics*, pp. 3–23. West Lafayette: Purdue University Press. Revised from "A Market Adjustment Mechanism." Institute Paper No. 1, School of Industrial Management, Purdue University, 1959.

Renshaw, E. 1988. "The Crash of October 19 in Retrospect." *Market Chronicle* 22:1.

Rigdon, M. 2002. "Efficiency Wages in an Experimental Labor Market." *Proceedings of the National Academy of Sciences* 99(20):13348–51.

Rizzello, S. 1999. *The Economics of the Mind.* Cheltenham: Edward Elgar.

Robinson, J. 1979. "What Are the Questions?" *Journal of Economic Literature* 15:1318–39.

Rosenthal, R. 1981. "Games of Perfect Information, Predatory Pricing, and the Chain Store Paradox." *Journal of Economic Theory* 25:92–100.

Roth, A., and M. Sotomayer. 1990. *Two-Sided Matching: A Study of Game Theoretic Modeling and Analysis.* Cambridge: Cambridge University Press.

Rothkopf, M. H., A. Pekec, and R. M. Harstad. 1998. "Computationally Manageable Combinational Auctions." *Management Science* 44(8):1131–47.

Samuelson, P., and W. Nordhaus. 1985. *Economics.* New York: McGraw-Hill.

Satterthwaite, M. 1987. "Strategy-Proof Allocation Mechanisms." In J. Eatwell, M. Milgate, and P. Newman, eds., *The New Palgrave*, vol. 4, pp. 518–20. London: Macmillan.

Schelling, T. 1960. *The Strategy of Conflict.* Cambridge: Harvard University Press.

Schmitt, P. M. 2003. "On Perceptions of Fairness: The Role of Valuations, Outside Options, and Information in Ultimatum Bargaining Games." *Experimental Economics* 7:49–73.

Schotter, A. and B. Sopher. 2006. "Trust and Trustworthiness in Games: An Experimental Study of Intergenerational Advice." *Experimental Economics*, 9:123–45.

Schotter, A., K. Wiegelt, and C. Wilson. 1994. "A Laboratory Investigation of Multiperson Rationality and Presentation Effects." *Games and Economic Behavior* 6:445–68.

Schultz, W. 2000. "Multiple Reward Signals in the Brain." *Nature Reviews: Neuroscience* 1:199–207.

———. 2002. "Getting Formal with Dopamine and Reward." *Neuron* 36:241–63.

Schwartz, T., and J. S. Ang. 1989. "Speculative Bubbles in the Asset Market: An Experimental Study." Paper presented at the American Finance Association meeting, Atlanta, December.

Scully, G. W. 1988. "The Institutional Framework and Economic Development." *Journal of Political Economy* 96(3):652–62.

Segal, N. 1999. *Entwined Lives.* New York: Plume.

Segrè, E. 1984. *From Falling Bodies to Radio Waves.* New York: Freeman.

Selten, R. 1973. "A Simple Model of Imperfect Competition, Where Four Are Few and Six Are Many." *International Journal of Game Theory* 2(2):141–201.

———. 1975. "Re-Examination of the Perfectness Concept for Equilibrium Points in Extensive Games." *International Journal of Game Theory* 4(1):25–55.

Shaked, Avner. 2005. "The Rhetoric of Inequity Aversion." Available online at http://www.wiwi.uni-bonn.de/shaked/rhetoric/.

Shepard, G. B. 1995. *Rejected.* Sun Lakes: Thomas Horton and Daughters.

Shubik, M. 1959. *Strategy and Market Structure.* New York: Wiley.

Siegel, S. 1959. "Theoretical Models of Choice and Strategy Behavior." *Psychometrika* 24:303–16.

———. 1961. "Decision Making and Learning under Varying Conditions of Reinforcement." *Annals of the New York Academy of Science* 89(5):766–83.

Siegel, S., and L. Fouraker. 1960. *Bargaining and Group Decision Making.* New York: McGraw-Hill.

Siegel, S., and D. L. Harnett. 1964. "Bargaining Behavior: A Comparison between Mature Industrial Personnel and College Students." *Operations Research* March: 334–43.

Simon, H. A. 1955. "A Behavioral Model of Rational Choice." *Quarterly Journal of Economics* 69:99–118

———. 1956. "A Comparison of Game Theory and Learning Theory." *Psychometrika* 3:267–72.

———. 1981; 1996. *The Sciences of the Artificial,* 3rd edition. Cambridge: MIT Press.

Singleton, S. 2000. "Will the Net Turn Car Dealers into Dinosaurs?" *Cato Institute Briefing Papers* 58.

Smith, A. 1759; 1982. *The Theory of Moral Sentiments.* Edited by D. Raaphaet and A. Mactie. Indianapolis: Liberty Fund.

———. 1776; 1981. *An Enquiry into the Nature and Causes of the Wealth of Nations,* vol 1. Edited by R. Campbell and A. Skinner. Indianapolis: Liberty Fund.

Smith, H. 2001. *Why Religion Matters.* New York: HarperCollins.

Smith, K., J. Dickhaut, K. McCabe, and J. Pardo. 2002. "Neuronal Substrates for Choice under Ambiguity, Risk, Gains, and Losses." *Management Science* 48:711–19.

Smith, V. L. 1962. "An Experimental Study of Competitive Market Behavior." *Journal of Political Economy* 70:111–37.

―――. 1967. "Experimental Studies of Discrimination versus Competition in Sealed-Bid Auction Markets." *Journal of Business* 70:56–84.

―――. 1976. "Experimental Economics: Induced Value Theory." *American Economic Review Proceedings* 66:274–9.

―――. 1980. "Relevance of Laboratory Experiments to Testing Resource Allocation Theory." In J. Kmenta and J. B. Ramsey, eds., *Evaluation of Econometric Models.* San Diego: Academic Press, pp. 345–77.

―――. 1982a. "Microeconomic Systems as an Experimental Science." *American Economic Review* 72:923–55.

―――. 1982b. "Markets as Economizers of Information: Experimental Examination of the 'Hayek Hypothesis'." *Economic Inquiry* 20:165–79.

―――. 1986. "Experimental Methods in the Political Economy of Exchange." *Science* 234:167–73.

―――. 1991. *Papers in Experimental Economics.* Cambridge: Cambridge University Press.

―――. 1992. "Game Theory and Economics: Beginnings and Early Influences." In E. Weintraub, ed., *Toward a History of Game Theory.* Durham: Duke University Press, pp. 241–82.

―――. 1997. "Experimental Economics: Behavioral Lessons for Microeconomic Theory and Policy." In D. P. Jacobs, E. Kalai, and M. I. Kamien, eds., *Frontiers of Research in Economic Theory: The Nancy L. Schwartz Memorial Lectures, 1983–1997.* Cambridge: Cambridge University Press, pp. 104–22.

―――. 1998. "The Two Faces of Adam Smith." *Southern Economic Journal* 65:1–19.

―――. 2000. *Bargaining and Market Behavior.* Cambridge: Cambridge University Press.

―――. 2002. "Method in Experiment: Rhetoric and Reality." *Experimental Economics* 5:91–110.

―――. 2003. "Constructivist and Ecological Rationality in Economics." *Les Prix Nobel.* Stockholm: Nobel Foundation. Reprinted in *American Economic Review* 93:465–508.

―――. 2005a. *From Starlight to Stockholm, A Memoir.* Manuscript in typescript.

―――. 2005b. "Sociality and Self Interest." *Behavioral and Brain Sciences* 28:833–4.

―――. 2007. *Discovery: Starlight to Stockholm* (title tentative). Forthcoming.

Smith, V. L., K. McCabe, and S. Rassenti. 1991. "Lakatos and Experimental Economics." In N. de Marchi and M. Blaug, eds., *Appraising Economic Theories.* London: Edward Elgar.

Smith, V. L., G. Suchanek, and A. Williams. 1988. "Bubbles, Crashes and Endogenous Expectations in Experimental Spot Asset Markets." *Econometrica* 56:1119–51.

Smith, V. L. and F. Szidarovszky. 2004. "Monetary Rewards and Decision Cost in Strategic Interactions." In M. Augier and J. March, eds., *Models of a Man: Essays in Memory of Herbert A. Simon.* Cambridge: MIT Press.

Smith, V. L., and J. Walker. 1993a. "Monetary Rewards and Decision Cost in Experimental Economics." *Economic Inquiry* April:245–61.

―――. 1993b. "Rewards, Experience and Decision Costs in First Price Auctions." *Economic Inquiry* April:237–44.

Smith, V. L., and A. Williams. 1982. "An Experimental Comparison of Alternative Rules for Competitive Market Exchange." In M. Shubik, ed., *Auctioning and Bidding.* New York: New York University Press, pp. 43–64.

Smith, V. L., A. Williams, K. Bratton, and M. Vannoni. 1982. "Competitive Market Institutions: Double Auctions vs. Sealed Bid-Offer Auctions." *American Economic Review* 72:58–77.

Smuts, B. 1999. "Multilevel Selection, Cooperation and Altruism: Reflections on Unto Others." *Human Nature* 10:156–61.

Sobel, J. 2005. "Interdependent Preferences and Reciprocity." *Journal of Economic Literature* 42:392–436.

Soberg, M. 2005. "The Duhem-Quine Thesis and Experimental Economics." *Journal of Economic Methodology* 12:581–97.

Sopher, B., and G. Gigliotti. 1993. "Intransitive Cycles: Rational Choice or Random Error." *Theory and Decision* 35:311–36.

Spence, M. 2002. "Signaling in Retrospect and the Informational Structure of Markets." *American Economic Review* 92(3):434–59.

Stahl, D. 1993. "Evolution of Smart Players." *Games and Economic Behavior* 5:604–17.

Stigler, G. 1957. "Perfect Competition, Historically Contemplated." *Journal of Political Economy* 65:1–17.

———. 1967. "Imperfections in the Capital Market." *Journal of Political Economy* June: 287–92.

Stiglitz, J. E. 2002. "Information and the Change in the Paradigm in Economics." Nobel Prize Lecture, December 8, 2001. In T. Frängsmyr, ed., *The Nobel Prizes 2001*. Stockholm: Nobel Foundation. Reprinted in *American Economic Review* 92(3): 460–501.

Strathern, P. 1996. *Wittgenstein*. Chicago: Ivan Dee.

Straub, P., and K. Murnigan. 1995. "An Experimental Investigation of Ultimatum Games: Information, Fairness, Expectations, and Lowest Acceptable Offers." *Journal of Economic Behavior and Organization* 27:345–64.

Sundar, S. 2004. "Market as Artifact: Aggregate Efficiency from Zero Intelligence Traders." In M. Augier and J. March, eds., *Models of a Man: Essays in Memory of Herbert A. Simon*. Cambridge: MIT Press.

Suppes, P. 1969. "Models of Data." In *Studies in the Methodology and Foundations of Science*. Dordrecht: Reidel, pp. 24–35.

Suroweicki, J. 2004. *The Wisdom of Crowds*. New York: Doubleday.

Thaler, R. 1980. "Toward a Positive Theory of Consumer Choice." *Journal of Economic Behavior and Organization* 1:39–60.

Thaler, R., and W. T. Ziemba. 1988. "Parimutuel Betting Markets: Racetracks and Lotteries." *Journal of Economic Perspectives* 2(2):161–74.

Thut, G., W. Schultz, U. Roelcke, et al. 1997. "Activation of the Human Brain by Monetary Reward." *NeuroReport* 8:1225–8.

Tremblay, L., and W. Schultz. 1999. "Relative Reward Preference in Primate Orbitofrontal Cortex." *Nature* 389:704–8.

Treynor, J. 1987. "Market Efficiency and the Bean Jar Experiment." *Financial Analysts Journal* 43:50–3.

Trivers, R. 2004. "Mutual Benefits at All Levels of Life." *Science* 304:964–5.

Tullock, G. 1985. "Adam Smith and the 'Prisoners' Dilemma." *Quarterly Journal of Economics* 100 Supplement:1073–81.

Tversky, A., and D. Kahneman. 1983. "Extensional versus Intuitive Reasoning: The Conjunction Fallacy in Probability Judgment." *Psychological Bulletin* 90(4):293–315.

_____. 1987. "Rational Choice and the Framing of Decisions." In R. M. Hogarth and M. W. Reder, eds., *Rational Choice: The Contrast between Economics and Psychology*. Chicago: University of Chicago Press.

Umbeck, J. 1977. "The California Gold Rush: A Study of Emerging Property Rights." *Explorations in Economic History* 14:197–226.

Van Boening, M., and N. Wilcox. 1996. "Avoidable Cost: Ride a Double Auction Roller Coaster." *American Economic Review* 86(3):461–77.

Vickery, W. 1961. "Counterspeculation, Auctions, and Competitive Sealed Tenders." *Journal of Finance* 16:8–37.

_____. 1976. "Auction Markets and Optimal Allocation." In Y. Amihud, ed., *Bidding and Auctioning for Procurement and Allocation*. New York: Columbia University Press, pp. 43–64.

Viner, J. 1991. "Adam Smith." In D. Irvin, ed., *Essays on the Intellectual History of Economics*. Princeton: Princeton University Press, pp. 248–61.

von Winterfeldt, D., and W. Edwards. 1986. *Decision Analysis and Behavioral Research*. Cambridge: Cambridge University Press.

Wagner, U., S. Gais, H. Haider, R. Verleger, and J. Born. 2004. "Sleep Inspires Insight." *Nature* 427:352–5.

Weber, R. A., and C. F. Camerer. 2006. "Behavioral Experiments in Economics." *Journal of Economic Behavior and Organization* 9:187–92.

Wierzbicka, A. 2006. *English: Meaning and Culture*. Oxford: Oxford University Press.

Williams, A. 1979. "Intertemporal Competitive Equilibrium: On Further Experimental Results." In V. L. Smith, ed., *Research in Experimental Economics*, vol. 1. Greenwich: JAI Press, pp. 225–78.

_____. 1980. "Computerized Double-Auction Markets: Some Initial Experimental Results." *Journal of Business* 53:235–57.

Williams, A., and V. L. Smith. 1986. "Simultaneous Trading in Two Competitive Markets." Manuscript, Indiana University, Department of Economics.

Williams, A., V. L. Smith, J. O. Ledyard, and S. Gjerstad. 2000. "Concurrent Trading in Two Experimental Markets with Demand Interdependence." *Journal of Economic Theory* 16:511–28.

Williamson, O. 1977. "Predatory Pricing: A Strategic and Welfare Analysis." *Yale Law Journal* December: 284–340.

Wilson, B. 2005. "Language Games of Reciprocity." Working Paper, Interdisciplinary Center of Economic Science, George Mason University.

Wilson, R. B. 1987. "On Equilibria of Bid-Ask Markets." In G. Feiwel, ed., *Arrow and the Ascent of Modern Economic Theory*. New York: NYU Press, pp. 375–414.

_____. 1992. "Strategic Analysis of Auctions." In R. Aumann and S. Hart, eds., *Handbook of Game Theory*. Amsterdam: North-Holland, pp. 227–79.

_____. 1993. "Design of Efficient Trading Procedures." In J. Friedman and J. Rust, eds., *The Double Auction Market: Institutions, Theories, and Evidence: Proceedings of the Workshop on Double Auction Markets Held June 1991 in Santa Fe, NM*. Santa Fe: Perseus, pp. 125–52.

_____. 2002. "Architecture of Power Markets." *Econometrica* 70:1299–1340.

Wittgenstein, L. 1963. *Philosophical Investigations*. Translated by G. E. M. Anscombe. Oxford: Basil Blackwell.

————. 1967. *Zettel.* Edited by G. Anscombe and G. von Wright. Translated by G. Anscombe. Berkeley: University of California Press, 1967.

Zajac, E. E. 1995. *Political Economy of Fairness.* Cambridge: MIT Press.

————. 2002. "What Fairness-and-Denial Research Could Have Told the Florida Supreme Court (and Can Tell the Rest of Us)." *Independent Review* 6(3):377–97.

Zeaman, D. 1949. *Response Latency as a Function of the Amount of Reinforcement.* Dissertation Thesis, Columbia University.

Index

For EU product safety concerns, contact us at Calle de José Abascal, 56–1°,
28003 Madrid, Spain or eugpsr@cambridge.org.

www.ingramcontent.com/pod-product-compliance
Ingram Content Group UK Ltd.
Pitfield, Milton Keynes, MK11 3LW, UK
UKHW040949090126
466816UK00019B/334